FUNDAMENTALS OF APERTURE ANTENNAS AND ARRAYS

FUNDAMENTALS OF APERTURE ANTENNAS AND ARRAYS

FROM THEORY TO DESIGN, FABRICATION AND TESTING

Trevor S. Bird PhD, FTSE

Principal Antengenuity, Adjunct Professor Macquarie University &
Honorary CSIRO Fellow, Australia

WILEY

Library of Congress Cataloging-in-Publication Data

Bird, Trevor S. 1949–
Fundamentals of aperture antennas and arrays : from theory to design, fabrication and testing / Trevor S. Bird.
   pages   cm
  Includes bibliographical references and index.
  ISBN 978-1-118-92356-6 (cloth : alk. paper)   1. Aperture antennas.   I. Title.
  TK7871.6.B49 2016
  621.3841′35–dc23

                                    2015021598

A catalogue record for this book is available from the British Library.

Set in 10/12pt Times by SPi Global, Pondicherry, India

1   2016

This book is dedicated to my mentors and collaborators, particularly
Peter J.B. Clarricoats
George A. Hockham
Peter L. Arlett
Geoffery T. Poulton
David G. Hurst

Contents

Preface

Aperture antennas are a class of antennas in common daily use and some have even become synonymous with areas of science and technology. Typical examples include reflectors, horns, lenses, waveguides, slits, and slots. Other antennas can be conveniently described by means of aperture concepts. Some of these include microstrip patches and reflectarrays. In this book we describe the underlying theory and application of these antennas as well as their use in arrays.

The history of aperture antennas is inextricably linked with historical developments in wireless and also the verification of Maxwell's equations. The very first waveguide was demonstrated by Lodge in 1894 and in 1895 Bose used circular waveguides as an antenna along with pyramidal horns for experiments on the polarization properties of crystals. About thirty years later a 10 m diameter reflector became the first radiotelescope when it detected emissions from electrons in interstellar space. In the 1960s aperture antennas accompanied the first humans on the moon and more recently they have contributed to the wireless revolution that is presently underway.

Aperture antennas are normally associated with directional beams and, indeed, this is their role in many applications. They can also occur on non-planar or curved surfaces such as on aircraft or ground-based vehicles. These antennas may consist of a single radiator or in arrays. In this form they are often used to provide directional or shaped beams.

Directional beams are needed in terrestrial and satellite microwave links to efficiently use the available power as well as to reduce interference and noise. Radar systems also require directional antennas to identify targets. As well, arrays of aperture antennas can produce almost omnidirectional radiation.

A limitation of a directional planar antenna is that when it is scanned from broadside the beam broadens and the pattern deteriorates. When the antenna is conformal to a convex surface, such as a cylinder or a cone, the beam can be scanned in discrete steps through an arc while maintaining a constant pattern. Of importance in the design of low sidelobe antenna arrays, both planar and conformal, is predicting the effect of mutual coupling between the array elements. Maximum performance is achieved from arrays when the coupling between elements is fully taken into account.

This book gives an introduction to the techniques that are used to design common aperture antennas as well as some approaches to their fabrication and testing. The intention is for it to be a single textbook for a course in antennas in the final year undergraduate or in a master's degree by coursework. It assumes that the reader has undertaken a course on Maxwell's equations, fields and waves. Some of these topics are reviewed in the early few chapters to provide continuity and background for the remainder of the book. The antennas covered in later chapters include horns, reflectors and arrays. Some examples are pyramidal and corrugated horns, parabolic and spherical reflectors, reflectarrays, planar lenses and coaxial waveguide array feeds. To provide more than a simplified treatment of arrays, the topic of mutual coupling is covered in more detail than most similar books on this topic. Also included is an introduction to sources and arrays on non-planar surfaces, which is of importance for applications involving aerodynamic surfaces and for making aperture antennas unobtrusive. A chapter is included on modern aperture antennas that extend the concepts introduced in earlier chapters. This is to show where advances have been made in the past and how they could be made in the future. Also included are some topics of a practical nature detailing some techniques for fabrication of aperture antennas and their measurement.

Acknowledgement

The author thanks his many colleagues who have provided comments on some of the material and for their contribution to the projects described herein. In particular for this book, thanks are extended to colleagues Drs. Stuart Hay, Doug Hayman, Nasiha Nikolic, Geoff James, Stephanie Smith and Andrew Weily who readily gave up time to providing comments on early drafts of the chapters. Thanks are extended to members of the editorial staff at Wiley, in particular Anna Smart, Sandra Grayson and Teresa Nezler, and also to the SPi publications content manager Shiji Sreejish. The kind permission to use pictures by CSIRO Australia, the Institution of Engineering and Technology (IET), Engineers Australia, and SES ASTRA is gratefully acknowledged.

The author also wishes to acknowledge in particular several folk who provided mentorship and gave great support and encouragement during his early research career. To mention a few significant mentors by name: Prof. Peter Clarricoats, FRS gave the author a great start in the area of aperture antennas and arrays; the late Dr. George Hockham FRS, inspired the author and demonstrated the importance of some of the techniques described herein for applications and how practical antennas could be realised; Dr. Geoff Poulton provided a number of important opportunities for the author and long-term friendship and the many technical discussions during our working careers; and finally, the late Mr. David Hurst helped the author in his first attempts in the wireless area graciously and shared his wide knowledge and practical expertise. The author wishes to thank Ms. Dallas Rolph for assisting greatly in editing the text as well as general assistance when they worked together at CSIRO. Finally, the author is very grateful to his wife, Val, for her love and strong support over 40 years when many of the topics discussed herein were investigated.

1

Introduction

The topic of this book covers a class of antenna in common use today as well as a way of describing many others. Examples include waveguides, horns, reflectors, lenses, slits, slots and printed antennas. Some examples are illustrated in Figure 1.1. In the following chapters, the background theory and application of some basic forms of these antennas are described as well as how they can be designed, fabricated and tested. Additionally, detail will be provided on some of the individual antennas pictured in Figure 1.1.

Aperture antennas are normally associated with directional radiation beams and, indeed, this is their purpose in many applications. They can also create other types of beams such as shaped or contoured beams either separately or combined as arrays as will be shown. Aperture antennas can also occur on non-planar or conformal surfaces such as on aircraft or missile bodies where airflow and aerodynamic performance are paramount. Conformal antennas can consist of a single radiator or arrays in the surface where they can be used to provide directional and shaped beams.

Aperture antennas can be used to produce omnidirectional radiation patterns, which are important if the antenna platform is unstable or the user direction is unknown, for all-round electronic surveillance and monitoring or where the location of another user cannot be guaranteed such as in mobile radio systems. A 360-degree coverage can be achieved with a conformal antenna or with electronic switching between planar elements.

Directional beams are required in terrestrial and satellite microwave links to efficiently use the available power as well as to reduce interference and noise. Directional antennas are also required in radar systems to identify targets. A limitation of a directional planar antenna is that when it is scanned from broadside (typically boresight) the beam broadens and the pattern deteriorates. When the antenna is conformal to a convex surface, such as a cylinder or a cone, the

Fundamentals of Aperture Antennas and Arrays: From Theory to Design, Fabrication and Testing,
First Edition. Trevor S. Bird.
© 2016 John Wiley & Sons, Ltd. Published 2016 by John Wiley & Sons, Ltd.
Companion website: www.wiley.com/go/bird448

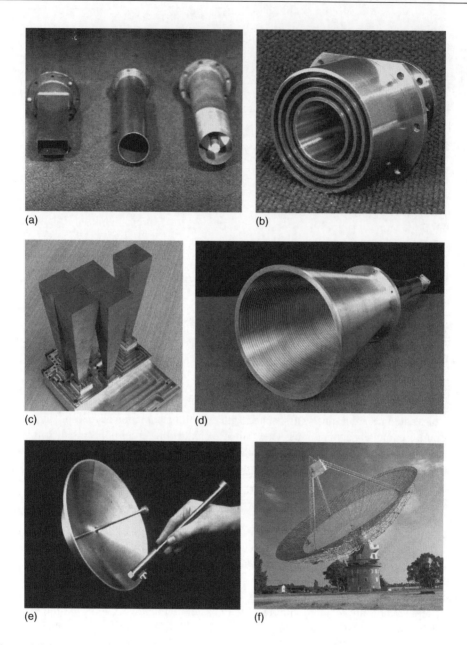

Figure 1.1 Examples of aperture antennas. (a) Open-ended waveguide antennas (right to left) coaxial, circular and rectangular. (b) Circular waveguide (diameter 32.7 mm) with three ring-slots designed for operation at 9 GHz. (c) Feed array of pyramidal horns for 12.25–12.75 GHz. (d) 11–14.5 GHz high-performance circular corrugated feed horn, diameter 273 mm, and flare angle 11.8°. (e) Small paraboloidal reflector and rear waveguide feed designed for a 15 GHz microwave link. (f) 64 m Parkes radio telescope is a front-fed paraboloid ($f/D = 0.408$). This versatile instrument has been used for frequencies from 30 MHz to >90 GHz. *Source*: Reproduced with permission from CSIRO (a–f)

Figure 1.1 (*continued*) (g) Two multibeam earth station antennas at Danish Radio's multimedia house in Ørestad in Copenhagen, Denmark, covering different segments of the geostationary satellite arc. (h) Multibeam feed system for the Parkes radio telescope. *Source*: Reproduced with permission from CSIRO. (i) On-board Ku-band satellite antennas under test on an outdoor test range prior to launch. (j) Dual-offset Cassegrain antenna with a waveguide array feed cluster under test in anechoic chamber (Bird & Boomars, 1980). (k) Series-fed microstrip patch array for a microwave landing system. *Source*: Reproduced from INTERSCAN International Ltd. (l) Conformal array of rectangular waveguides (22.86 × 10.16 mm) on a cylinder of radius 126.24 mm. *Source*: Picture courtesy of Plessey Electronic Systems

beam can be scanned in discrete steps through an arc while maintaining a constant pattern. Recent developments in microwave and optical components have simplified the design of feed networks, thereby making conformal antennas and arrays attractive alternatives for directive applications as well as for scanned beam and in ultra-low sidelobe antennas. Of importance in the design of the latter, both planar and conformal antenna arrays are often employed, and in this application predicting the effect of mutual coupling between the array elements should be undertaken. Maximum performance is achieved from arrays when the effects of coupling are known and included in the design. Otherwise, the full potential of the array flexibility may not be realized.

Aperture antennas may be analysed in much the same way as the conceptually simpler wire antennas. First, the designer needs to find the currents on the conductors or in other materials from which the antenna is constructed. To do this exactly is usually impossible except in a few idealized cases, and numerical methods are required to obtain approximate solutions. After the currents are known, the radiated fields are obtained from Maxwell's equations. Sometimes, however, adequate design information may be obtained from simplified approximations to the current, similar in some regards to adopting a sinusoidal current approximation on a linear wire antenna. This approach is especially valuable for analysing the far-field radiation characteristics, which are relatively insensitive to second-order variations in the current distribution. However, for more detailed information or quantities such as the input impedance, reflection coefficient at the input of horns or the effects of mutual coupling from nearby antennas, an accurate representation of the currents is usually required to properly take account of the current variations and near-field behaviour.

The representation of actual currents on the antenna structure may be difficult, or impossible, to achieve analytically because of the geometry and materials involved. It is convenient, and also physically allowable, to replace the actual sources by equivalent sources at the radiating surface, the antenna 'aperture', which need not lie on the actual antenna surface but on another often fictitious surface close by. For example, the aperture of a paraboloid reflector may be the projection of the rim onto a suitable plane. These equivalent sources are used in the same way as actual sources to find the radiated fields. Once these fields are known, an assessment of the antenna's performance can be made.

For the engineer wishing to specialize in the area of communications systems, some knowledge is needed of the theory and design of aperture antennas. The intention of this book is to provide some of this basic information. Today, compared with prior to the 1980s and even earlier, a variety of full wave computer solvers are now available and are particularly valuable for final design and analysis. The fundamental material available in this book is important as a starting point and for understanding the physical nature of the antenna structure before more detailed design is undertaken. It is intended that readers should be able to move from the present material to more specialized topics and to the research literature. In addition, the details provided herein should help the non-specialist in antennas to critically assess aperture antenna specifications. Where possible, useful design information has also been included. An underlying assumption is that the reader is familiar with the basic concepts of electromagnetic fields, waves and radiation, as presented, in a variety of excellent textbooks (Harrington, 1961; Jones, 1964; Jordan & Balmain, 1968; Kraus & Carver, 1973; Johnk, 1975). Some topics of a more advanced nature have also been included here, beyond those of a typical introductory course. These are indicated by an asterisk (∗) after the section heading. They have been included as possible extensions from standard material for more specialized courses, research or possibly part of a project.

The material included here is based on notes for several courses in antennas given to fourth year students in Electrical Engineering at James Cook University of North Queensland and also at the University of Queensland in the 1980s. At that time there was no suitable modern textbook available on antennas for undergraduate teaching. Since then, several excellent textbooks have appeared (Balanis, 1982). In addition, the notes were found useful over the years by members of my research group at CSIRO. Other relevant material had been developed on mutual coupling for presentation at several symposia held in the 1990s, and some of this information has been included here. As might be anticipated, practical topics of relevance that were encountered during my research career have been included as well.

The purpose of this book is to provide a stand-alone textbook for a course in antennas, possibly in the final undergraduate years or in a master's degree by coursework. It should also be useful for Ph.D. candidates and practising engineers. For continuity, some background electromagnetics, fields and waves are included.

The antennas described in detail include horns, reflectors, lenses, patch radiators and arrays of some of these antennas. Because of its importance and to provide more than a superficial treatment of arrays, the topic of mutual coupling is covered in greater detail than most similar books in the area. Also included is an introduction to sources and arrays on non-planar surfaces, which is important for applications involving aerodynamic surfaces and for making aperture antennas unobtrusive. An introduction to the fabrication and test of aperture antennas is included as well as some recent examples of them.

The theory needed for analysing aperture antennas is given in Chapter 3. Material is also included for handling conformal aperture antennas. Starting with the concept of equivalent sources, the equations for radiation from an aperture are developed from the fields radiated by a small electric dipole and a small loop of current. The basic theory that is needed for more detailed development is also provided. This includes details of the far-field radiation from uniformly illuminated rectangular and circular apertures and also how phase aberrations on the aperture impact the far-fields. The radiation from waveguide and horn aperture antennas are described in Chapter 4, and material is included for the radiation from rectangular waveguide antenna. This model is used as a basis for detailed description of the pyramidal horn. The radiation properties of circular waveguides and horns are reviewed in this chapter and details are provided on the corrugated horn. A simple model of the microstrip patch antenna is given in Chapter 5 along with details of the radiation properties of these antennas. The purpose is to describe another form of aperture antenna and as background for reflectarrays. The properties of reflector antennas in common use are described in Chapter 6, including the paraboloid the Cassegrain, and spheroid geometries as well as some offset counterparts. Planar arrays of aperture antennas and mutual coupling in arrays are detailed in Chapter 7. This is followed in Chapter 8 by similar details for apertures on conformal surfaces. The areas of arrays and reflectors come together in the reflectarray antenna, which is introduced in Chapter 9. This chapter also includes details of some other aperture antennas not treated elsewhere, in particular, lenses, and the Fabry-Pérot cavity antennas. Finally, some possible approaches for the fabrication and testing of aperture antennas are described in Chapter 10. In addition it includes examples of some aperture antennas that make use of many of the techniques covered earlier in the book. At all times, the intention is an emphasis on fundamentals and, where possible, practical information for design is also included.

References

Balanis, C.A. (1982): 'Antenna theory: analysis and design', Harper and Row, New York.

Bird, T.S. and Boomars, J.L. (1980): 'Evaluation of focal fields and radiation characteristics of a dual-offset reflector antenna', IEE Proc. (Pt. H): Microwav. Optics Antennas, Vol. **127**, pp. 209–218. Erratum: IEE Proc. (Pt. H), Vol. 128, 1981, p. 68.

Harrington, R.F. (1961): 'Time-harmonic electromagnetic fields', McGraw-Hill, New York.

Johnk, C.T.A. (1975): 'Engineering electromagnetic fields and waves', John Wiley & Sons, Inc., New York.

Jones, D.S. (1964): 'The theory of electromagnetism', Pergamon Press, London, UK.

Jordan, E.C. and Balmain, K.G. (1968): 'Electromagnetic waves and radiating systems', 2nd ed., Prentice-Hall, Eaglewood Cliffs, NJ.

Kraus, J.D. and Carver, K.R. (1973): 'Electromagnetics', 2nd ed., McGraw-Hill, International Student Edition, Kagakuska Ltd., Tokyo, Japan.

2

Background Theory

In this chapter, some background theory is provided and notation is introduced in preparation for use throughout the remainder of this text. The equations that were devised by Maxwell and placed in differential form by Heaviside and Hertz are introduced. Throughout this book, all field and sources are assumed to be time harmonic and the formulation of the field equations and their consequences will be explored under this limitation. The important concepts of field duality, equivalent sources and image theory are summarized. Finally, radiation from elementary sources is investigated, and this allows a description of some basic radiation parameters as well as an introduction to mutual coupling.

2.1 Maxwell's Equations for Time-Harmonic Fields

The instantaneous vector field quantity $\mathcal{A}(\mathbf{r},t)$ may be expressed in terms of a complex vector field, $\mathbf{A}(\mathbf{r})$, where all fields and sources have a time-harmonic dependence, as follows:

$$\mathcal{A}(\mathbf{r},t) = \mathrm{Re}\{\mathbf{A}(\mathbf{r})\exp(j\omega t)\}, \tag{2.1}$$

where bold type face indicates vector quantities, $\omega = 2\pi f$ is the angular frequency (rad/s), t denotes time (s) and f is the frequency (Hz) of the harmonic oscillation.

Field and source quantities are defined as follows (MKS units given in square brackets):

$\mathbf{E}(\mathbf{r}) = $ Electric field intensity $[\mathrm{V\ m^{-1}}]$
$\mathbf{H}(\mathbf{r}) = $ Magnetic field intensity $[\mathrm{A\ m^{-1}}]$
$\mathbf{J}(\mathbf{r}) = $ Electric current density $[\mathrm{A\ m^{-2}}]$

Fundamentals of Aperture Antennas and Arrays: From Theory to Design, Fabrication and Testing,
First Edition. Trevor S. Bird.
© 2016 John Wiley & Sons, Ltd. Published 2016 by John Wiley & Sons, Ltd.
Companion website: www.wiley.com/go/bird448

$\mathbf{M(r)}$ = Magnetic current density [V m^{-2}]
$\rho_e(r)$ = Electric charge density [C m^{-3}]
$\rho_m(r)$ = Magnetic charge density [Wb m^{-3}],

where V is the volt, A is the Ampere, m is metre, Wb is Weber and C is the Coulomb.

The equations governing the interaction of these fields and sources are known as Maxwell's equations, after James Clerk Maxwell (1831–1879), who first presented them in component form and in terms of potentials from the earlier results of Faraday, Öersted, Ampere, Weber and others. Oliver Heaviside (1850–1925), and independently Heinrich Hertz (1857–1894), reduced these 20 equations to the four vector field equations that are essentially used today (Sarkar et al., 2006). For Heaviside, the concepts of fields, symmetry and vector notation were vital. With the present assumption that fields and sources vary harmonically with time, Maxwell's equations are expressed as follows:

$$\nabla \times \mathbf{E} = \mathbf{M} - j\omega\mu\mathbf{H} \tag{2.2a}$$

$$\nabla \times \mathbf{H} = \mathbf{J} + j\omega\varepsilon\mathbf{E} \tag{2.2b}$$

$$\nabla \cdot \mathbf{E} = \frac{\rho_e}{\varepsilon} \tag{2.2c}$$

$$\nabla \cdot \mathbf{H} = \frac{\rho_m}{\mu}, \tag{2.2d}$$

where ∇ is the gradient operator, $\nabla \times$ is the curl operation, $\nabla \cdot$ denotes divergence, ε is the electric permittivity [F m^{-1}] and μ is the magnetic permeability [H m^{-1}].

A general field may be considered as the superposition of the fields due to two types of sources, respectively, electric (e) and magnetic (m) as follows:

$$\mathbf{E} = \mathbf{E}_e + \mathbf{E}_m \quad \text{and} \quad \mathbf{H} = \mathbf{H}_e + \mathbf{H}_m.$$

The partial field pairs, $(\mathbf{E}_e, \mathbf{H}_e)$ and $(\mathbf{E}_m, \mathbf{H}_m)$ satisfy separate sets of Maxwell's equations as shown in Table 2.1 and originate from electric or magnetic sources. The former is due to physical electric currents and charges, while the latter is due to magnetic currents and charges, which are of an equivalent type and were introduced to maintain the symmetry of Maxwell's equations (Harrington, 1961; Jones, 1964). More will be said about equivalent sources in the following sections.

Table 2.1   Maxwell's equations for electric and magnetic sources

Electric: $\mathbf{J} \neq 0$ $\mathbf{M} = 0$	Magnetic: $\mathbf{J} = 0$ $\mathbf{M} \neq 0$
$\nabla \times \mathbf{E}_e = -j\omega\mu\mathbf{H}_e$	$\nabla \times \mathbf{H}_m = j\omega\varepsilon\mathbf{E}_m$
$\nabla \times \mathbf{H}_E = \mathbf{J} + j\omega\varepsilon\mathbf{E}_e$	$\nabla \times \mathbf{E}_m = \mathbf{M} - j\omega\mu\mathbf{H}_m$
$\nabla \cdot \mathbf{E}_e = \rho_e/\varepsilon$	$\nabla \cdot \mathbf{H}_m = \rho_m/\mu$
$\nabla \cdot \mathbf{H}_e = 0$	$\nabla \cdot \mathbf{E}_m = 0$

2.1.1   Field Representation in Terms of Axial Field Components in Source-Free Regions

In problems involving sections of uniform structures that guide electromagnetic waves such as waveguides and transmission lines when **J** and **M** are absent, it is convenient to represent all field components in terms of the field components in the direction of propagation, that is, in the direction of uniformity. By convention this direction is usually taken as the z-component in a cylindrical co-ordinate system with directions denoted by (u,v,z). It is recognized that the field components $E_z(u,v,z)$ and $H_z(u,v,z)$ satisfy Helmholtz wave equations, where (u,v) are transverse co-ordinates. The guiding structures are assumed to exhibit reflection symmetry and, therefore, it is sufficient to represent the total field as the superposition of forward and reverse travelling wave solutions in the z-direction. For time harmonic fields of the type defined by Eq. 2.1, a forward travelling wave (in the +z-direction) has the following fields:

$$\mathbf{E}(u,v,z) = (\mathbf{E}_t(u,v) + \hat{z}E_z(u,v))\exp(-j\gamma z) \tag{2.3a}$$

$$\mathbf{H}(u,v,z) = (\mathbf{H}_t(u,v) + \hat{z}H_z(u,v))\exp(-j\gamma z), \tag{2.3b}$$

where $\mathbf{E}_t(u, v)$ and $\mathbf{H}_t(u, v)$ are the transverse electric and magnetic field vectors and $\gamma = \beta - j\alpha$ is the complex propagation constant. β is the phase shift per unit length and α is the attenuation constant. For lossless structures $\alpha = 0$. With the field represented by Eqs. 2.3, the transverse field components can be obtained from Maxwell's equations in the following form:

$$k_z^2\mathbf{E}_t(u,v) = j\omega\mu\hat{z} \times \nabla_t H_z - j\gamma\nabla_t E_z \tag{2.4a}$$

$$k_z^2\mathbf{H}_t(u,v) = -j\omega\varepsilon\hat{z} \times \nabla_t E_z - j\gamma\nabla_t H_z, \tag{2.4b}$$

where $k_z^2 = \omega^2\varepsilon\mu - \gamma^2$ is the axial wave number, and ∇_t is the transverse gradient operator. For homogeneous materials, the permittivity and permeability are $\varepsilon = \varepsilon_r\varepsilon_0$ and $\mu = \mu_r\mu_0$, respectively, where ε_r is the relative permittivity, μ_r is the relative permeability, $\varepsilon_0 = 8.854 \times 10^{-12}$ F/m and $\mu_0 = 4\pi \times 10^{-7}$ H/m are the permittivity and permeability of free-space. The first term on the right-hand side of k_z^2, namely, $k = \omega\sqrt{\varepsilon\mu}$, is the propagation constant of a plane wave in the homogeneous medium. In free-space $\varepsilon_r = 1$ and $k = 2\pi/\lambda = k_0 = \omega\sqrt{\mu_0\varepsilon_0} = \omega/c$ where $c = 1/\sqrt{\varepsilon_0\mu_0}$ is the free-space wave velocity and equals $c = 2.99859 \times 10^8$ ms^{-1}. Ratios of components of **E** and **H** in Eqs. 2.4 have dimensions of impedance and are referred to as the wave impedance. In a general medium, the intrinsic impedance is $\eta = \sqrt{\mu/\varepsilon}$ Ω. By substituting Eqs. 2.4 into Eqs. 2.2b and 2.2d, it can be shown that the axial field components satisfy the following scalar wave equations:

$$\left(\nabla_t^2 + k_z^2\right)\begin{cases} E_z(u,v) \\ H_z(u,v) \end{cases} = 0. \tag{2.5}$$

It is seen from Eq. 2.5 that k_z^2 is also the transverse wavenumber. If the co-ordinates u and v are separable there will be separation constants in these directions as well. For fields that are TE to the propagation direction, $E_z = 0$, and the simultaneous pair of wave equations simplify to a solution of the wave equation in H_z only and Eqs. 2.4a and 2.4b become as follows:

$$k_z^2 \mathbf{E}_t(u,v) = j\omega\mu\hat{z} \times \nabla_t H_z \tag{2.6a}$$

$$k_z^2 \mathbf{H}_t(u,v) = -j\gamma\nabla_t H_z, \tag{2.6b}$$

where now k_z^2 is constant for a fixed geometry and is the cut-off wavenumber of the guiding structure. For example, for the TE modes of circular cylindrical waveguide of radius a, k_z is a root of the derivative of the Bessel function (see Appendix B), namely, $J_n'(k_z a) = 0$, where n is the azimuthal period that arises in the solution of the wave equation in the azimuthal co-ordinate Φ in the transverse plane. Similarly, a field that is transverse magnetic (TM) to the propagation direction is obtained by setting $H_z = 0$ and an equivalent simplification occurs in Eqs. 2.4, namely,

$$k_z^2 \mathbf{E}_t(u,v) = -j\gamma\nabla_t E_z \tag{2.7a}$$

$$k_z^2 \mathbf{H}_t(u,v) = -j\omega\varepsilon\hat{z} \times \nabla_t E_z. \tag{2.7b}$$

2.1.2   Boundary Conditions

Consider a volume that is divided into two regions 1 and 2 by a surface S as shown in Figure 2.1. There are currents on this surface, namely, an electric surface current $\mathbf{J}_s$ [A m^{-1}] and a magnetic surface current $\mathbf{M}_s$ [V m^{-1}]. On either side of a surface of discontinuity the field pairs $(\mathbf{E}_1, \mathbf{H}_1)$ and $(\mathbf{E}_2, \mathbf{H}_2)$ satisfy the following continuity conditions:

$$\mathbf{J}_s = \hat{n} \times (\mathbf{H}_2 - \mathbf{H}_1) \tag{2.8a}$$

$$\mathbf{M}_s = -\hat{n} \times (\mathbf{E}_2 - \mathbf{E}_1). \tag{2.8b}$$

Thus the tangential components are discontinuous by an amount equal to the current at the surface. The associated boundary conditions for the normal components to the surface are as follows:

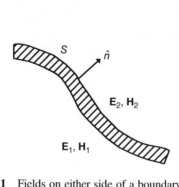

Figure 2.1   Fields on either side of a boundary surface S

$$\rho_{es} = \hat{n} \cdot (\varepsilon_2 \mathbf{E}_2 - \varepsilon_1 \mathbf{E}_1) \tag{2.9a}$$

$$\rho_{ms} = \hat{n} \cdot \left(\frac{\mathbf{H}_2}{\mu_2} - \frac{\mathbf{H}_1}{\mu_1} \right), \tag{2.9b}$$

where ρ_{es} is the electric surface charge in [C m^{-2}] and ρ_{ms} is the magnetic surface charge [Wb m^{-2}]. In many physical problems $\rho_{es} = 0$ and so $\hat{n} \cdot (\varepsilon_2 \mathbf{E}_2 - \varepsilon_1 \mathbf{E}_1) = 0$. No physical magnetic surface charges have been detected and, therefore, Eq. 2.9b is usually expressed $\hat{n} \cdot (\mathbf{H}_2/\mu_2 - \mathbf{H}_1/\mu_1) = 0$.

2.1.3 Poynting's Theorem

The time-averaged conservation of energy in the electromagnetic field is given by Poynting's theorem. In an isotropic medium of volume V with permeability μ, permittivity ε and conductivity σ, this is

$$\frac{1}{2} \oiint_\Sigma \mathbf{E} \times \mathbf{H}^* \cdot \hat{n} dS = \frac{1}{2} \iiint_V j\omega((\mu \mathbf{H} \cdot \mathbf{H}^* - \varepsilon \mathbf{E} \cdot \mathbf{E}^*) + \sigma \mathbf{E} \cdot \mathbf{E}^*) dV, \tag{2.10}$$

where Σ is the surface bounding V. On left-hand side, $\hat{n}$ is the normal to the surface and is directed into Σ. The quantity $\bar{\mathbf{P}} = 1/2\ \mathbf{E} \times \mathbf{H}^*$ is the power density entering V that is called the complex Poynting vector. The integral of this vector over the closed surface Σ is the power input, P_I. On the right-hand side, the three terms are related to, from left to right, the energy stored in the magnetic field, W_m, the energy stored in the electric field, W_e, and the power lost due to conduction loss, P_L. Expressed succinctly, Eq. 2.10 is $P_I = 2j\omega(W_m - W_e) + P_L$, where

$$W_m = \frac{1}{4} \mathrm{Re}\left\{ \iiint_V \mu \mathbf{H} \cdot \mathbf{H}^* dV \right\},$$

$$W_e = \frac{1}{4} \mathrm{Re}\left\{ \iiint_V \varepsilon \mathbf{E} \cdot \mathbf{E}^* dV \right\} \quad \text{and}$$

$$P_L = \frac{1}{2} \sigma \iiint_V \mu \mathbf{E} \cdot \mathbf{E}^* dV.$$

In an ideal lossless medium, $\sigma = 0$ and, as a consequence, $P_L = 0$. Therefore, $P_I = 2j\omega (W_m - W_e)$. This says that the input power converts totally to energy in the fields, which is totally reactive, and is the difference of the energies stored in the magnetic and electric fields.

2.1.4 Reciprocity

Of importance in all types of antenna systems is the relationship between the receiving and transmitting fields. In more general terms, the response in the vicinity of one source due to fields from a second source and the relationships when the roles are reversed are of particular significance. Suppose there are two sources in a region denoted by a and b. Thus the source pairs are $\left(\mathbf{J}_s^a, \mathbf{M}_s^a\right)$ and $\left(\mathbf{J}_s^b, \mathbf{M}_s^b\right)$. These produce fields $(\mathbf{E}^a, \mathbf{H}^a)$ and $(\mathbf{E}^b, \mathbf{H}^b)$, respectively, that satisfy their own sets of Maxwell's equations as shown in Table 2.1. Making use of the vector identity

$$\nabla \cdot \left(\mathbf{E}^a \times \mathbf{H}^b\right) = \mathbf{H}^b \cdot \nabla \times \mathbf{E}^a - \mathbf{E}^a \cdot \nabla \times \mathbf{H}^b$$

and the Maxwell curl relations, it follows that

$$\nabla \cdot \left(\mathbf{E}^a \times \mathbf{H}^b\right) = \mathbf{H}^b \cdot \mathbf{M}_s^a - j\omega\mu \mathbf{H}^a \cdot \mathbf{H}^b - \mathbf{E}^a \cdot \mathbf{J}_s^b - j\omega\varepsilon \mathbf{E}^a \cdot \mathbf{E}^b. \tag{2.11a}$$

Similarly,

$$\nabla \cdot \left(\mathbf{E}^b \times \mathbf{H}^a\right) = \mathbf{H}^a \cdot \mathbf{M}_s^b - j\omega\mu \mathbf{H}^a \cdot \mathbf{H}^b - \mathbf{E}^b \cdot \mathbf{J}_s^a - j\omega\varepsilon \mathbf{E}^a \cdot \mathbf{E}^b. \tag{2.11b}$$

Subtraction of Eq. 2.11b from Eq. 2.11a results in

$$\nabla \cdot \left(\mathbf{E}^a \times \mathbf{H}^b - \mathbf{E}^b \times \mathbf{H}^a\right) = -\mathbf{E}^a \cdot \mathbf{J}_s^b + \mathbf{E}^b \cdot \mathbf{J}_s^a + \mathbf{H}^a \cdot \mathbf{M}_s^b - \mathbf{H}^b \cdot \mathbf{M}_s^a.$$

At any point within the region where the sources are not present, the right-hand side is zero:

$$\nabla \cdot \left(\mathbf{E}^a \times \mathbf{H}^b - \mathbf{E}^b \times \mathbf{H}^a\right) = 0 \tag{2.12}$$

This result is called the Lorentz reciprocity theorem. When Eq. 2.12 is integrated throughout the source-free region Σ, the divergence theorem allows it to be expressed as follows:

$$\oiint_{\Sigma} \left(\mathbf{E}^a \times \mathbf{H}^b - \mathbf{E}^b \times \mathbf{H}^a\right) \cdot \hat{n}\, dS = 0, \tag{2.13}$$

where the integral sign refers to a closed surface with volume V. When sources are contained within the surface the result is

$$\oiint_{\Sigma} \left(\mathbf{E}^a \times \mathbf{H}^b - \mathbf{E}^b \times \mathbf{H}^a\right) \cdot \hat{n}\, dS = \iiint_{V} \left(-\mathbf{E}^a \cdot \mathbf{J}_s^b + \mathbf{E}^b \cdot \mathbf{J}_s^a + \mathbf{H}^a \cdot \mathbf{M}_s^b - \mathbf{H}^b \cdot \mathbf{M}_s^a\right) dV. \tag{2.14}$$

If the surface is a sphere with a very large radius, as the fields decay as $1/r$, the integral on the left-side limits to zero. As a result, the right-side of Eq. 2.14 gives

$$\iiint_{V} \left(\mathbf{E}^a \cdot \mathbf{J}_s^b - \mathbf{H}^a \cdot \mathbf{M}_s^b\right) dV = \iiint_{V} \left(\mathbf{E}^b \cdot \mathbf{J}_s^a - \mathbf{H}^b \cdot \mathbf{M}_s^a\right) dV, \tag{2.15}$$

where V is now all space. The two integrals on the left and right side of Eq. 2.15 are termed reaction integrals. Eq. 2.15 is sometimes expressed as the reaction of field a on source b is the same as the reaction of field b on source a. When a and b are the same the integral is called self-reaction. Although not immediately obvious, Eq. 2.13 is also applicable when the volume, V, contains all sources.

Table 2.2 Field duality

Electric dipole		Magnetic dipole
E	$\leftrightarrow$	H
H	$\leftrightarrow$	$-E$
k	$\leftrightarrow$	k
ε	$\leftrightarrow$	μ
μ	$\leftrightarrow$	ε
η	$\leftrightarrow$	$1/\eta$
J	$\leftrightarrow$	M

2.1.5   Duality

The symmetry of Maxwell's equations as summarized in Table 2.1 indicates that mathematically there may be equivalence between the extension of Faraday's law and Maxwell's extension of Ampere's equation and similarly between the divergence equations arising from Gauss's laws. This is, in fact, the case if there were physically a magnetic current and a magnetic charge density, which like the electric charge and divergence of the electric current, are related through a magnetic current continuity equation. This correspondence between Maxwell's equations and the field sources is referred to as duality and is summarized in Table 2.2. There will be occasions when a magnetic current is adopted, although physically it is fictitious, as it can simplify some of working and produce fields as if such a source or to construct field solutions as if these sources were actually present.

2.1.6   Method of Images

Adjacent to plane electric and magnetic conductors, the boundary conditions (Eqs. 2.8 and 2.9) on the electric and magnetic fields imply the presence of an 'image' field on the other side of the conductor. A summary of image theory is illustrated in Figure 2.2.

An electric field that is perpendicular to a perfect conductor has an image, which is parallel to the original field. On the other hand, an electric field that is parallel to the conductor has an image that is oppositely directed. A magnetic field that is perpendicular to a perfect conductor has an image that is anti-parallel to the original field, while a parallel field creates a parallel image. For electric and magnetic fields above a perfect magnetic conductor the roles reverse as shown in Figure 2.2b.

2.1.7   Geometric Optics

The basis of geometric optics is that the wavefronts of incident waves are equiphase level surfaces represented by the function $L(x,y,z)$. In an inhomogeneous medium with a refractive index $n(x,y,z)$ these surfaces satisfy the eikonal equation (Born & Wolf, 1965), which is expressed by

$$|\nabla L| = n = \frac{c}{v_p},\tag{2.16}$$

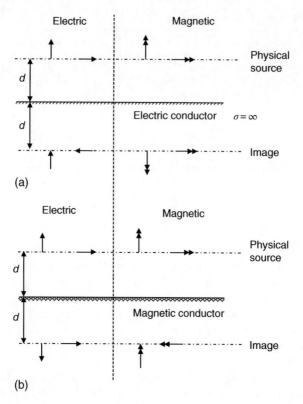

Figure 2.2 Field components and their images. Above a perfect (a) electric conductor and (b) magnetic conductor

where v_p is the phase velocity. The eikonal equation can be used to determine the ray paths for a given refractive index as will now be shown.

Suppose $\hat{s}$ is a unit vector tangent to the ray path and is, therefore, normal to the wavefront. Consequently,

$$\hat{s} = \frac{\nabla L}{|\nabla L|} = \frac{\nabla L}{n}. \tag{2.17}$$

From differential geometry, the curvature of this unit vector is (by Frenet's formula),

$$\frac{d\hat{s}}{ds} = -\hat{s} \times (\nabla \times \hat{s}) = \frac{\hat{n}_\perp}{\rho}, \tag{2.18}$$

where $\hat{n}_\perp$ is the principal unit-normal vector and ρ is the radius of curvature. Therefore, by means of Eq. 2.17

$$\frac{d\hat{s}}{ds} = -\hat{s} \times \left(\nabla \times \frac{\nabla L}{n} \right) = -\hat{s} \times \left[\nabla \left(\frac{1}{n} \right) \times \nabla L \right] = \hat{s} \times [\nabla (\ln n) \times \hat{s}].$$

Thus,

$$\frac{\hat{n}_\perp}{\rho} \cdot \hat{n}_\perp = \frac{d\hat{s}}{ds} \cdot \hat{n}_\perp = \hat{n}_\perp \cdot \nabla(\ln n) = \hat{n}_\perp \cdot \frac{\nabla n}{n}.$$

That is,

$$\frac{1}{\rho} = \hat{n}_\perp \cdot \frac{\nabla n}{n}. \tag{2.19}$$

In a homogenous medium, $n(x, y, z)$ is a constant and, therefore, the ray curvature of the ray path is zero and the ray paths are straight lines. However, the ray paths in inhomogeneous media are generally curved.

2.2 Equivalent Sources

Suppose the fields $\mathbf{E}_1, \mathbf{H}_1$ are produced by electric and magnetic current sources $\mathbf{J}_1$ and $\mathbf{M}_1$, respectively. Now surround these sources by a surface S to form a volume V as shown in Figure 2.3a. Outside S, in the volume V_1, the fields are unchanged. Now replace the original fields and sources in V with fields $\mathbf{E}_2, \mathbf{H}_2$ and also introduce surface currents $\mathbf{J}_s$ and $\mathbf{M}_s$ in V_1 as shown in Figure 2.3b. For continuity, surface currents must exist on S; otherwise, the boundary conditions would require a null field everywhere. These surface currents are given by

$$\mathbf{J}_s = \hat{n} \times (\mathbf{H}_1 - \mathbf{H}_2) \tag{2.20a}$$

$$\mathbf{M}_s = -\hat{n} \times (\mathbf{E}_1 - \mathbf{E}_2), \tag{2.20b}$$

where $\hat{n}$ is the outward normal to the surface S. The replacement of a set of fields and sources by another equal set of fields and sources is known as the field equivalence principle. Sources produced by this technique are called equivalent sources.

Several special cases of the equivalent problem can be devised. These are illustrated in Figure 2.4.

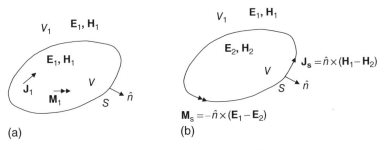

(a)   (b)

Figure 2.3 Equivalent sources. (a) Original fields and sources. (b) Equivalent sources to maintain the same external fields

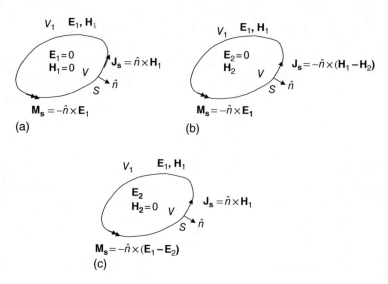

Figure 2.4 Special cases of equivalent sources and fields. (a) Null internal field to S. (b) Zero internal electric field. (c) Zero internal magnetic field

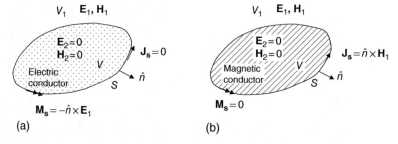

Figure 2.5 Null field internal to S with introduced media. (a) Electric conductor ($\sigma = \infty$) internal to S. (b) Internal magnetic conductor

When there is a null field inside S (as shown in Figure 2.4a), the contents of the medium in V can be changed without altering the field inside. There are two particular cases of interest, and these are illustrated in Figure 2.5. The first case shown in Figure 2.5a is useful as it can apply to many aperture antennas. A perfect electric conductor ($\sigma = 0$) is introduced into V without affect due to the null field. However, at the surface, the currents are affected because the conductor shorts out the electric surface current leaving only the magnetic surface current. The problem of finding the fields $\mathbf{E}_1, \mathbf{H}_1$ is modified now to determine $\mathbf{M_s}$ in the presence of a perfect electric conductor, the solution to which may be just as elusive as the original problem (Figure 2.3a).

The dual problem to the one in Figure 2.5a is illustrated in Figure 2.5b where a magnetic conductor is introduced into V. This shorts out the magnetic surface current leaving only an electric current acting in the presence of the magnetic conductor.

2.2.1   Aperture in a Ground Plane

Suppose that the surface chosen for the equivalent sources is an infinite plane (Figure 2.6a). As in Figure 2.5a, let a perfect electric conductor be introduced into the space V in which there is a null field. There is now a plane sheet of magnetic current acting near a perfect conductor. Image theory tells us that a magnetic source induces an identical image source, as shown in Figure 2.6b, in the conductor. The field produced in V_1 is the one due to $\mathbf{M}_s$ and its image source, which is also $\mathbf{M}_s$. That is, the field produced is due to an equivalent source of twice the strength of the original source as illustrated in Figure 2.6c.

2.2.2   Conformal Surfaces

It is common for antennas and sources to be located on or near non-planar surfaces. A special case is when the source is on the surface and conformal to it. The simplifications found for planar surfaces do not arise for curved surfaces, either concave or convex ones. Many of the other principles described earlier, such as equivalent sources, are still valid although the geometry for conformal surfaces is more complex. To demonstrate this, consider two examples

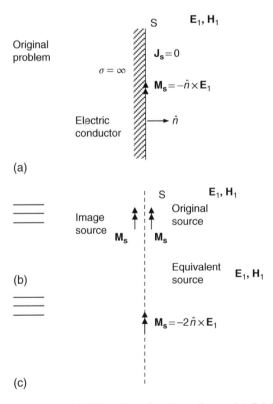

Figure 2.6   Magnetic source near an infinite plane electric conductor. (a) Original problem. (b) Equivalent problem. (c) Image replaced with equivalent source

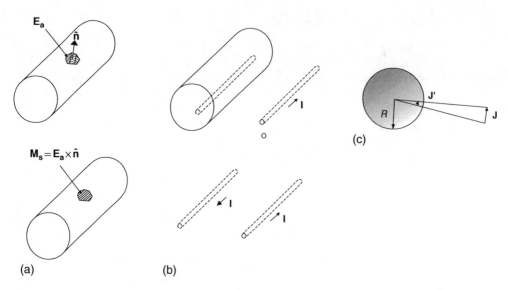

Figure 2.7 Equivalent sources on convex surfaces. (a) Magnetic current source. (b) Electric line source. (c) Image source on sphere

of sources on convex surfaces as shown in Figure 2.7. The first is an aperture in a cylinder in Figure 2.7a. Modes are excited in the aperture, and the radiated field is equivalent to the radiation from a magnetic current source on the cylinder. The second is a line source that is parallel to a conducting cylinder in Figure 2.7b. A cylinder does not create images from point sources as occurring on a plane, but it produces images for line sources. When the line source is parallel to a cylinder as in Figure 2.7b, an image line is produced inside the cylinder. It does this in such a way that the cylinder surface is an equiphase surface for the image. Finally, a sphere produces images from point sources as illustrated in Figure 2.7c. These are special cases but often problems with a complicated geometry can be replaced by means of the method of images to a simpler problem, which may be more amenable to detailed analysis.

2.3   Radiation

Consider a very short wire of length dz that is excited by a time-harmonic electric current as shown in Figure 2.8 in a homogeneous medium. It is convenient to express the electric and magnetic fields due to this current element in terms of its magnetic vector potential, $\mathbf{A}$, as follows:

$$\mathbf{E} = -j\omega\mathbf{A} + \frac{1}{j\omega\mu\varepsilon}\nabla(\nabla\cdot\mathbf{A}) \qquad (2.21a)$$

$$\mathbf{H} = \frac{1}{\mu}\nabla\times\mathbf{A}, \qquad (2.21b)$$

where

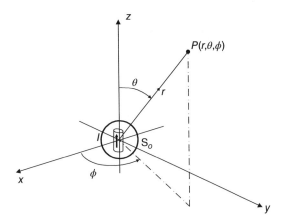

Figure 2.8   Radiation from an electric current element

$$A = \hat{z}\frac{\mu I dz}{4\pi r}e^{-jkr}.$$

The resulting non-zero electric field components are given by

$$E_r = \frac{\mu I dz}{2\pi}e^{-jkr}\cos\theta\frac{1}{r^2}\left[1 + \frac{1}{jkr}\right] \tag{2.22a}$$

$$E_\theta = \frac{\mu I dz}{4\pi}e^{-jkr}\sin\theta\frac{1}{r^2}\left[jkr + 1 + \frac{1}{jkr}\right]. \tag{2.22b}$$

The related magnetic field can be obtained from Eq. 2.21b from which the only non-zero component is

$$H_\phi = \frac{I dz}{4\pi}e^{-jkr}\sin\theta\frac{1}{r^2}[jkr + 1]. \tag{2.22c}$$

Observations on Eqs. 2.22 are as follows:

a. The instantaneous fields are found from Eq. 2.1. Contour plots of the instantaneous fields given by Eqs. 2.22 with $\eta I d\ell/4\pi = 1$ have been made in the vicinity of a current element at instants of times $t = 0$, $T/8$, $T/4$, $3\ T/8$, where $T = 2\pi/\omega$ is the period of the source and these are shown in Figure 2.9. Because of symmetry, only one quadrant is shown in Figures 2.9 for $0 < \theta < \pi/2$, with $0 < kz < 15$.
b. It can be seen that $E_\theta = 0$ and $H_\phi = 0$ in the plane of the element, while E_θ and H_ϕ are maximum in the plane perpendicular to the element.
c. The radial field component vanishes, that is, $E_r = 0$, in the plane perpendicular to the element and it is maximum in the plane of the element.
d. All non-zero field components of Eqs. 2.22 contain terms involving powers of $1/r$.

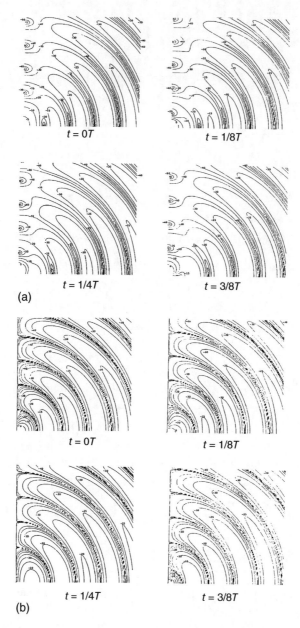

Figure 2.9 (a) Instantaneous electric field intensity in the vicinity of a short dipole at time instants $0\,T$, $1/8\,T$, $1/4\,T$ and $3/8\,T$, where T is the period of the oscillation. 3 dB contours are plotted in a single quadrant to a distance from the dipole $kr = 15$. (b) Instantaneous magnetic field intensity in the vicinity of a short dipole (vertical) under same conditions as (a).

2.3.1   Near-Field

The near-field region is defined by $kr \ll 1$, which implies that $r \ll \lambda$. Neglecting all terms but the highest power of r in Eq. 2.22 results in

$$E_r = -\frac{j\eta Idz}{2\pi kr^3}\cos\theta \qquad (2.23a)$$

$$E_\theta = -\frac{j\eta Idz}{4\pi kr^3}\sin\theta. \qquad (2.23b)$$

$$H_\phi = \frac{Idz}{4\pi kr^2}\sin\theta. \qquad (2.23c)$$

where $\mu = \eta/k$.
It is noted that:

a. H_ϕ is identical to the field produced by a short wire carrying a constant current. Also, as the electric field contains terms proportional to $1/r^3$, the near-field is predominantly electric in nature, and is the gradient of a scalar quantity. Thus,

$$\mathbf{E} = -\nabla\left(-\frac{jIdz}{4\pi\varepsilon\omega r^2}\cos\theta\right) \qquad (2.24)$$

$$= -\nabla\Phi,$$

where

$$\Phi = -\frac{jIdz}{4\pi\varepsilon\omega r^2}\cos\theta.$$

This scalar is the potential due to equal and opposite charges a distance dz apart, that is, a dipole, which is oscillating at a frequency ω.
b. The electric and magnetic fields are out of phase as they are with a standing wave. As a result the average power flow/unit area is zero. However, since the complex Poynting vector is non-zero, the near-field stores energy and is reactive.

2.3.2   Far-Field

At distances very far from the current element, in the far-field region, $kr \gg 1$. That is, $r \gg \lambda$. In this case, Eqs. 2.22 reduce to

$$E_r = 0 \qquad (2.25a)$$

$$E_\theta = \frac{j\eta kIdz}{4\pi r}e^{-jkr}\sin\theta \qquad (2.25b)$$

$$H_\phi = \frac{jkIdz}{4\pi r} e^{-jkr} \sin\theta. \qquad (2.25c)$$

Regarding Eqs. 2.25, it is observed that:

a. As E_r is negligibly small the far-field is predominantly a spherical wave. The remaining field components, E_θ and H_ϕ, are tangential to the surface of this radiation sphere of radius r and, hence, both are perpendicular to the direction of propagation.
b. The ratio of the two non-zero field components is

$$\frac{E_\theta}{H_\phi} = \eta = \sqrt{\frac{\mu}{\varepsilon}}. \qquad (2.26)$$

For a current element radiating in free-space, the wave impedance is $\eta = \eta_o = 376.73 \approx 120\pi$ ohms, is called the free-space wave impedance. In the light of comment (a), Eq. 2.26 is generalized to

$$\mathbf{H} = \frac{1}{\eta}\hat{r} \times \mathbf{E}. \qquad (2.27)$$

c. As the non-zero field components are in phase, the far-field has a non-zero power density. The power density for time harmonic far-fields is given by

$$\bar{\mathbf{P}} = \frac{1}{2} \operatorname{Re}\{\mathbf{E} \times \mathbf{H}^*\}$$

$$= \frac{1}{2} \operatorname{Re}\left\{\mathbf{E} \times \frac{1}{\eta}(\hat{r} \times \mathbf{E}^*)\right\} \qquad (2.28)$$

$$= \hat{r} \frac{\mathbf{E} \cdot \mathbf{E}^*}{2\eta} .$$

Eq. 2.28 is a general result for the far-field radiation, and it shows that the power density is in the radial direction, which is normal to the surface of the propagating spherical wave. In the present case,

$$\bar{\mathbf{P}} = \hat{r} \frac{|E_\theta|^2}{2\eta}$$

$$= \hat{r} \frac{\eta}{32} \left(\frac{k|I|dz}{\pi r} \sin\theta\right)^2 .$$

The units of power density are in watts/m^2.

The fields due to a magnetic current element of length dz can be obtained in the same way. However, instead of following a similar development, use is made of duality given in Table 2.2.

As a result, the fields due to a magnetic current element, Mdz, are obtained directly from the electric current element results. Therefore, by means of Table 2.2 and Eqs. 2.25, the far-fields due to a magnetic dipole are expressed as follows:

$$H_r = 0 \tag{2.29a}$$

$$H_\theta = \frac{jYkMdz}{4\pi r} e^{-jkr} \sin\theta \tag{2.29b}$$

$$E_\phi = -\frac{jkMdz}{4\pi r} e^{-jkr} \sin\theta \tag{2.29c}$$

where $Y = 1/\eta$. Similar, to Eq. 2.26, for this dual problem, Eq. 2.26 gives $E_\phi/H_\theta = -\eta$.

2.3.3   Mutual Coupling Between Infinitesimal Current Elements

Two or more current elements interact with each other depending on their orientation. This interaction is referred to as mutual coupling. To provide an initial insight into mutual coupling and its effects, consider two infinitesimally short electric dipoles of length dl_1 and dl_2 that are located in free-space and are supporting time harmonic currents of amplitude I_1 and I_2 with angular frequency ω. Dipole 2 is rotated at an angle α in the same plane (z–y plane) relative to dipole 1, as shown in Figure 2.10 The theory of a short electric dipole given by Eqs. 2.22 allows the elemental electric and magnetic fields of dipole 1 to be expressed as follows:

$$\mathbf{dE_1} = \frac{\eta_o I_1 dl_1}{4\pi} e^{-jkr_1} \left[\hat{\boldsymbol{\theta}}_1 \sin\theta_1 \left(\frac{jk}{r_1} + \frac{1}{r_1^2} \right) + \hat{\mathbf{r}}_1 \cos\theta_1 \frac{2}{r_1^2} \right] \tag{2.30a}$$

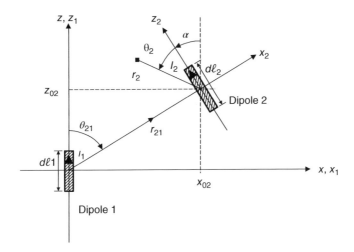

Figure 2.10   Geometry for coupling between two short electric dipoles

$$\mathbf{dH_1} = \frac{1}{\eta_o} \hat{\mathbf{r}}_1 \times \mathbf{dE_1}, \tag{2.30b}$$

where (r_1, θ_1, ϕ_1) are the spherical polar co-ordinates at the centre of dipole 1. Similar expressions apply to dipole 2 where the subscript 2 replaces the subscript 1 in the above equations. Now let the electric field produced on dipole 2 due to dipole 1 be given by $\mathbf{dE_{21}}$. In turn this field induces an electromotive force (emf) across the element in the following form (Schelkunoff & Friis, 1952):

$$dV_{21} = -\mathbf{dE_{21}} \cdot \hat{\mathbf{z}}_2 d\ell_2. \tag{2.31}$$

Since I_1 is the current producing this emf, the mutual impedance of element 2 due to element 1 is defined as follows:

$$Z_{21} = \frac{dV_{21}}{I_1} = -\frac{\mathbf{dE_{21}} \cdot \hat{\mathbf{z}}_2 d\ell_2}{I_1}. \tag{2.32}$$

Similarly, the current I_2 induces an emf in dipole 1 allowing the mutual impedance at dipole 1 due to dipole 2 to be given by

$$Z_{12} = \frac{dV_{12}}{I_2} = -\frac{\mathbf{dE_{12}} \cdot \hat{\mathbf{z}}_1 d\ell_1}{I_2}. \tag{2.33}$$

A relationship between these two mutual impedances is found by applying Lorentz's reciprocity theorem Eq. 2.12 to the two sets of fields and sources. This theorem results in

$$I_1 dV_{12} = I_2 dV_{21}, \tag{2.34}$$

which is the reciprocity theorem for elementary dipoles. Furthermore, Eq. 2.34 along with Eqs. 2.28 and 2.29 requires that

$$Z_{12} = Z_{21}. \tag{2.35}$$

Extending this result, when there a number of elements, reciprocity requires the mutual impedance matrix for these elements to be symmetric.

A formal expression for the mutual impedance can be obtained from Eqs. 2.30a and 2.33. Using some vector identities, this mutual impedance of the dipoles is given by

$$Z_{21} = \frac{\eta_o d\ell_1 d\ell_2}{4\pi} e^{-jkr_{21}} \left[\frac{jk}{r_{21}} \sin\theta_{21} \sin(\theta_{21} - \alpha) + \frac{1}{r_{21}^2} \left(\cos\alpha \left(1 - 3\cos^2\theta_{21}\right) \right. \right.$$
$$\left. \left. + \sin\alpha \sin\theta_{21} \cos\theta_{21} \left(2\cos\theta_{21} - 1\right) \right) \right]. \tag{2.36}$$

As shown in Figure 2.10, θ_{21} is the angle subtended at dipole 1 by dipole 2 and r_{21} is the distance between the dipole's centres. In the special case of parallel dipoles ($\alpha = 0$), that is in a broadside arrangement, $\theta_{21} = 90°$ (i.e. in the H-plane), Eq. 2.36 simplifies to

$$Z_{21} = \frac{\eta_0 d\ell_1 d\ell_2}{4\pi} e^{-jkr_{21}} \left(\frac{jk}{r_{21}} + \frac{1}{r_{21}^2} \right). \tag{2.37}$$

When the dipoles are in an echelon arrangement, that is end-to-end ($\theta_{21} = 0°$ and in E-plane), the mutual impedance is

$$Z_{21} = -\frac{\eta_0 d\ell_1 d\ell_2}{2\pi r_{21}^2} e^{-jkr_{21}}. \tag{2.38}$$

Therefore, for a broadside configuration (H-plane coupling) of dipoles, the mutual imped-ance is inversely proportional to the distance between the dipoles, while in the echelon config-uration (E-plane coupling), the distance dependence is inverse square.

In the same way, the coupling of short magnetic dipoles can be studied. However, the fields due to electric and magnetic dipoles are duals of each other and, therefore, the corresponding results for the magnetic dipole may be obtained by inspection from the above results. For mag-netic dipoles the dipole moments are, respectively, $\mathbf{dm_1} = \hat{z}_1 V_1 d\ell_1$ and $\mathbf{dm_2} = \hat{z}_2 V_2 d\ell_2$, where V_1 and V_2 are the applied voltages. Now from the duality summarized in Table 2.1, the mutual admittance of elemental magnetic dipoles is

$$Y_{21} = \frac{dI_{21}}{V_1} = -\frac{\mathbf{dH_{21}} \cdot \hat{z}_2 d\ell_2}{V_1}. \tag{2.39}$$

Therefore,

$$Y_{21} = \frac{Y_0 d\ell_1 d\ell_2}{4\pi} e^{-jkr_{21}} \left[\frac{jk}{r_{12}} \sin\theta_{21} \sin(\theta_{21}-\alpha) + \frac{1}{r_{21}^2} \left(\cos\alpha \left(1 - 3\cos^2\theta_{21} \right) \right. \right.$$
$$\left. \left. + \sin\alpha \sin\theta_{21} \cos\theta_{21} (2\cos\phi_{21} - 1) \right) \right]. \tag{2.40}$$

Equation 2.40 shows that when magnetic dipoles are arranged broadside to each other (i.e. E-plane), the mutual admittance varies inversely with the distance between them while, in an echelon arrangement (i.e. H-plane), the dependence is as the square of the distance.

There is a general relationship between the mutual impedance and admittances of electric and magnetic dipoles that finds widespread use. Let the mutual impedance for electric dipoles in free-space, Eq. 2.33, be written as Z_{21}^{ef} and the admittance for magnetic dipoles also in free-space, given by Eq. 2.39, be Y_{21}^{mf}. The ratio of these quantities is

$$\frac{Z_{21}^{ef}}{Y_{21}^{ef}} = \eta_0^2. \tag{2.41}$$

Eq. 2.41 is similar to Booker's relation for complementary antennas in free space (Booker, 1946). Similar expressions can be found for other arrangements such as for dipoles backed by plane conducting sheets. For example, the mutual admittance of a magnetic dipole located adja-cent to a plane electric conductor can be shown to be $Y_{21}^{me} = 2Y_{21}^{mf}$ where the superscript 'e' on the right side refers to an electric wall. Booker's relation between this mutual admittance and the mutual impedance of an electric dipole in free space is, therefore,

$$\frac{Z_{21}^{ef}}{Y_{21}^{me}} = \frac{\eta_o^2}{2}. \qquad (2.42)$$

The more common form of Booker's formula (Kraus & Carver, 1973) gives the relation between the input impedance of a half-wave dipole in free-space and its complementary structure, which is a slot in a ground plane. For that case, the right side of Eq. 2.42 is further divided by 2 to give $Z_{11}^{dipole}/Y_{11}^{slot} = \eta_o^2/4$.

From the simple theory given in this section, four principles may be stated for assessing, in a qualitative fashion, the likely impact of mutual coupling between antennas. These are:

a. Mutual coupling is a function of distance between the antennas. Although there is a general downward trend in the level of coupling with increasing distance, this dependence is not monotonic. The level goes through a series of maxima and minima in the same manner as the radiation pattern.
b. Coupling depends on the radiation pattern of the elements.
c. Coupling depends on the antennas' polarization. Highest coupling occurs when the radiated fields have the same polarization and are aligned, for example, for electric dipoles in echelon. If the interaction is predominantly electric dipole related, strongest coupling occurs in the H-plane, while for antennas that are predominantly magnetic dipole type, strongest coupling occurs in the E-plane.
d. Booker's relation can be used to convert an unknown coupling problem to one that may have a simpler, or known, solution.

2.4   Problems

P2.1 Two different current sheets overlay a plane aperture with dimensions $a \times b$ on an interface located at $z = 0$. The current sheets are $\mathbf{J}_s = \hat{x}/\eta_o$ and $\mathbf{M}_s = \hat{y}\cos(\pi y/b)$. If a plane wave in region 1, given by $\mathbf{E}_1 = \hat{x}\exp(-jkz)$, is incident on the interface, determine the fields in region 2 on the other side of this interface $(z > 0)$.

P2.2 A thin dielectric layer of thickness d where $d \ll \lambda$ and relative permittivity ε_r covers a metal plate (or large cylinder). A plane wave is normally incident on this plate. Verify that the field at the top surface of the dielectric has a reflection coefficient approximately given by $\approx -1\exp(2jkd)$ and the total field ≈ 0.

P2.3 A narrow slot antenna of length L resonates in its fundamental mode. Show that the equivalent magnet current is the dual of the electric current excited on a thin dipole also of length L.

P2.4 An infinite conducting plane separates a half-space (region 1) with constitutive parameters ε_1, μ_1 from a metallic enclosed region (region 2) with constitutive parameters ε_2, μ_2. Harmonic sources $\mathbf{J}_i, \mathbf{M}_i$ exist in each region $i = 1,2$. The two regions are coupled by a common aperture.

    a. Use the equivalence theorem to divide the problem into two separate parts.

    b. Obtain expressions for the tangential magnetic and electric fields to ensure continuity across the aperture.

P2.5 A coaxial aperture with inner and outer radii b and a is located in a cylinder that extends a height g above an infinite ground plane as shown in Figure P2.1. The aperture is excited in the TEM mode only.

    a. Obtain equivalent sources over the coax.

    b. Use the sources in (a) to represent a possible approach to a solution using equivalence.

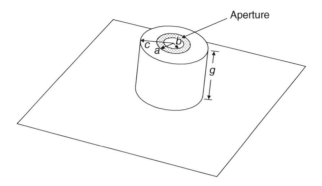

Figure P2.1   Protruding coaxial aperture in a ground plane

P2.6 Use Booker's relation and the input impedance a half-wave dipole in free-space, to obtain an expression for the input admittance of the complementary structure, which is a slot in a ground plane.

References

Booker, H.G. (1946): 'Slot aerials and their relation to complementary wire aerials (Babinet's principle)', J. Inst. Elect. Eng. Pt. IIIA, Vol. **93**, pp. 620–626.

Born, M. and Wolf, E. (1965): 'Principles of optics', 3rd ed., Pergamon Press, London, UK.

Harrington, R.F. (1961): 'Time-harmonic electromagnetic fields', McGraw-Hill, New York.

Jones, D.S. (1964): 'The theory of electromagnetism', Pergamon Press, London, UK.

Kraus, J.D. and Carver, K.R. (1973): 'Electromagnetics', 2nd ed., McGraw-Hill, International Student Edition, Kagakuska Ltd., Tokyo, Japan.

Sarkar, T.K., Mailloux, R.J., Oliner, A.A., Salazar-Palma, M. and Sengupta, D.L. (2006); 'History of wireless', John Wiley & Sons, Inc., Hoboken, NJ.

Schelkunoff, S.A. and Friis, H.T. (1952): 'Antennas: theory and practice', John Wiley & Sons, Inc., New York.

3

Fields Radiated by an Aperture

The previous chapter has provided the background for examining radiation from an aperture. Special cases of interest to the topic will be examined, and parameters related to radiation that are used to describe the characteristics of radiation will be defined.

3.1 Radiation Equations

Suppose the aim is to determine the fields at a point P arising from fields excited on an aperture A. It is convenient to do this in a spherical co-ordinate system as shown in Figure 3.1. In the previous section, it was shown that fields $\mathbf{E}_a$, $\mathbf{H}_a$, on a surface, may be replaced by equivalent sources

$$\mathbf{J}_s = \hat{n} \times \mathbf{H}_a \tag{3.1a}$$

$$\mathbf{M}_s = -\hat{n} \times \mathbf{E}_a. \tag{3.1b}$$

where $\hat{n}$ is the normal to the surface.

These equivalent currents imply the situation described in Section 2.2 where there is now a null field inside A. Our aim is to find the fields radiated by these sources using a simplified model.

Initially consider an infinitesimally small element dS' of the surface A. On this surface element, suppose there are electric and magnetic dipole sources with electric and magnetic dipole moments **dp** and **dm**, respectively. These dipole moments are related to the surface currents as follows:

$$\mathbf{dp} = \mathbf{J}_s dS' \tag{3.2a}$$

$$\mathbf{dm} = \mathbf{M}_s dS'. \tag{3.2b}$$

Fundamentals of Aperture Antennas and Arrays: From Theory to Design, Fabrication and Testing,
First Edition. Trevor S. Bird.
© 2016 John Wiley & Sons, Ltd. Published 2016 by John Wiley & Sons, Ltd.
Companion website: www.wiley.com/go/bird448

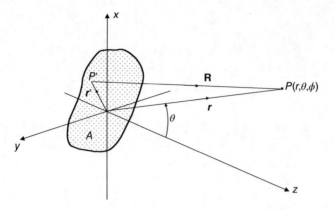

Figure 3.1   Geometry for fields radiated by an aperture A. P is an observation or field point; P' is a source point

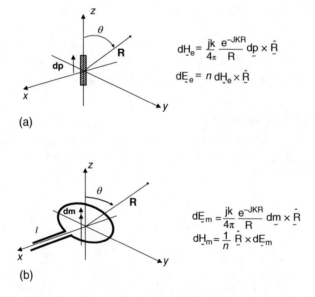

Figure 3.2   Models of electric and magnetic dipoles. (a) Elementary electric dipole. (b) Elementary magnetic dipole

Representations of both dipoles are shown in Figure 3.2. A short wire of length $d\ell'$ directed along the z-axis (Figure 3.2a) carrying a uniform current I along its length has an electric dipole moment $\mathbf{dp} = \mathbf{z}Id\ell'$. At distance R such that $kR \gg 1$ from the wire, the non-zero field components are given by Eqs. 2.25. In vector form, these fields are expressed as follows:

$$\mathbf{dH_e} = \frac{jk}{4\pi}\frac{e^{-jkR}}{R}\,\mathbf{dp} \times \hat{R} \tag{3.3a}$$

$$\mathbf{dE_e} = \eta_0 \mathbf{dH_e} \times \hat{R}, \tag{3.3b}$$

where the identity $\hat{z} \times \hat{R} = \hat{\phi} \sin \theta$ has been used. The subscript e is introduced on the fields in Eqs. 3.3 to indicate they are due to an electric dipole. For a magnetic dipole the fields may be obtained from Eqs. 2.29. The fields due to **dm** are, therefore,

$$\mathbf{dE_m} = -\frac{jk}{4\pi} \frac{e^{-jkR}}{R} \mathbf{dm} \times \hat{R} \tag{3.4a}$$

$$\mathbf{dM_m} = \frac{1}{\eta_o} \mathbf{dE_m} \times \hat{R}. \tag{3.4b}$$

The subscript m in Eqs. 3.4 indicates a magnetic dipole source. Eq. 3.4 is equivalent to the fields due to a wire loop of cross section dS' supporting a harmonic current, I, as depicted in Figure 3.2b. The loop has a magnetic dipole moment

$$\mathbf{dm} = \hat{z} j\omega\mu_o I dS'. \tag{3.5}$$

Thus, a model of the sources at each point P' on aperture A is a small electric dipole acting in conjunction with a small loop of current. The electric field at P due to both dipole sources is

$$\mathbf{dE} = \mathbf{dE_e} + \mathbf{dE_m} \tag{3.6a}$$

and

$$\mathbf{dH} = \mathbf{dH_e} + \mathbf{dH_m}. \tag{3.6b}$$

Thus, the total electric field due to both sources is

$$\begin{aligned} \mathbf{dE} &= \frac{jk}{4\pi} \frac{e^{-jkR}}{R} \left[-\mathbf{dm} \times \hat{R} + \eta_o \left(\mathbf{dp} \times \hat{R} \right) \times \hat{R} \right] \\ &= \frac{jk}{4\pi} \frac{e^{-jkR}}{R} \left[-\mathbf{M_s} \times \hat{R} + \eta_o \left(\mathbf{J_s} \times \hat{R} \right) \times \hat{R} \right] dS'. \end{aligned} \tag{3.7}$$

Adding all contributions from such sources on aperture A through integration results in the electric field

$$\mathbf{E} = \frac{jk}{4\pi} \int_A \frac{e^{-jkR}}{R} \left[\left(\hat{R} \times \mathbf{M_s} \right) + \eta_o \left(\mathbf{J_s} \times \hat{R} \right) \times \hat{R} \right] dS', \tag{3.8a}$$

where $\mathbf{R} = \mathbf{r} - \mathbf{r}'$ as shown in Figure 3.1. The integral in Eq. 3.8a is with respect to the primed co-ordinates, that is the source co-ordinates on aperture A. The magnetic field is obtained similarly and is

$$\mathbf{H} = \frac{jk}{4\pi} \int_A \frac{e^{-jkR}}{R} \left[\left(\mathbf{J_s} \times \hat{R} \right) + \frac{1}{\eta_o} \left(\mathbf{M_s} \times \hat{R} \right) \times \hat{R} \right] dS'. \tag{3.8b}$$

As a consequence of the assumptions leading to Eqs. 3.3 and 3.4, Eqs. 3.8 are valid for all field points P such that $kR \gg 1$ and thus they are applicable at intermediate distances from the aperture as well as in the far-field, that is, at distances that are comparable in size to the aperture itself and beyond (Silver, 1946).

3.2 Near-Field Region

Consider the case when the distance r is close to the aperture where r is comparable to the largest dimension of A. No simplifying approximations can be made to Eqs. 3.8. What is required is a return to the basic sources and to make approximations that are valid close to the source. A summation is still required across all infinitesimal sources although there is little to be gained from this compared with using Eqs. 3.8 directly. In the near-field region, the field differs little from that in the aperture field itself although there are fluctuations due to diffraction from any nearby edges or rim and the phase will have become slightly non-uniform. The field rapidly falls away from the edge in the aperture plane. If required, this field decay can be predicted by means of Eqs. 3.8. The field is typically concentrated in the region of the normal to the aperture.

3.3 Fresnel Zone

At greater distances from the aperture, the intermediate-field region is encountered, which is also called the Fresnel zone after a similar region in optics. For the Fresnel zone approximation to apply, the distance r should be $> \sqrt[3]{D^4/\lambda}/3$. Several simplifying approximations can be introduced into Eqs. 3.8 by virtue that the distance r is now assumed to be larger than the largest dimension in the region A. With this assumption, $R = |\mathbf{r} - \mathbf{r}'|$ can be approximated by the binomial series in r. The first three terms of this series are $(1+x)^n \approx 1 + nx + ((n(n-1))/2!)x^2$. This allows simplifications to be made to the integrals in Eqs. 3.8. Thus,

$$R = |\mathbf{r} - \mathbf{r}'|$$
$$= \sqrt{r^2 - 2\mathbf{r} \cdot \mathbf{r}' + (\mathbf{r}' \cdot \mathbf{r}')^2}$$
$$\approx r\left(1 - \frac{(\mathbf{r} \cdot \mathbf{r}')}{r^2} + \frac{1}{2}\frac{(\mathbf{r}' \cdot \mathbf{r}')^2}{r^2}\right)$$

(3.9)

This approximation is used in the exponential phase function and

$$\frac{1}{R} \approx \frac{1}{r} \quad \text{and} \quad \hat{R} \approx \hat{r}$$

in the amplitude of the integral. Therefore, in the Fresnel zone the electric field can be expressed as follows:

$$\mathbf{E} \approx \frac{jk}{4\pi}\frac{e^{-jkr}}{r}\int_A e^{-jkF}\left[(\hat{r} \times \mathbf{M_s}) + \eta_0(\mathbf{J_s} \times \hat{r}) \times \hat{r}\right]dS'$$

(3.10)

where $F = \hat{r} \cdot \mathbf{r}' + (1/2)\left((\mathbf{r}' \cdot \mathbf{r}')^2/r\right)$ comes about from the Fresnel phase approximation. When $\mathbf{r}$ is orthogonal to $\mathbf{r}'$ as in the case of Figure 3.1 when $\theta = 0°$, $F = (1/2)\left(r'^2/r\right)$. The exponential factor inside the integral contributes to a phase variation across the aperture, which is significant. The main impact of this is to introduce a quadratic phase error across the aperture, which broadens the central beam and fills in the sidelobe nulls. Computation with Eq. 3.10 is quite feasible and, therefore, is often chosen for applications such as reflector antennas where feeds and subreflectors can be in the Fresnel zone of each other. Quadratic phase error will be discussed in greater detail in Section 3.6

3.4   Far-Field Region

Attention is now turned to distances, r, that are large compared with the dimensions of the aperture. A minimum distance commonly specified is the Rayleigh distance criterion. According to this criterion,

$$r > \frac{2(\text{largest aperture dimension})^2}{\text{wavelength}}. \tag{3.11}$$

At the Rayleigh distance a field illuminating the antenna produces a maximum phase variation of $\pm \pi/8$ radian across the largest dimension. This distance is sufficient for measuring the first few sidelobes down to about -30 dB of the main beam. A greater distance is required for accurately measuring lower level near-in sidelobes. When the distances from the aperture are large $|\mathbf{r}| >> |\mathbf{r}'|$, the distance R can be approximated as follows:

$$\begin{aligned} R &= |\mathbf{r} - \mathbf{r}'| \\ &= \sqrt{r^2 + r'^2 - 2\mathbf{r} \cdot \mathbf{r}'} \\ &\approx r - \hat{r} \cdot \mathbf{r}', \end{aligned} \tag{3.12}$$

where use has been made of the first two terms of the binomial series expansion. The term $\hat{r} \cdot \mathbf{r}'$ accounts for the path-length difference from the source to the far-field point and from the origin to the same point.

On rectangular apertures, the following applies:

$$\hat{r} \cdot \mathbf{r}' = \sin\theta(x'\cos\phi + y'\sin\phi) + z'\cos\theta, \tag{3.13}$$

where the primed co-ordinates refer to the source point while on circular apertures it is

$$\hat{r} \cdot \mathbf{r}' = \rho'\sin\theta\cos(\phi - \phi') + z'\cos\theta. \tag{3.14}$$

To obtain approximations to Eqs. 3.8 in the far-field, the exponential is far more sensitive to approximation than the complex amplitude of the integrand. Therefore, Eq. 3.12 is used in the exponential, but in the amplitude let

$$\frac{1}{R} \approx \frac{1}{r} \quad \text{and} \quad \hat{R} \approx \hat{r} \tag{3.15}$$

Making these approximations, Eqs. 3.8 and 3.9 give

$$\mathbf{E} \approx \frac{jk}{4\pi} \frac{e^{-jkr}}{r} \hat{r} \times \int_A [\mathbf{M_s} - \eta_o(\mathbf{J_s} \times \hat{r})] \exp(jk\hat{r} \cdot \mathbf{r'}) dS' \tag{3.16a}$$

and

$$\mathbf{H} = \frac{1}{\eta_o} \mathbf{r} \times \mathbf{E}. \tag{3.16b}$$

Eq. 3.16a implies that the field radiated by the aperture is a wave with a spherical wavefront. The field is polarized tangential to the sphere, there being no radial components (i.e. $E_r = 0 = H_r$).

For a plane aperture situated in the x–y plane, the surface currents are

$$\mathbf{J_s} = \hat{n} \times \mathbf{H_a} \text{ and } \mathbf{M_s} = -\hat{n} \times \mathbf{E_a}. \tag{3.17}$$

From Eq. 3.16a, the field is

$$\mathbf{E} \approx -\frac{jk}{4\pi} \frac{e^{-jkr}}{r} \hat{r} \times \int_A [\hat{z} \times \mathbf{E_a} + \eta_o(\hat{z} \times \mathbf{H_a}) \times \hat{r}] \exp(jk\hat{r} \cdot \mathbf{r'}) dS'. \tag{3.18}$$

Now making use of

$$\hat{x} = \hat{r} \sin\theta \cos\phi + \hat{\theta} \cos\theta \cos\phi - \hat{\phi} \sin\phi \tag{3.19a}$$

$$\hat{y} = \hat{r} \sin\theta \sin\phi + \hat{\theta} \cos\theta \sin\phi + \hat{\phi} \cos\phi \tag{3.19b}$$

$$\hat{z} = \hat{r} \cos\theta - \hat{\theta} \sin\theta, \tag{3.19c}$$

the spherical components of Eq. 3.18 are found to be

$$E_r = 0 \tag{3.20a}$$

$$E_\theta \approx \frac{jk}{4\pi} \frac{e^{-jkr}}{r} \left[(N_x \cos\phi + N_y \sin\phi) + \eta_o \cos\theta(-L_x \sin\phi + L_y \cos\phi) \right] \tag{3.20b}$$

$$E_\phi \approx \frac{jk}{4\pi} \frac{e^{-jkr}}{r} \left[\cos\theta(-N_x \sin\phi + N_y \cos\phi) - \eta_o(L_x \cos\phi + L_y \sin\phi) \right], \tag{3.20c}$$

where N_x, N_y and L_x, L_y are rectangular components of the following transforms:

$$\mathbf{N}(\theta, \phi, \lambda) = \int_A \mathbf{E_a} \exp(jk\hat{r} \cdot r') dS' \tag{3.21}$$

$$\mathbf{L}(\theta, \phi, \lambda) = \int_A \mathbf{H_a} \exp(jk\hat{r} \cdot r') dS'. \tag{3.22}$$

To help interpret these vectors, consider a plane aperture situated at $z' = 0$. Also let

$$u = \frac{1}{\lambda} \sin \theta \cos \phi \tag{3.23a}$$

$$v = \frac{1}{\lambda} \sin \theta \sin \phi \tag{3.23b}$$

so that $(jk\hat{r}\cdot\mathbf{r}') = 2\pi(ux' + vy')$. As a result define

$$\mathbf{N}(u,v) = \int_A \mathbf{E_a}(x',y') \exp(j2\pi(ux' + vy')) dx' dy' \tag{3.24a}$$

$$\mathbf{L}(u,v) = \int_A \mathbf{H_a}(x',y') \exp(j2\pi(ux' + vy')) dx' dy'. \tag{3.24b}$$

It is observed from Eqs. 3.24 that the components of $\mathbf{N}$ and $\mathbf{L}$ are two-dimensional Fourier transforms of the aperture field components. Therefore, the far-zone electric and magnetic fields, via Eqs. 3.20 and 3.16b, are proportional to Fourier transforms of the aperture field distributions. Conversely, the aperture fields are related to inverse Fourier transforms of the far-fields. This relationship is particularly useful as results in later sections can be interpreted from a knowledge of Fourier transforms (Oppenheim & Schafer, 1975). The transforms in Eqs. 3.24 can also be evaluated numerically by means of the fast Fourier transform (FFT) (Brigham, 1974).

Two special cases of practical importance are now considered. The first is when the electric current is located close to a magnetic ground plane. In this case, $\mathbf{J_s} = 2\hat{n} \times \mathbf{H_a}$ and $\mathbf{M_s} = 0$. It is easy to show that the far-field components are now

$$E_\theta \approx \frac{jk\eta_o}{2\pi} \frac{e^{-jkr}}{r} \cos\theta \left(-L_x \sin\phi + L_y \cos\phi\right) \tag{3.25a}$$

and

$$E_\phi \approx -\frac{jk\eta_o}{2\pi} \frac{e^{-jkr}}{r} \left(L_x \cos\phi + L_y \sin\phi\right). \tag{3.25b}$$

The second case is when the magnetic current is above a conducting surface, $\mathbf{M_s} = -2\hat{n} \times \mathbf{E_a}$ and $\mathbf{J_s} = 0$. Therefore,

$$E_\theta \approx \frac{jk}{2\pi} \frac{e^{-jkr}}{r} \left(N_x \cos\phi + N_y \sin\phi\right) \tag{3.26a}$$

and

$$E_\phi \approx \frac{jk}{2\pi} \frac{e^{-jkr}}{r} \cos\theta \left(-N_x \sin\phi + N_y \cos\phi\right). \tag{3.26b}$$

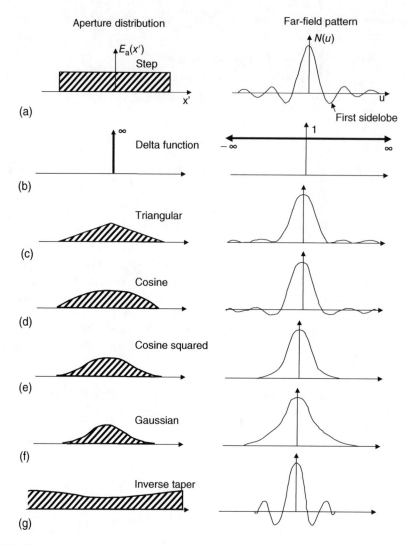

Figure 3.3   Fourier transform relationship between the aperture field distribution and the far-field pattern. (a) Step. (b) Delta function. (c) Triangular. (d) Cosine. (e) Cosine squared. (f) Gaussian. (g) Inverse taper

In Figure 3.3, several Fourier transforms of one-dimensional aperture distributions are illustrated. For apertures with a separable co-ordinate system (e.g. rectangular co-ordinates), the two-dimensional Fourier transform is a product of two one-dimensional transforms and, therefore, Figure 3.3 can be used as a guide in the general case as well. In particular, it is noted that a uniform aperture distribution gives rise to a more directive far-field pattern than a tapered distribution, for example, the cosine or cosine-squared distributions. However, the sidelobe levels are higher for a uniform distribution than for a tapered distribution. Also, the sharper the taper the broader the main lobe but the lower the sidelobes. Transforms of three aperture distributions of particular interest in the sections to follow are plotted in Figure 3.4. These correspond to a

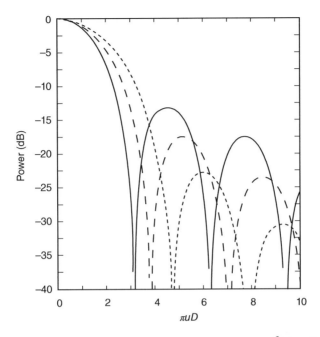

Figure 3.4 Transforms of some aperture illuminations. ─────── $S^2(x)$ uniformly illuminated rectangular aperture; - - - - - - - - - $C^2(x)$ cosine illuminated rectangular aperture; – – – – – – – $[2(J_1(x)/x)]^2$ uniformly illuminated circular aperture

uniform aperture distribution on a rectangular aperture, a cosine distribution on a rectangular aperture and a uniform distribution on a circular aperture. The transforms of these respective distributions on apertures of dimension D are as follows:

Uniform distribution on rectangular aperture

$$\int_{-D/2}^{D/2} \exp(j2\pi u x')dx' = DS(\pi u D). \tag{3.27}$$

Cosine taper on rectangular aperture

$$\int_{-D/2}^{D/2} \cos\left(\frac{\pi x'}{D}\right)\exp(j2\pi u x')dx' = \frac{2D}{\pi}C(\pi u D). \tag{3.28}$$

Uniform distribution on circular aperture

$$\int_{0}^{2\pi} d\phi' \int_{0}^{D/2} d\rho' \exp(jk\rho' \sin\theta \cos(\phi-\phi')) = 2\pi \int_{0}^{D/2} d\rho' J_0(k\rho' \sin\theta)$$

$$= \pi\left(\frac{D}{2}\right)^2 2\frac{J_1(w)}{w}, \tag{3.29}$$

where $w = (\pi D/\lambda) \sin \theta$ and the identity Eq. B.3 in Appendix B has been used. The functions J_0 and J_1 are Bessel functions of order 0 and 1, respectively. The new functions involved are defined as follows:

$$S(x) = \frac{\sin x}{x} \qquad (3.30)$$

$$C(x) = \frac{\cos x}{1 - (2x/\pi)^2}. \qquad (3.31)$$

Plots of the square of the functions $S(x)$, $C(x)$ and $2J_1(x)/x$ are shown in Figure 3.4. The first sidelobe level relative to the peak in each case is, respectively, -13.3, -23.0 and -17.6 dB. The 3 dB points occur at $x = 1.39$, 1.88 and 1.60, respectively (Silver, 1946). Hence the half-power beamwidths (HPBWs) are approximately $0.88\lambda/D$ for the uniformly illuminated rectangular aperture, $1.2\lambda/D$ for a cosine distribution on a rectangular aperture and $1.02\lambda/D$ for the uniformly illuminated circular aperture.

The power radiated in any of the three regions identified above is obtained using the Poynting vector. The power traversing an area Σ is given by

$$P = \frac{1}{2}\mathrm{Re}\left\{\iint_\Sigma \mathbf{E} \times \mathbf{H}^* \cdot \hat{n} dS\right\}, \qquad (3.32)$$

where $\hat{n}$ is the normal to Σ. In the Fresnel and far-field regions, the magnetic field is given by Eq. 3.16b. Therefore, the Poynting vector is

$$\bar{\mathbf{P}} = \frac{1}{2\eta_0}\mathbf{E} \cdot (\mathbf{r} \times \mathbf{E})^*.$$

In the far-field $E_r = 0$ and, therefore,

$$\bar{\mathbf{P}} = \frac{\hat{r}}{2\eta_0}\mathbf{E} \cdot \mathbf{E}^*$$

which indicates the power radiates in a radial direction.

3.4.1   Example of a Uniformly Illuminated Rectangular Aperture

The fields radiated by the rectangular aperture shown in Figure 3.5 are to be determined when uniform electric and magnetic fields in the aperture are assumed. Outside the aperture in the aperture plane the field is zero. From these fields find the maximum power gain.

Referring to Figure 3.5, let

$$\mathbf{E_a} = \hat{x}E_o; \quad -\frac{a}{2} < x' < \frac{a}{2}; \quad -\frac{b}{2} < y' < \frac{b}{2}$$

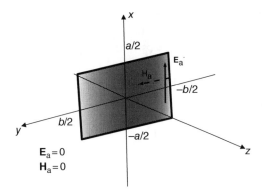

Figure 3.5 Rectangular aperture

$$\mathbf{H_a} = \frac{1}{\eta_o}\hat{z} \times \mathbf{E_a} = \frac{1}{\eta_o}\hat{y}E_o,$$

where E_o is a constant. The fields are zero elsewhere on the aperture plane. The radiated fields are calculated from Eqs. 3.20 and 3.24 with the assumed aperture field, which is uniform in both amplitude and phase.

The only non-zero component of Eq. 3.24a in this example is

$$N_x(u,v) = E_o \int_{-b/2}^{b/2} dx' \int_{-a/2}^{a/2} dy' \exp(j2\pi(ux' + vy'))$$

$$= abE_o S(\pi ua)S(\pi vb),$$

(3.33)

where S is the sinc function defined by Eq. 3.30. Also

$$L_y(u,v) = \frac{1}{\eta_o}N_x(u,v).$$

(3.34)

The radiated fields are, therefore, given by

$$E_\theta \approx \frac{jkabE_o}{4\pi}\frac{e^{-jkr}}{r}(1 + \cos\theta)S(\pi ua)S(\pi vb)\cos\phi$$

(3.35a)

$$E_\phi \approx -\frac{jkabE_o}{4\pi}\frac{e^{-jkr}}{r}(1 + \cos\theta)S(\pi ua)S(\pi vb)\sin\phi.$$

(3.35b)

The S function (which is called the sinc function) is plotted in Figure 3.4. If the aperture is large in terms of wavelengths (a, $b \gg \lambda$), S varies much faster than the term $(1 + \cos\theta) \approx 2$ when θ is small, and hence S predicts the pattern close to the normal of the aperture, the boresight direction.

Of particular importance is the power radiated in a particular direction and especially in the boresight direction $\theta = 0°$ where the radiated fields are maximum. The total power radiated by the uniform aperture field is found by means of Poynting's theorem and is given by

$$P_T = \frac{1}{2} \iint_A E \times H* \cdot \hat{z} dS'$$

$$= \frac{1}{2} \int_{-a/2}^{a/2} dy' \int_{-b/2}^{b/2} dx' \frac{E_o^2}{\eta_o} \tag{3.36}$$

$$= \frac{ab}{2\eta_o} E_o^2.$$

On the other hand, by means of Eq. 2.28, the maximum power density at $\theta = 0°$ is

$$P_r|_{\theta=0} = \frac{1}{\eta_o r^2} \left[\frac{kabE_0}{4\pi} \right]^2. \tag{3.37}$$

Now suppose it is possible to radiate this density over a sphere of radius r. This means Eq. 3.37 should be multiplied by the area of the sphere, that is, $4\pi r^2$. The maximum gain of the radiating aperture is defined as the ratio of this apparent power evaluated at the peak of the beam and the total power available. That is,

$$G_{max} = \frac{4\pi r^2 P_r|_{\theta=0}}{P_T}. \tag{3.38}$$

Substituting in the quantities for the uniformly illuminated rectangular aperture given by Eqs. 3.36 and 3.37, then it follows

$$G_{max} = 4\pi \frac{ab}{\lambda^2}. \tag{3.39}$$

In general, the maximum gain of a uniformly illuminated aperture with physical area A is

$$G_o = 4\pi \frac{A}{\lambda^2}. \tag{3.40}$$

An approximate expression for maximum gain is discussed in more detail in the next section. Eq. 3.37 can be generalized in terms of system quantities as $P_r = GP_{in}/L$ where G is the antenna gain, P_{in} is the power input and $L = (4\pi r/\lambda)^2$ is the free-space loss factor.

3.5 Radiation Characteristics

The performance of an antenna is usually described in terms of its far-field radiation characteristics and its terminal impedance. Some terms have been introduced in the previous sections. However, this section will detail most of the important terms applied to antennas. A typical radiation pattern of an aperture antenna is illustrated in Figure 3.6.

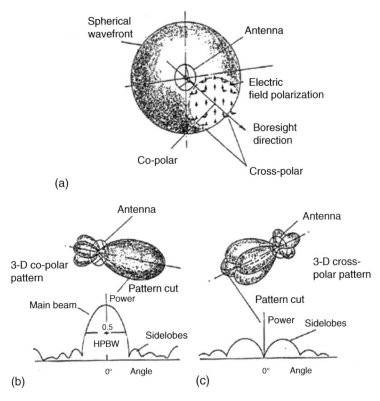

Figure 3.6 Antenna radiation patterns. (a) Radiation sphere enclosing an antenna showing the (linear) field polarization on a cap about the direction of the main beam (boresight). (b) Co-polar radiation pattern cut. (c) Cross-polar radiation pattern cut.

3.5.1 Radiation Pattern

In the far-field or radiation zone of an antenna, the amplitude of electromagnetic fields are proportional to $1/r$, where r is the distance from the antenna. Plots of the magnitude of the electric and magnetic fields at a constant distance are called field strength patterns. Plots of the radiated power at a constant distance or radius r are called radiation power patterns. These are illustrated in Figure 3.6. The power pattern is defined:

$$P(\theta,\phi) = P_r(r,\theta,\phi)r^2 = \text{power density per unit solid angle.} \tag{3.41}$$

Power density (W/m^2) is the radial component of the Poynting vector which is

$$P_r(r,\theta,\phi) = \frac{1}{2}\text{Re}[\mathbf{E}\times\mathbf{H}^*]$$

$$= \frac{1}{2\eta_o}|\mathbf{E}|^2, \tag{3.42}$$

where $\mathbf{E}$ is the electric field intensity (V/m), $\mathbf{H}$ is the magnetic field intensity (A/m) transverse to the direction of wave propagation in the radial (r) direction and $\eta_o = \sqrt{\mu_o/\varepsilon_o}$ is the wave impedance of free-space where ε_o and μ_o are the permittivity and permeability of free-space. Equation 3.42 assumes that the fields are in the far-field of the source of radiation.

Usually the power pattern is normalized relative to its maximum value P_{max}. The normalized power pattern is

$$P_n(\theta,\phi) = \frac{P(\theta,\phi)}{P_{max}}. \tag{3.43}$$

For example, the normalized pattern of a small current element is $P_n(\theta,\phi) = \sin^2\theta$.

3.5.2   Half-Power Beamwidth

The HPBW is the angle subtended at the -3 dB points of the normalized power pattern (see Figure 3.6b). The beamwidth between first nulls (BWFN) and the tenth-power beamwidth are also used, for example, the HPBW of an elemental dipole is $90°$ and for a half-wave dipole it is approximately $78°$. The HPBW of a uniformly illuminated aperture of width a can be shown to be $\sim 1.2\lambda/a$ radians.

3.5.3   Front-to-Back Ratio

The front-to-back ratio (FTBR) gives a measure of the isolation provided by a directional antenna from or to sources in the direction opposite the direction of maximum gain θ_{max}. Expressed in dB, the FTBR is

$$\mathrm{FTBR} = 10\log_{10}\left[\frac{P_n(\theta_{max},\phi)}{P_n(\theta_{max} \pm \pi,\phi)}\right]$$
$$= -10\log_{10}[P_n(\theta_{max} \pm \pi,\phi)]. \tag{3.44}$$

For example, a half-wave dipole has FTBR of 0 dB. A reflector antenna typically has a FTBR > 20 dB.

3.5.4   Polarization

At distances far from the antenna, the radiated fields are tangential to the surface of a sphere centred on the antenna (see Figure 3.6a). In general, the field on the sphere has components in both the $\hat{\theta}$ and $\hat{\phi}$ directions. It is linearly polarized if the components E_θ and E_ϕ are in-phase everywhere, and if they are $\pm 90°$ out of phase, the field is circularly polarized. The field is elliptically polarized for an arbitrary phase difference.

Radiation patterns measured in the two principal planes of linearly polarized antennas are referred to as the E- and H-plane patterns. As shown in Figure 3.7, the E-plane pattern is the cut taken parallel to the electric field, and the H-plane pattern is the cut taken perpendicular to the electric field.

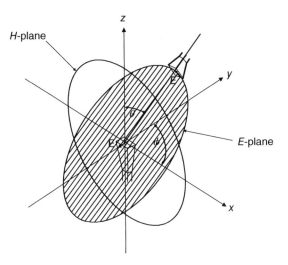

Figure 3.7   Principal plane radiation pattern cuts

The polarization generally varies over the surface of the sphere, and as a consequence, there is also an unwanted field component polarized in the opposite direction. For example, although an antenna is required to produce right-hand circularly polarized (RHCP) radiation, a practical antenna also produces a small amount of the opposite polarization (LHCP) as well. This orthogonally polarized field is called cross-polarization. If $\mathbf{E}$ is the radiated field, $\mathbf{p}$ is a unit vector in the direction of the reference polarization, or the co-polarized component, and $\mathbf{q}$ is a unit vector in the cross-polarized direction, then

$\mathbf{E} \cdot \mathbf{p}$ is the co-polar component of the electric field.
$\mathbf{E} \cdot \mathbf{q}$ is the cross-polar component of the electric field.

The choice of vectors $\mathbf{p}$ and $\mathbf{q}$ is somewhat arbitrary. For predominantly linearly polarized fields, however, one definition is preferred. With this definition, the co-polar antenna field component is found by conventional far-field measurement with the polarization of the distant source antenna initially aligned with the test antenna on boresight. Maintaining this alignment, the test antenna is rotated about a chosen origin, the phase centre. The signal received by the test antenna is the co-polar radiation pattern. If the polarization of the distant source antenna is now rotated through 90 degrees and the radiation pattern measurement repeated, the received signal is the cross-polar radiation pattern. If the antenna has its principal electric field vector parallel to the x-axis, as in Figure 3.7, in the E-plane, the co-polar component is $\mathbf{E}_\theta$, while the cross-polar component is $\mathbf{E}_\phi$. In the H-plane, $\mathbf{E}_\phi$ is the co-polar component, and $\mathbf{E}_\theta$ is the cross-polar component. In general, with the electric field polarized in the ϕ_o direction, the co-polar and cross-polar components in the ϕ direction are given by

$$\begin{bmatrix} E_p(\theta,\phi) \\ E_q(\theta,\phi) \end{bmatrix} = \begin{bmatrix} \cos(\phi-\phi_o) & \sin(\phi-\phi_o) \\ \sin(\phi-\phi_o) & -\cos(\phi-\phi_o) \end{bmatrix} \cdot \begin{bmatrix} E_\theta(\theta,\phi) \\ E_\phi(\theta,\phi) \end{bmatrix}. \qquad (3.45)$$

For example, if the reference polarization is at $\phi_o = 0°$ as in Figure 3.7, the co-polar and cross-polar components of the field in the $\phi = 45°$ plane are

$$E_p\left(\theta, \frac{\pi}{4}\right) = \frac{1}{\sqrt{2}}\left[E_\theta\left(\theta, \frac{\pi}{4}\right) + E_\phi\left(\theta, \frac{\pi}{4}\right)\right]$$

$$E_q\left(\theta, \frac{\pi}{4}\right) = \frac{1}{\sqrt{2}}\left[E_\theta\left(\theta, \frac{\pi}{4}\right) - E_\phi\left(\theta, \frac{\pi}{4}\right)\right].$$

In addition, the E-plane corresponds to the $\phi = 0°$ plane where

$$E_p(\theta, 0) = E_\theta(\theta, 0) \quad \text{and} \quad E_q(\theta, 0) = -E_\phi(\theta, 0)$$

and the H-plane occurs in $\phi = 90°$ plane where

$$E_p\left(\theta, \frac{\pi}{2}\right) = F_\phi\left(\theta, \frac{\pi}{2}\right) \quad \text{and} \quad E_q\left(\theta, \frac{\pi}{2}\right) = E_\theta\left(0, \frac{\pi}{2}\right).$$

3.5.5 Phase Centre

The phase centre of an antenna is the apparent location of emanating spherical waves when it is transmitting.

One approach is to estimate the phase centre from the two-dimensional discrete patterns using the method of least squares (Fröberg, 1974). It can be shown that in the p-th azimuth plane, the phase centre for a symmetric radiator in the plane p ($= 1, ..., NP$) is approximately given by

$$kd_p = \frac{b_1 - a_1 b_0}{a_2 - a_1^2}, \tag{3.46}$$

where $k = 2\pi/\lambda$, $a_q = \left(\sum_{i=1}^{M} \cos^q \theta_i\right)/M$ and $b_q = \left(\sum_{i=1}^{M} \Phi(\theta_i, \phi_p) \cos^q \theta_i\right)/M$, $q = 0, 1, 2, \theta_i$ is angle i ($= 1, ..., M$) in the n-th pattern cut through the plane ϕ_p which is symmetric about boresight, M is the number of angular directions in the pattern and $\Phi(\theta_i, \phi_p)$ is the continuous (unwrapped) phase function expressed in radians. An improved estimate of phase centre is obtained by averaging several pattern cuts.

3.5.6 Antenna Gain and Directivity

The power gain of an antenna in a given direction (θ, ϕ) may be defined as the ratio of power intercepted by a sphere enclosing the antenna if the same power density at (θ, ϕ) is radiated isotropically and the total radiated power. The power density at (θ, ϕ) is given by Eq. 3.42. If this power were radiated isotropically, the power that would be radiated at a large sphere of radius r is $P_r \times 4\pi r^2$. The total power radiated, P_T, is given by

$$P_T = \frac{1}{2} \text{Re} \iint_S (\mathbf{E} \times \mathbf{H}^*) dS. \qquad (3.47)$$

S may be any surface enclosing the field. For aperture antennas it is common to let $S = A$ where A is the aperture surface because it is easier to integrate over this surface than over the far-zone sphere. From these definitions the power gain function is

$$G(\theta, \phi) = \frac{4\pi r^2 P_r(r, \theta, \phi)}{P_T} \qquad (3.48)$$

in the far-field and using Eq. 3.42,

$$G(\theta, \phi) = \frac{2\pi r^2}{\eta_0} \frac{|\mathbf{E}(r, \theta, \phi)|^2}{P_T}.$$

The maximum value of gain is a parameter often used to describe the performance of an antenna. However, sometimes when the term 'gain' is used, it is the maximum power gain that is being referred to. Gain is referred to another antenna with the same input power. Here the reference is the isotropic radiator although any convenient reference antenna may be used. For example, the maximum gain of a short dipole relative to an isotropic radiator is 3/2 or 1.76 dBi, where dBi indicates the gain is in decibels above the gain of an isotropic radiator, which has a gain of unity.

The gain of an antenna with respect to the gain of a uniformly illuminated aperture of the same dimensions is known as the aperture efficiency. This is defined as

$$\eta_a = \frac{G_{\text{max}}}{G_o}, \qquad (3.49)$$

where $G_o = (4\pi A)/\lambda^2$ is the gain of a uniformly illuminated aperture, in both amplitude and phase, over an area A. A useful rough approximation for the maximum gain of an aperture with HPBWs θ_E radians in the E-plane and θ_H radians in the H-plane is

$$G_{\text{max}} \approx \frac{4\pi}{\theta_E \theta_H}$$

$$\approx \frac{41253}{\theta_E(\text{deg}) \theta_H(\text{deg})},$$

where $\theta_{E,H}(\text{deg})$ refer to the HPBW expressed in degrees.

A related quantity to gain is the directivity D. This is defined as

$$D = \frac{\text{peak radiated power}}{\text{power in radiated field (W)}}$$

$$= \frac{4\pi r^2 P_r(r, \theta, \phi)|_{\text{peak}}}{\iint\limits_{4\pi} P_r(r, \theta, \phi) r^2 d\theta d\phi}. \qquad (3.50)$$

The directivity always exceeds unity, that is, $D \geq 1$. The directivity of an ideal short dipole is the same as its gain. In general, gain is related to directivity through various conversion factors. These include the efficiency of feeding and efficiency of conversion of power to radiation.

The gain of a uniformly illuminated aperture is not the maximum gain that is physically possible. Bouwkamp & de Bruijn (1946) and also Riblet (1948) showed that there is no theoretical limit to directivity, for an antenna of given size, if the current distribution is unconstrained, that is, the amplitude and phase is non-uniform. As a result, efficiencies in excess of 100% are achievable. Translating this to aperture antennas means that the effective radiating aperture is greater than the physical area. This effect is called a supergain and it occurs when the gain is greater than that produced by an aperture distribution which is uniform in both amplitude and phase. It has been shown (Bird & Granet, 2013) that for rectangular and circular apertures, for example, it is possible to achieve efficiencies close to 100%. Thus, as the number of modes in the aperture goes to infinity,

$$\text{Rectangular aperture}: \eta_{a,\max} \approx \frac{8}{\pi^2} \sum_{n=1,3,\ldots}^{N} \frac{1}{n^2} \Rightarrow 1 \quad \text{as} \quad N \to \infty$$

$$\text{Circular aperture}: \eta_{a.\max} \approx \varepsilon_{0m} \sum_{n=1}^{N} \frac{1}{(\alpha_{mn}^2 - m^2)} \Rightarrow \left(\frac{1}{8(2 - \varepsilon_{0m}) + m} \right) \quad \text{as} \quad N \to \infty,$$

where $\alpha_{mn} = k_{c,mn}a$, a is the radius, $\varepsilon_{0m} = 1$ if $m = 0$ and is 2 otherwise, $k_{c,mn}$ is the cut-off wavenumber of the TE_{mn} mode and the subscript n runs over the number of modes in the aperture where N is the largest integer satisfying $\alpha_{mN} < ka$, that is, all modes that propagate. It has been shown that only the TE modes contribute directly to the above summations. Also the above assumes that there no mode coupling or mismatch at the aperture. For example, apertures that contain TE modes with a single period ($m = 1$) (i.e. TE_{1n}) can achieve a maximum efficiency of 100%, while for the axisymmetric TE modes ($m = 0$), the maximum efficiency is limited to 12.5% and for the double period TE modes ($m = 2$), it is 50%. However, a combination of modes in the aperture can produce higher maximum efficiencies. How such an efficiency could be achieved is a subject for further design. One approach will be described in Section 4.5.3.

3.5.7 Effective Aperture

A receiving antenna is characterized by the equivalent area over which it collects energy from an incident wave. The receiving cross section of the antenna is defined as follows:

$$A_r(\theta, \phi) = \frac{\text{received power}}{\text{power density of incident wave}}. \tag{3.51}$$

It is possible to show that when the receiving antenna is oriented to receive maximum signal and the antenna is matched to the terminating load, then

$$A_r(\theta, \phi) = \frac{\lambda^2}{4\pi} G(\theta, \phi). \tag{3.52}$$

The maximum aperture is known as the effective aperture, A_e, and it is given by

$$A_e = \frac{\lambda^2}{4\pi} G_{\max}, \tag{3.53}$$

where $G_{\max}$ is the maximum value of gain (Eq. 3.48). For example, the effective aperture of a short dipole is $A_e = 3\lambda^2/8\pi$.

3.5.8 Radiation Resistance

Assuming no conduction losses, the total power radiated by an antenna is equal to the total power at the input terminals. If I_0 is the peak value of the current, then

$$P_T = \frac{1}{2}[I_0]^2 R_r,$$

where R_r is the radiation resistance associated with the power that is radiated. That is,

$$R_r = \cdot \frac{2P_T}{[I_0]^2} \tag{3.54}$$

For an example, the radiation resistance of a short dipole of length $d\ell$ is

$$R_r = \frac{2\pi}{3}\eta_o \left(\frac{d\ell}{\lambda}\right)^2.$$

3.5.9 Input Impedance

The input impedance of an antenna is the impedance presented to the feeder. It is usually a complex quantity. The real part is due to the energy losses associated with the antenna. For practical antennas these losses are not only due to power transfer into the far-field but also due to loss mechanisms such as lossy ground and finite conductivity of the antenna structure.

The reactive part of the impedance is due to near-field energy storage, and its value is highly dependent upon the antenna geometry.

Assuming a lossless antenna radiating into free-space, the power transferred through a surface Σ enclosing the antenna and very close to its surface is

$$P = \iint_\Sigma \bar{\mathbf{P}} \cdot d\mathbf{S} = P_T + jQ_T, \tag{3.55}$$

where $\bar{\mathbf{P}} = 1/2\,\mathbf{E} \times \mathbf{H}^*$ is the complex Poynting vector. At very large distances, the reactive power Q_T becomes very small, but close to the antenna it makes a significant contribution and its size depends on the antenna. The input impedance is defined as follows:

$$Z_{in} = \frac{2P}{\left\lfloor I_{in} \right\rfloor^2} = R_{in} + jX_{in}, \tag{3.56}$$

where I_{in} is the current at the input terminals of the antenna, for example, for a dipole antenna of length L the input current is $I_{in} = I_o \sin^2(\beta L/2)$ where I_o is the peak value of the current. Therefore, for a $\lambda/2$ dipole $R_{in} = R_r = 73.1$ ohms.

3.5.10   Antenna as a Receiver

Reciprocity and in particular Eq. 2.13 can be used to show the behaviour of an antenna when used as a receiver in terms of its characteristics as a transmitter. To do this, in Eq. 2.13, the fields of $(\mathbf{E}^a, \mathbf{H}^a)$ that are taken are those emitted when the antenna transmits and $(\mathbf{E}^b, \mathbf{H}^b)$ are the fields of a plane wave at the same frequency which is incident on the antenna. If $\mathbf{F}(\theta, \phi)$ is the transmitting radiation pattern where $\mathbf{E}_{rad} = \mathbf{F}(\theta, \phi) \exp(-jkr)/r$ is the radiated electric field on a large sphere of radius r, then the received field in the feeder is proportional to

$$\mathbf{E}_{rec} \propto \mathbf{E}_o \cdot \mathbf{F}(\theta', \phi'),$$

where $\mathbf{E}_o$ is a constant vector giving the direction and magnitude of the incident plane wave and the primed co-ordinates refer to receiver direction.

3.6   Aberrations

The radiation pattern of an aperture antenna is sensitively dependent upon the phase distribution of the aperture field. Variation from the ideal uniform phase distribution causes phase aberrations. These aberrations occur inadvertently during antenna design and manufacture. Sometimes they arise intentionally when, for example, beam shaping or for beam steering. Due to aberrations, the aperture field, $\mathbf{E_a}$, can be considered to have a phase distribution superimposed on it; thus,

$$\mathbf{E_a} \exp(j\Phi(x', y')),$$

where $\Phi(x', y')$ is the aberration function. To simplify the discussion of aberrations consider a circular aperture of unit radius. Then it is convenient to define polar co-ordinates, so that $x' = t \cos \xi$ and $y' = t \sin \xi$. Without loss of generality, the aberration function is assumed to be an even function of ξ. Moreover, it may be shown (Born and Wolf, 1959) that the aberration function can be represented as follows:

$$\Phi(t, \xi) = \sum_{n, m=0}^{\infty} \Delta_{nm} t^n \cos m\xi, \tag{3.57}$$

where Δ_{nm} are the aberration coefficients and are non-zero when $m + n$ **is** even.

It is possible to see the effect of each term in Eq. 3.57 by letting all coefficients apart from one be zero. The primary aberrations are illustrated in Figure 3.8. They are referred to as linear $(n = 1, \ m = 1)$, quadratic $(n = 2, \ m = 0)$, coma $(n = 3, \ m = 1)$, astigmatic $(n = 2, \ m = 2)$ and

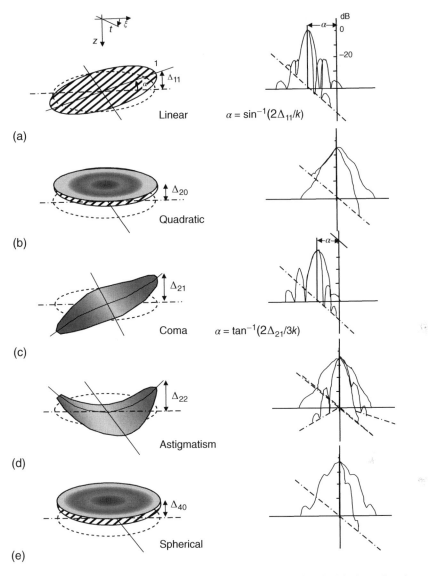

Figure 3.8 Phase distributions (left) and radiation patterns (right) associated with aberrations in a circular aperture of unit radius. (a) Linear $\Phi = \Delta_{11} t$. (b) Quadratic $\Phi = \Delta_{20} t^2$. (c) Coma $\Phi = \Delta_{31} t^3 \cos \xi$. (d) Astigmatic $\Phi = \Delta_{22} t^2 \cos 2\xi$. (e) Spherical aberrations $\Phi = \Delta_{40} t^4$ (after Masterman, 1973)

spherical ($n = 4$, $m = 0$) aberration. Linear aberration in Figure 3.8a shifts the direction of the main beam by the angle $\alpha = \sin^{-1}(\Delta_{11}/k)$ without changing the structure of the beam. Quadratic phase error in Figure 3.8b causes a reduction in antenna gain and increases both the beamwidth and sidelobe level. Another effect is that the nulls in the pattern are filled in. Cubic phase error, or coma, in Figure 3.8c causes the beam to shift an angle $\alpha = \arcsin(2\Delta_{31} a^2/3k)$ and also reduces gain. In addition, the pattern is asymmetrical in the plane containing the shifted beam and the central axis (the plane of 'scan'). The sidelobes closest to the central axis are lower than the sidelobes without coma, and those in the direction of scan are higher than without coma. The effect of astigmatism in Figure 3.8d is similar to a quadratic phase error. When astigmatism and quadratic phase error occur together, the width of the main beam and the sidelobes are different in the two principal planes. Finally, in Figure 3.6e, spherical aberration produces a symmetrical distortion of the radiation pattern with an effect similar to quadratic phase error.

The gain of a uniformly illuminated aperture affected by aberration is approximately given by (Bracewell, 1961)

$$G_a = \frac{1}{1 + \kappa \Delta_e^2} G, \qquad (3.58)$$

where G_a is the gain with aberration, G is the gain without aberration and Δ_e is the phase error in radians at the edge of the aperture. κ is a constant that depends on the type of aberration and equals 0 for linear, 1/12 for quadratic, 1/72 for coma, 1/6 for astigmatism and 4/45 for spherical aberration.

In any practical antenna, all types of aberrations can occur, some to a greater extent than others. For example, a method commonly used to scan the beam of a reflector antenna is to displace the feed laterally from the reflector axis in order to produce a linear phase shift across the aperture. In this situation, as well as the desired linear aberration (beam shift), coma is strongly represented in the radiation pattern. Astigmatism is also produced but it is of lesser importance than coma for small lateral shifts. When the feed is moved from the focus in the axial direction, either towards or away from the reflector vertex, quadratic and spherical aberrations are created. This is often used to improve the pattern of the reflector when the feed has a diffuse phase centre.

3.7   Power Coupling Theorem[*]

A corollary of Lorentz reciprocity that finds use in aperture antennas is the power coupling theorem (Robieux, 1959, Wood, 1980). Suppose an antenna is defined by surfaces $S_1 = S_1' + S_1''$ and S_2. Power is coupled into the antenna through surface S_2 as shown in Figure 3.9. Inside the antenna, the surface S_1' covers the receiving port.

Other parts of the inside surface, indicated as S_1'', are perfect conductors. Let $(\mathbf{E}^a, \mathbf{H}^a)$ be the fields on S_1 and S_2 when the antenna transmits. Also let $(\mathbf{E}^b, \mathbf{H}^b)$ be the fields on these surfaces when the antenna receives. Under these conditions, with no sources being present in V, the reciprocity theorem, Eq. 2.13, gives

$$\iint_{S_1} \left(\mathbf{E}^a \times \mathbf{H}^b - \mathbf{E}^b \times \mathbf{H}^a \right) \cdot \hat{n} \, dS = - \iint_{S_2} \left(\mathbf{E}^a \times \mathbf{H}^b - \mathbf{E}^b \times \mathbf{H}^a \right) \cdot \hat{n} \, dS. \qquad (3.59)$$

On S_1 the fields are zero except on S_1'. This is because tangential components of the electric field are zero on S_1''. Assuming that the impedance of the load in S_1' in the receiving case is the complex conjugate of the impedance seen in the transmitting situation by virtue of the fields propagating in opposite directions, the field on S_1' must be related as follows: $\mathbf{E}^b = c\mathbf{E}^{a*}$ and $\mathbf{H}^b = -c\mathbf{H}^{a*}$, where c is a complex constant. Introducing these into Eq. 3.59, the result is

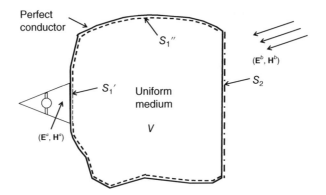

Figure 3.9 Power coupling theorem geometry

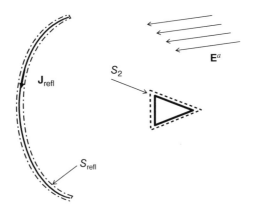

Figure 3.10 Power coupling theorem for a reflector and feed

$$c \iint_{S_1'} \left(\mathbf{E}^a \times \mathbf{H}^{a*} + \mathbf{E}^{a*} \times \mathbf{H}^a \right) \cdot \hat{n} \, dS = - \iint_{S_2} \left(\mathbf{E}^a \times \mathbf{H}^b - \mathbf{E}^b \times \mathbf{H}^a \right) \cdot \hat{n} \, dS. \qquad (3.60)$$

The integral on the left is a real quantity, and the right-hand integral is complex. Multiplying together Eq. 3.60 and its complex conjugate, it follows that

$$cc* \left(\iint_{S_1'} \left(\mathbf{E}^a \times \mathbf{H}^{a*} + \mathbf{E}^{a*} \times \mathbf{H}^a \right) \cdot \hat{n} \, dS \right)^2 = \left| \iint_{S_2} \left(\mathbf{E}^a \times \mathbf{H}^b - \mathbf{E}^b \times \mathbf{H}^a \right) \cdot \hat{n} \, dS \right|^2. \qquad (3.61)$$

The received power is, therefore, given by (Figure 3.10)

$$P_{\text{rec}} = \frac{1}{4} \iint_{S_1'} \left(\mathbf{E}^b \times \mathbf{H}^{b*} + \mathbf{E}^{b*} \times \mathbf{H}^b \right) \cdot \hat{n} \, dS$$

$$= \frac{cc*}{4} \iint_{S_1'} \left(\mathbf{E}^a \times \mathbf{H}^{a*} + \mathbf{E}^{a*} \times \mathbf{H}^a \right) \cdot \hat{n} \, dS.$$

Substituting the received power into Eq. 3.61 results in

$$P_{\text{rec}} = \frac{1}{8} \frac{\left| \iint_{S_2} (\mathbf{E}^a \times \mathbf{H}^b - \mathbf{E}^b \times \mathbf{H}^a) \cdot \hat{n} \, dS \right|^2}{\text{Re} \left\{ \iint_{S_2} (\mathbf{E}^a \times \mathbf{H}^{a*}) \cdot \hat{n} \, dS \right\}}. \tag{3.62}$$

Eq. 3.62 is one form of the power coupling theorem, which is also called field correlation. There are several variations of this theorem, which are useful in specific applications (Poulton et al., 1972; Wood, 1980). For example, if field a is due to a feed antenna (which could also be the field scattered by a subreflector) and b is due to an incident wave on a large reflector, the power coupled into the feed is from Eq. 3.62:

$$P_{\text{rec}} = \frac{1}{4} \frac{\left| \iint_{S_{\text{refl}}} (\mathbf{E}^a \times \mathbf{H}^b) \cdot \hat{n} \, dS \right|^2}{\text{Re} \left\{ \iint_{S_2} (\mathbf{E}^a \times \mathbf{H}^{a*}) \cdot \hat{n} \, dS \right\}}.$$

Letting $\mathbf{J}_{\text{refl}} = 2\hat{n} \times \mathbf{H}^b$ be the current induced on the reflector due to the incident wave (shown as b in Figure 3.9) and P_T be the total power transmitted by the feed, then

$$P_{\text{rec}} = \frac{\left| \iint_{S_{\text{refl}}} (\mathbf{E}^a \cdot \mathbf{J}_{\text{refl}}(\theta, \phi)) \, dS \right|^2}{16 P_T}. \tag{3.63}$$

The efficiency of the reflector system is then

$$\frac{P_{\text{rec}}}{P_b} = \eta(\theta, \phi) = \frac{\left| \iint_{S_{\text{refl}}} (\mathbf{E}^a \cdot \mathbf{J}_{\text{refl}}(\theta, \phi)) \, dS \right|^2}{16 P_T P_b}, \tag{3.64}$$

where P_b is the power contained in the incident wave.

The radiation pattern of the antenna can be calculated from Eq. 3.64 by changing the angle of incidence (θ, ϕ) of the plane wave. This has the advantage in calculations, such as for a complex antenna made up of several reflectors, as the field from the feed through the reflector system except the final one. The integration of this field with the induced current on the final reflector usually need only be computed once by numerical integration by employing techniques such as Simpson's rule or Gaussian quadrature (Fröberg, 1974).

3.8 Field Analysis by High-Frequency Methods*

In many situations involving aperture antennas, the size of the aperture and distance to the radiated field are large in terms of wavelengths. This means that approximate high frequency methods can be adopted and will often yield accurate results. Two main techniques for aperture antennas are described in this section. One technique involves a direct approximation of the physical optics integrals and the method limits in the extreme of very large apertures and distances, which is called asymptotic physical optics (APO) (James, 1986). The second approach uses techniques from geometric optics as well as particular high-frequency solutions to particular or canonical problems. These can derive from APO or as limits of known mathematical solutions such as electromagnetic scattering from a metallic wedge.

3.8.1 Asymptotic Physical Optics*

The radiation from line or ring sources is expressed in terms of the single integral (van Kampen, 1949)

$$I = \int_a^b d\xi f(\xi) \exp[jkg(\xi)], \qquad (3.65)$$

where $g(\xi)$ is the argument of the phase function on the aperture domain $D : \xi = (a, b)$. When the phase function $\exp(jkg)$ varies rapidly over the integral domain, integrals of this type can be evaluated asymptotically, which means the result applies at very high frequencies or the aperture dimensions are large compared with the wavelength. In that case, the value of the integral is given in the vicinity of the stationary points of the function g. That is, the integrals are given at the points of stationary phase. Near the stationary phase points, the function f and the exponential are expanded in Taylor series and these series can be arranged in descending powers of kR where R is related to the aperture dimensions. When kR is large, sufficient accuracy for practical applications is possible by taking only the leading terms of the asymptotic expansion.

Of the possible stationary points yielding contributions to an asymptotic expansion of Eq. 3.65, there are critical points of the first kind defined as

$$g_\xi(\xi) = 0, \qquad (3.66)$$

where $g_\xi = \partial g(\xi, \psi)/\partial \xi$. Critical points of the second kind, or edge points, are defined by (van Kampen, 1949)

$$g_\xi(a) = 0 \quad \text{or} \quad g_\xi(b) = 0. \qquad (3.67)$$

The critical points of the first kind produce a geometric optics type of contribution to the integral, while those of the second kind represent diffraction from the edge of the aperture. Now let ξ_o denote a critical point in general. In the vicinity of this point the functions g and f may be represented by their Taylor expansions

$$g(\xi) = g_0 + g_\xi u + \frac{1}{2} g_{\xi\xi} u^2 + \cdots$$

and

$$f(\xi) = f_0 + f_\xi u + \frac{1}{2} f_{\xi\xi} u^2 + \cdots,$$

where $u = \xi - \xi_o$, $g_o = g(\xi_o)$ and $f_o = f(\xi_o)$. If g and f are assumed to be single-valued in the domain D, the limits of the integrals can be extended infinitely as the remainder of the integral cancels out by virtue of rapid oscillation with argument. Also at any critical point it is possible to reverse the order of integration. The integrals in Eq. 3.65 become

$$I = \int_{-\infty}^{\infty} d\xi \cdots - \int_{b}^{\infty} d\xi \cdots - \int_{-a}^{\infty} d\xi f(-\xi) \exp(jkg(-\xi))$$

$$= I_\infty - I_b - I_{-a}.$$

Consider the integral of infinite extent, I_∞, if the interval contains a critical point of the first kind, then

$$g(\xi) \approx g_o + \frac{1}{2} g_{\xi\xi} u^2 \quad \text{and} \quad f(\xi) \approx f_o$$

and

$$I_\infty \sim f_o \exp(jkg_o) \int_{-\infty}^{\infty} du \exp\left[jkg_{\xi\xi}(\xi_o) \frac{u^2}{2} \right]. \tag{3.68}$$

The integral in Eq. 3.68 can be expressed in terms of a complex Fresnel integral, $\mathcal{K}(z)$, with zero argument, which has a simple value (see Appendix E). Thus,

$$\int_{-\infty}^{\infty} du \exp\left[jkg_{\xi\xi}(\xi_o) \frac{u^2}{2} \right] = \sqrt{\frac{2\pi}{k|g_{\xi\xi}(\xi_o)|}} \exp\left(j \operatorname{sgn}(g_{\xi\xi}) \frac{\pi}{4} \right),$$

where use has been made use of the symmetry of the integral and the identity

$$\int_{0}^{\infty} dt \exp\left[\pm jt^2 \right] = \frac{\sqrt{\pi}}{2} \exp\left(\pm j \frac{\pi}{4} \right).$$

Note that if f_o or $g_{\xi\xi}(\xi_o)$ are zero then the next terms in the Taylor series should be taken. The former involves little extension from the above. However, the latter involves the use of Airy functions. For this extension, the reader is directed to the references (James, 1986; Felsen & Marcuvitz, 1973).

The remaining two integrals are evaluated similarly. If the stationary points are not at the end points a or b, a first-order approximation for large arguments can be obtained by integration by parts. Thus, consider integral I_b:

$$I_b = \int_{b}^{\infty} d\xi\, f(\xi) \exp[jk\, g(\xi)] = \frac{1}{jk} \int_{b}^{\infty} d\xi\, \frac{f(\xi)}{g_\xi} (jkg_\xi) \exp[jk\, g(\xi)]$$

$$\sim \frac{1}{jk} \frac{f(b)}{g_\xi(b)} \exp[jk\, g(b)]. \tag{3.69}$$

A stronger result that applies whether the stationary point is at the end point or not is (James, 1986)

$$I_b \approx H(-\delta')I_\infty + \delta' f(b) \exp\left[jkg(b) - \mu jv^2\right] \sqrt{\frac{2}{k|g_{\xi\xi}(b)|}} F_\mu(v), \qquad (3.70)$$

where $\mu = \mathrm{sgn}(g_{\xi\xi}(b))$, $\delta = \mathrm{sgn}(b-\xi_o)$, $H(x)$ is Heaviside step function, $F_\mu(x)$ is a Fresnel integral, which has upward extending limits, and a positive or negative argument depending on the sign of μ as defined in Appendix E, and

$$v = \sqrt{\frac{k}{2|g_{\xi\xi}(b)|}} |g_\xi(b)|.$$

3.8.1.1   Example: Scattering Radiation from Large Conducting Wire Loop*

A large circular wire loop is illuminated by a plane wave that is polarized in the x–z plane as shown in Figure 3.11.
   The loop has a large radius $R \gg \lambda$ and small cross section $\tau \gg \lambda$ to the incident wave. The incident wave direction is

$$\hat{s}_I = -\hat{x}\sin\theta_I - \hat{z}\cos\theta_I$$

and the incident field

$$\mathbf{E}_I = E_o(\hat{x}\cos\theta_I - \hat{z}\sin\theta_I)\exp(-jks_I).$$

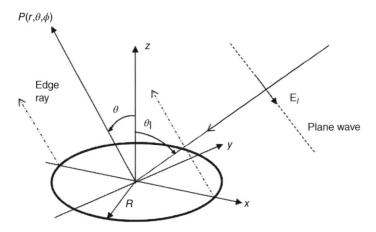

Figure 3.11   Circular loop illuminated by a plane wave

The current induced on loop is

$$\mathbf{J}_s = 2\hat{n} \times \mathbf{H}_I = \frac{2}{\eta_o}\hat{z} \times (\hat{s}_I \times \mathbf{E}_I)\exp(-jks_I) = \frac{2}{\eta_o}\hat{x}E_o\exp(-jks_I).$$

The radiated field can be calculated from

$$L_x = \int_0^{2\pi} d\phi' \int_{R-\tau/2}^{R+\tau/2} \left(\frac{2E_o}{\eta_o}\right)\exp(-jk(\boldsymbol{\rho}'\cdot\hat{r})-s_I)\rho'd\rho'$$

where $\boldsymbol{\rho}' = R(\hat{x}\cos\phi' + \hat{y}\sin\phi')$ and $\tau \ll R$. Therefore,

$$L_x = \frac{2E_o}{\eta_o}\exp(-jks_I)\int_0^{2\pi} d\phi' \int_{R-\tau/2}^{R+\tau/2}\exp(-jkR\cos(\phi'-\phi))\rho'd\rho'$$

$$\approx \frac{E_oR\tau}{\eta_o}\exp(-jks_I)\int_0^{2\pi} d\phi'\exp(-jkR\cos(\phi'-\phi)). \tag{3.71}$$

The integral in Eq. 3.71 can be evaluated in closed form and the result is

$$L_x = \frac{E_oR\tau}{\eta_o}\exp(-jks_I)(2\pi J_0(kR)). \tag{3.72}$$

Therefore, the scattered fields are expressed as

$$E_\theta \approx -jE_okR\tau\frac{\exp(-jkr-jks_I)}{r}J_0(kR)\sin\phi \tag{3.73a}$$

$$E_\phi \approx jE_okR\tau\frac{\exp(-jkr-jks_I)}{r}J_0(kR)\cos\phi. \tag{3.73b}$$

Alternatively, the integral in Eq. 3.71 could be evaluated asymptotically, and this provides a physical interpretation of the result. This approach is equivalent to finding the critical points of the second kind on the periphery because of the narrow width of the loop. Thus,

$$L_x = \frac{E_oR\tau}{\eta_o}\exp(-jks_I)\int_0^{2\pi} d\phi'\exp(jkg(\phi')),$$

where $g(\phi') = -R\cos(\phi'-\phi)$.
Now,

$$g_{\phi'}(\phi') = R\sin(\phi'-\phi) \quad \text{and} \quad g_{\phi'}(\phi') = 0.$$

When $\phi' = \phi$ and $\phi' = \phi + \pi$. Also $g_{\phi'\phi'}(\phi') = R\cos(\phi' - \phi)$. Therefore, there are two stationary point contributions to the integral. As well, because of the periodicity of the integral extend the region over an infinite domain as follows:

$$L_x \sim \frac{E_o R\tau}{\eta_o} \exp(-jks_I) \int_{-\infty}^{\infty} d\phi' \exp(jkg(\phi')). \tag{3.74}$$

There are two contributions to the integral of the type Eq. 3.65, one obtained in the vicinity of $\phi' = \phi$ and the other at $\phi' = \phi + \pi$. Thus,

$$L_x \sim \frac{E_o R\tau}{\eta_o} \exp(-jks_I) \left[\sqrt{\frac{2\pi}{kR}} \exp\left(-jkR + j\frac{\pi}{4}\right) + \sqrt{\frac{2\pi}{kR}} \exp\left(jkR - j\frac{\pi}{4}\right) \right]$$

$$= \frac{E_o \tau}{\eta_o} \sqrt{\frac{2\pi}{kR}} \exp(-jks_I) \left[\exp\left(-jkR + j\frac{\pi}{4}\right) + \exp\left(jkR - j\frac{\pi}{4}\right) \right] \tag{3.75}$$

$$= \frac{E_o \tau}{\eta_o} \sqrt{\frac{2\pi}{kR}} \exp(-jks_I) 2\cos\left(kR - \frac{\pi}{4}\right).$$

This is equivalent to expressing J_0 in Eq. 3.73 asymptotically as

$$J_0(z) \sim \sqrt{\frac{2}{\pi z}} \cos\left(z - \frac{\pi}{4}\right),$$

which is a standard large argument approximation to this function (Abramowitz & Stegun, 1965) (see Appendix B). The physical interpretation of this radiated field can therefore be given as consisting of two edge ray contributions that are 180° apart as illustrated in Figure 3.11.

3.8.1.2 Special Case: APO in Two Dimensions*

Many of the integrals involved in calculating radiation from circular apertures such as reflectors can be expressed (Jones & Kline, 1958)

$$I = \int_0^{2\pi} d\xi \int_0^{\psi_c} d\psi \, f(\xi, \psi) \exp[jk\,g(\xi, \psi)]. \tag{3.76}$$

The quantity $\psi = \psi_c$ defines the upper rim of the aperture. Let (ψ_o, ξ_o) denote a critical point on the surface. In the vicinity of this point, the functions g and f are expanded in their Taylor series once again as follows:

$$g(\xi, \psi) = g_o + g_\xi u + g_\psi v + \frac{1}{2}\left(g_{\xi\xi} u^2 + 2g_{\xi\psi} uv + g_{\psi\psi} v^2\right) + \cdots \tag{3.77}$$

and

$$f(\xi,\psi)=f_o+f_\xi u+f_\psi v+\frac{1}{2}\left(f_{\xi\xi}u^2+2f_{\xi\psi}uv+f_{\psi\psi}v^2\right)+\cdots,\tag{3.78}$$

where $u=\xi-\xi_o$, $v=\psi-\psi_o$, $g_o=g(\xi_o,\psi_o)$ and $f_o=f(\xi_o,\psi_o)$. The integral is now re-written to take advantage of the various critical points. Note that at any critical point it is possible to reverse the order of integration. The integrals in Eq. 3.76 become

$$\int_0^{2\pi}d\xi\int_0^{\psi_c}d\psi\cdots=\int_{-\infty}^{\infty}d\xi\int_{-\infty}^{\infty}d\psi\cdots-\int_{\psi_c}^{\infty}d\psi\int_{-\infty}^{\infty}d\xi\cdots.$$

Also Eq. 3.77 is re-expressed as

$$g(\xi,\psi)=g_o+\frac{g_{\xi\xi}}{2}\left(u+\left(\frac{g_\xi+g_{\xi\psi}v}{g_{\xi\xi}}\right)\right)^2+v\alpha+\frac{1}{2}v^2\beta-\frac{1}{2}\frac{(g_\xi)^2}{g_{\xi\xi}},$$

where $\alpha=g_\psi-g_\xi g_{\xi\psi}/g_{\xi\xi}$ and $\beta=\Delta/g_{\xi\xi}$ where

$$\Delta=\begin{bmatrix}g_{\xi\xi}&g_{\xi\psi}\\g_{\xi\psi}&g_{\psi\psi}\end{bmatrix}.$$

Completing the square in v gives

$$g(\xi,\psi)=g_o+\frac{g_{\xi\xi}}{2}\left(u+\left(\frac{g_\xi+g_{\xi\psi}v}{g_{\xi\xi}}\right)\right)^2+\frac{\beta}{2}\left(v+\frac{\alpha}{\beta}\right)^2-\frac{1}{2}\frac{\alpha^2}{\beta}-\frac{1}{2}\frac{(g_\xi)^2}{g_{\xi\xi}}.$$

Hence, the argument of the integral is

$$f(\xi,\psi)\exp[jkg(\xi,\psi)]\sim\left(f_o+f_\xi u+f_\psi v\right).\exp(jkg_o)\cdot\exp\left[\frac{-jk}{2}\left(\frac{\alpha^2}{\beta}+\frac{(g_\xi)^2}{g_{\xi\xi}}\right)\right]$$

$$\times\exp\left[\frac{jkg_{\xi\xi}}{2}\left(u+\left(\frac{g_\xi+g_{\xi\psi}v}{g_{\xi\xi}}\right)\right)^2\right]\cdot\exp\left[\frac{jk\beta}{2}\left(v+\frac{\alpha}{\beta}\right)^2\right].$$

If f neither varies too rapidly in the vicinity of the critical point nor vanishes there, as it would for zero edge illumination, the first term of its Taylor series is usually sufficiently accurate. Thus,

$$I\sim f_o.\exp(jkg_o)\cdot\exp\left[\frac{-jk}{2}\left(\frac{\alpha^2}{\beta}+\frac{(g_\xi)^2}{g_{\xi\xi}}\right)\right]\cdot\left(\int_{-\infty}^{\infty}d\xi\int_{-\infty}^{\infty}d\psi-\int_{\psi_c}^{\infty}d\psi\int_{-\infty}^{\infty}d\xi\right)$$

$$\times\exp\left[\frac{jkg_{\xi\xi}}{2}\left(u+\left(\frac{g_\xi+g_{\xi\psi}v}{g_{\xi\xi}}\right)\right)^2\right]\cdot\exp\left[\frac{jk\beta}{2}\left(v+\frac{\alpha}{\beta}\right)^2\right].\tag{3.79}$$

Considerable simplification results if the integral with respect to ξ is evaluated first. With the substitution

$$t = \sqrt{\frac{k|g_{\xi\xi}|}{2}} \left[u + \left(\frac{g_\xi + g_{\xi\psi} v}{g_{\xi\xi}} \right) \right],$$

the integral in ξ is expressed as

$$\int_{-\infty}^{\infty} d\xi \exp\left[\frac{jkg_{\xi\xi}}{2} \left(u + \left(\frac{g_\xi + g_{\xi\psi} v}{g_{\xi\xi}} \right) \right)^2 \right] = \sqrt{\frac{2}{k|g_{\xi\xi}|}} \int_{-\infty}^{\infty} dt \exp\left[j\,\text{sgn}(g_{\xi\xi}) t^2 \right]$$

$$= \sqrt{\frac{2\pi}{k|g_{\xi\xi}|}} \exp\left[j\sigma\frac{\pi}{4} \right],$$

(3.80)

where $\sigma = \text{sgn}(g_{\xi\xi})$. Therefore,

$$I \sim \sqrt{\frac{2\pi}{k|g_{\xi\xi}|}} \exp\left[j\sigma\frac{\pi}{4} \right] f_0 \cdot \exp(jkg_0) \cdot \exp\left[\frac{-jk}{2} \left(\frac{\alpha^2}{\beta} + \frac{g_\xi^2}{g_{\xi\xi}} \right) \right].$$

$$\times \left(\int_{-\infty}^{\infty} - \int_{\psi_c}^{\infty} \right) d\psi \exp\left[\frac{jk\beta}{2} \left(v + \frac{\alpha}{\beta} \right)^2 \right].$$

The integral with respect to ψ is now evaluated with the substitution

$$s = \sqrt{\frac{k|\beta|}{2}} \left(v + \frac{\alpha}{\beta} \right).$$

The result is dependent upon the region of integration. When the domain is infinite, as it is for the first integral, the evaluation is similar to the integral in ξ. Thus,

$$\int_{-\infty}^{\infty} d\psi \exp\left[\frac{jk\beta}{2} \left(v + \frac{\alpha}{\beta} \right)^2 \right] = \sqrt{\frac{2}{k|\beta|}} \int_{-\infty}^{\infty} \exp(j\mu s^2)$$

$$= \sqrt{\frac{2\pi}{k|\beta|}} \exp\left(j\mu\frac{\pi}{4} \right),$$

(3.81)

where $\mu = \text{sgn}(\beta)$. The second integral with respect to ψ is bounded on one side and this leads to a functionally different result. Making use of the same substitution, the integral simplifies to

$$\int_{\psi_c}^{\infty} d\psi \exp\left[\frac{jk\beta}{2} \left(v + \frac{\alpha}{\beta} \right)^2 \right] = \sqrt{\frac{2}{k|\beta|}} \int_{s_c}^{\infty} ds \exp\left[j\mu s^2 \right],$$

(3.82)

where $s_c = \sqrt{k|\beta|/2}\,(\psi_c - \psi_o + \alpha/\beta)$. Finally, combining all the above results, the initial expression in Eq. 3.79 is expressed asymptotically as

$$I \sim \sqrt{\frac{2}{k|g_{\xi\xi}|}}\sqrt{\frac{2}{k|\beta|}}\exp\left(j\sigma\frac{\pi}{4}\right)f_o\cdot\exp(jkg_o)\cdot\exp\left[\frac{-jk}{2}\left(\beta + \frac{(g_\xi)^2}{g_{\xi\xi}}\right)\right] \times$$

$$\left\{\sqrt{\pi}\exp\left(j\mu\frac{\pi}{4}\right) - F_\mu(s_c)\right\}, \tag{3.83}$$

where the function $F_\mu(x)$ is a Fresnel integral (refer to Appendix E) with sign μ on the exponential in the integrand.

Four special cases can be identified for the term in the curly braces of Eq. 3.83 for the instance when the critical point lies on the boundary or rim of the surface, that is, $\psi_o = \psi_c$ and $s_c = \alpha/\beta\sqrt{k|\beta|/2}$. The result depends on the signs of α and β, which in turn are functions of the derivatives of the phase function g. To do this use is made of the identify Eq. E.5 in Appendix E.

Case 1. $\alpha > 0$ and $\beta < 0$

$$\left\{\sqrt{\pi}\exp\left(j\mu\frac{\pi}{4}\right) - F_\mu(s_c)\right\} = F_-\left(\alpha\sqrt{\frac{k|\beta|}{2}}\right). \tag{3.84}$$

Case 2. $\alpha < 0$ and $\beta < 0$

$$\left\{\sqrt{\pi}\exp\left(j\mu\frac{\pi}{4}\right) - F_\mu(s_c)\right\} = \sqrt{\pi}\exp\left(-j\frac{\pi}{4}\right) - F_-\left(|\alpha|\sqrt{\frac{k}{2|\beta|}}\right). \tag{3.85}$$

Case 3. $\alpha < 0$ and $\beta > 0$

$$\left\{\sqrt{\pi}\exp\left(j\mu\frac{\pi}{4}\right) - F_\mu(s_c)\right\} = F_+\left(|\alpha|\sqrt{\frac{k}{2\beta}}\right). \tag{3.86}$$

Case 4. $\alpha > 0$ and $\beta > 0$

$$\left\{\sqrt{\pi}\exp\left(j\mu\frac{\pi}{4}\right) - F_\mu(s_c)\right\} = \sqrt{\pi}\exp\left(j\frac{\pi}{4}\right) - F_+\left(\alpha\sqrt{\frac{k}{2\beta}}\right). \tag{3.87}$$

Each case given by Eqs. 3.84–3.87 can be interpreted as a diffracted ray contribution emanating from a point on the boundary or rim of a surface not unlike the loop example in Figure 3.11. An extension of the ray diffraction interpretation is given in the following section.

3.8.2 Geometrical Theory of Diffraction*

The method of geometrical theory of diffraction (GTD) is a ray-based method of analysis, which assumes a high operating frequency compared with the size of the objects in the analysis. The wavefront and surfaces are fully described through geometric optics. The diffraction from other objects such as wedges or corners can be represented approximately in a ray-based system. The joining together of geometric optics and such canonical solutions is the essence of GTD. Initially consider the representation of a propagating wave in geometric optics. Suppose the direction of propagation is given by the vector s_I. The rays and ray paths are subject to Fermat's principle. This states that the path taken by rays from a source to an observation point is stationary with respect to small variations in that path, that is, with respect to a neighbouring path, the chosen path has a maximum or minimum value.

Around each ray as shown in Figure 3.12, there is a bundle of rays called a ray pencil. Consider the wavefronts (1) and (2) of the incident ray. Let there be an element of area dA_I around the incident ray. The wavefront will generally be curved and described by two radii of curvature ρ_{I1} and ρ_{I2}. For this element $dA_I = \alpha \rho_1 \rho_2$ where α is a constant for a propagating wave. As the wave propagates, the power flow is entirely through the ray pencil. From position A to B, conservation of energy requires

$$|\mathbf{E}_{IA}|^2 dA_{IA} = |\mathbf{E}_{IB}|^2 dA_{IB}, \tag{3.88}$$

where $\mathbf{E}_{IA}$ and $\mathbf{E}_{IB}$ are the vector field values at A and B with wavefront areas dA_{IA} and dA_{IB}, respectively. If s is the distance between the two wavefronts as the wave propagates along s_I, then the wavefront areas are related by

$$\frac{dA_{IB}}{dA_{IA}} = \frac{(\rho_{I1}+s)\,(\rho_{I2}+s)}{\rho_{I1}\rho_{I2}}. \tag{3.89}$$

From Eq. 3.88, this gives

$$|\mathbf{E}_{IB}| = \sqrt{\frac{\rho_{I1}\rho_{I2}}{(\rho_{I1}+s)\,(\rho_{I2}+s)}}|\mathbf{E}_{AB}|$$

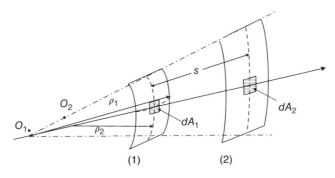

Figure 3.12 Representation of wavefront on same bundle of rays

and the application of the phase factor from position A to B gives

$$\mathbf{E}_{IB} = \sqrt{\frac{\rho_{I1}\rho_{I2}}{(\rho_{I1}+s)(\rho_{I2}+s)}}\mathbf{E}_{AB}\exp(-jk_1 s). \tag{3.90}$$

A reflected field could be given by Eq. 3.90 multiplied by the reflection matrix $\mathbf{R}$ as the ray pencil is modified by the interface. Similarly, the transmitted field is given by Eq. 3.90 multiplied by a transmission matrix $\mathbf{T}$. These matrices apply at the reflection/transmission point and are given by

$$\mathbf{R} = \begin{bmatrix} R^{\mathrm{e}} & 0 \\ 0 & R^{\mathrm{m}} \end{bmatrix} \tag{3.91a}$$

and

$$\mathbf{T} = \begin{bmatrix} T^{\mathrm{e}} & 0 \\ 0 & T^{\mathrm{m}} \end{bmatrix}, \tag{3.91b}$$

where R^{e} and R^{m} are the reflection coefficients for electric and magnetically polarized reflected fields and similarly T^{e} and T^{m} for the transmitted field. These are standard expressions for a plane interface (Kraus & Carver, 1973).

Reflection and refraction at a plane dielectric interface are shown in Figure 3.13. Medium 1 has a relative permittivity (dielectric constant) ε_{r1} and a relative permeability μ_{r1} and similarly for medium 2. The refractive index of medium i is defined as $n_i = \sqrt{\mu_{ri}\varepsilon_{ri}}$. Tables of relative permittivity and permeability of various materials are found listed in the references (Harrington, 1961; Bodnar, 2007). Implementation of Fresnel's principle results in two laws each for reflection and refraction. These laws are as follows:

1. The incident and reflected rays lie in the same plane as the normal to the interface, that is, $\hat{n}\cdot(\mathbf{s}_I \times \mathbf{s}_R) = 0$. From Figures 3.13 and 3.14, it is seen $\hat{n}$ is the normal to the surface while $\mathbf{s}_I$ and $\mathbf{s}_R$ are vectors in the incident and reflected ray directions.

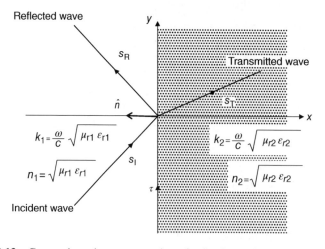

Figure 3.13 Geometric optics representation of reflection and refraction at an interface

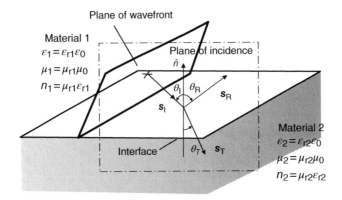

Figure 3.14   Reflection and refraction at a plane interface

2. The incident and reflected rays are at equal angles to the normal, that is, $(\mathbf{s}_I + \mathbf{s}_R) \cdot \hat{n} = 0$.
3. The refracted ray is diverted closer to the normal of the surface as the refractive index of the output media increases, that is, $(n_1 \mathbf{s}_I - n_2 \mathbf{s}_T) \times \hat{n} = 0$. This latter equation is a vector form of Snell's law. It indicates also that at a dielectric interface the vectors perpendicular to the plane of refraction are continuous on either side of the boundary.
4. The result in reflection law 2 applies for refraction as well.

The reflection coefficient for a wave incident on the interface depends on whether the E-field is parallel ($\parallel$) or perpendicular ($\perp$) to the plane of incidence and also on the material properties. From a consideration of a plane wave incident on the interface of lossy isotropic media, it can be shown (Jackson, 1999) that the reflection coefficient for the E-field parallel to the interface (i.e. $\parallel$) is given by

$$\Gamma^{\parallel} = \frac{(\mu_{r1}/\mu_{r2}) n_2^2 \cos\theta_1 - n_1\sqrt{n_2^2 - n_1^2 \sin^2\theta_1}}{(\mu_{r1}/\mu_{r2}) n_2^2 \cos\theta_1 + n_2\sqrt{n_2^2 - n_1^2 \sin^2\theta_1}}. \tag{3.92a}$$

In the case when the E-field perpendicular to the interface (i.e. $\perp$) the reflection coefficient is

$$\Gamma^{\perp} = \frac{n_1 \cos\theta_1 - (\mu_{r1}/\mu_{r2})\sqrt{n_2^2 - n_1^2 \sin^2\theta_1}}{n_1 \cos\theta_1 + (\mu_{r1}/\mu_{r2})\sqrt{n_2^2 - n_1^2 \sin^2\theta_1}}. \tag{3.92b}$$

θ_1 is the angle from the normal direction into region 1. The corresponding transmission coefficients are

$$T^{\parallel} = \left(1 + \Gamma^{\parallel}\right) \frac{\mu_{r2} n_1}{\mu_{r1} n_2} \tag{3.93a}$$

and

$$T^{\perp} = 1 + \Gamma^{\perp},$$ (3.93b)

where θ_2 is the angle from the normal direction in region 2. It can be shown that since s_I, s_R and $\hat{n}$ lie in the same plane $s_R = -s_I - 2\hat{n} \times (\hat{n} \times s_I) = s_I - 2(\hat{n} \cdot s_I)\hat{n}$. Also $n_2 s_T = n_1 s_I$. For example, suppose the interface of a dielectric with refractive index lies in the $x-z$ plane and $\hat{n} = \hat{y}$. A wave that is incident from air with a direction $s_I = -0.67\hat{x} - 0.5\hat{y}$, then $s_R = -0.67\hat{x} + 0.5\hat{y}$ and $s_T = 0.5(-0.67\hat{x} + 0.5\hat{y})$. Eqs. 3.93 are known collectively as the Fresnel equations for isotropic media. When the materials are lossless and non-magnetic ($\mu_{r1} = 1 = \mu_{r2}$), the reflection coefficients simplify to

$$\Gamma^{\parallel} = \frac{(\varepsilon_{r2}/\varepsilon_{r1})\cos\theta_1 - \sqrt{(\varepsilon_{r2}/\varepsilon_{r1}) - \sin^2\theta_1}}{(\varepsilon_{r2}/\varepsilon_{r1})\cos\theta_1 + \sqrt{(\varepsilon_{r2}/\varepsilon_{r1}) - \sin^2\theta_1}}$$

and

$$\Gamma^{\perp} = \frac{\cos\theta_1 - \sqrt{(\varepsilon_{r2}/\varepsilon_{r1}) - \sin^2\theta_1}}{\cos\theta_1 + \sqrt{(\varepsilon_{r2}/\varepsilon_{r1}) - \sin^2\theta_1}}.$$

Continuing with the description of the wavefront, Eq. 3.90 is the field of a spherical wave. However, when one of the wave's radii of curvature is large, say, $\rho_{12} \to \infty$, the wavefront is cylindrical. A cylindrical wave is represented by

$$\mathbf{E}_{IB} = \sqrt{\frac{\rho_{I1}}{(\rho_{I1} + s)}} \mathbf{E}_{AB} \exp(-jk_1 s).$$ (3.94)

When a wave is incident on a metallic wedge as shown in Figure 3.15, there is a diffracted ray depending on the observation position (ρ, ϕ). The type of field produced depends on whether the observation angle is inside the reflection boundary, that is, $\phi \le \pi - \phi_o$, or is beyond the shadow boundary, that is, $\phi \ge \phi_o + \pi$. In the same way as for reflection and refraction, Fresnel's principle provides two laws for diffraction. Once again the incident and diffracted rays lie in the same plane and also the optical path length from the source to the observation point is stationary with respect to small variations in the path. The ray bundle can be represented in the same way as described above for the previous cases. Thus, the diffracted field is represented as

$$\mathbf{E}_{DB} = \mathbf{D}\mathbf{E}_{IA} \sqrt{\frac{\rho_{I1}}{(\rho_{I1} + s)}} \exp(-jks),$$ (3.95)

where $\mathbf{D}$ is the edge diffraction matrix. The elements of the matrix are obtained from rigorous solutions for fields produced by electric or magnetic oriented sources in the form

$$\mathbf{D} = \begin{bmatrix} D^e & 0 \\ 0 & D^m \end{bmatrix}.$$

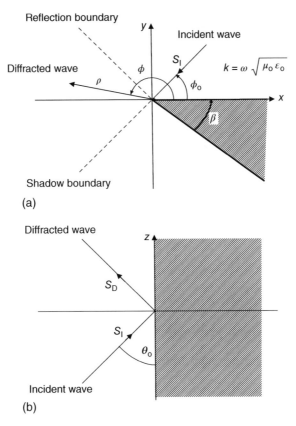

Figure 3.15   Plane wave illumination of a metallic wedge. (a) Incidence in plane of wedge angle β; and (b) incidence in plane of edge.

One of the first rigorous solutions for the wedge was obtained by Macdonald (1902), and the first rigorous uniform asymptotic series solution was obtained by Pauli (1938) of quantum exclusion principle fame. A field solution can be expressed as the sum of four terms: incident and reflected geometric optics fields and also a diffracted field associated with each optical term. As a result, the elements of the diffraction matrix for the metallic wedge can be expressed as (James, 1986)

$$D^{e,m} = \left[h(\Phi^i) + h(-\Phi^i)\right] \mp \left[h(\Phi^r) + h(-\Phi^r)\right], \tag{3.96}$$

where $\Phi^{i,r} = \phi \mp \phi_o$, $h(\Phi^{i,r}) = -\varepsilon^{i,r}\sqrt{\sigma^{i,r}} M_-(v^{i,r})\Lambda^{i,r}$, $\sigma^i = \sigma^r = \rho$ for a straight edge, $M_-(x)$ is a modified Fresnel integral (see Appendix E) and $v^{i,r} = \sqrt{k\sigma^{i,r}}|a^{i,r}|\sin\theta_o$ where θ_o is the incident angle to the edge. The modified Fresnel integral is given by (refer to Appendix E)

$$M_\pm(x) = \frac{\exp(\mp j(x^2 + \pi/4))}{\sqrt{\pi}} \int_x^\infty \exp(\pm jxt^2)\, dt. \tag{3.97}$$

The step function

$$\varepsilon^{i,r} = \mathrm{sgn}\left(a^{i,r}\right) = \begin{cases} +1 & \text{in source region} \\ -1 & \text{in shadow region} \end{cases}.$$

In addition,

$$a^{i,r} = \sqrt{2}\cos\left(\frac{\Phi^{i,r} + 2p\pi N}{2}\right) \tag{3.98}$$

with $N = (2\pi - \beta)/\pi$. Eq. 3.98 is independent of p for source and observation points removed from the optical boundaries $\phi = \pi \pm \phi_0$. On these boundaries, p is chosen to satisfy the following conditions:

$$\left|\Phi^{i,r} + 2p\pi N\right| = \pi \tag{3.99a}$$

where

$$\Lambda^{i,r} = 1 \tag{3.99b}$$

$$\Lambda^{i,r} = \frac{a^{i,r}}{\sqrt{2}N}\cot\left(\frac{\Phi^{i,r} + \pi}{2N}\right). \tag{3.99c}$$

The expressions in Eqs. 3.99 enable continuity to be achieved with the geometric optics field. They show that the correct behaviour across the two reflection boundaries is obtained by setting $p = 0$ in $h(-\Phi^r)$ and $p = -1$ in $h(\Phi^r)$. When a shadow boundary falls in visible space, the function $h\left(\pm\Phi^i\right)\big|_{p=0}$ is used with a value that depends on whether $\Phi^i = \mp\pi$ lies on the shadow boundary. The remaining term in Eq. 3.96 will be $h\left(\mp\Phi^i\right)\big|_{p=-1}$.

For large arguments of the modified Fresnel integral, $v^{i,r}$, in Eq. 3.96, the leading term of the asymptotic expansion given in Appendix E results in

$$h\left(\Phi^{i,r}\right) \sim \frac{-\csc\theta_0\cot\left(\left(\pi + \Phi^{i,r}\right)/2N\right)}{N\sqrt{8j\pi k}}. \tag{3.100}$$

In the special case of a metallic half-plane where $N = 2$, the diffraction coefficients Eq. 3.96 simplify to

$$D^{e,m} = -\left(\varepsilon^i\sqrt{\sigma^i}\right)M_-\left(v^i\right) \mp \left(\varepsilon^r\sqrt{\sigma^r}\right)M_-\left(v^r\right).$$

Diffraction coefficients can be derived for other geometries such as a dielectric wedge, a corner or a curved surface. It remains to say that the solution summarized here is sufficiently useful for applications described here such as diffraction by the edge of ground plane or the rim of a reflector antenna.

3.9 Problems

P3.1 a. Using field equivalence and the method of images show that when A is an aperture in a perfectly conducting ground plane containing the fields E, H (see Figure P3.1) the radiated electric field is given by

$$\mathbf{E}(r,\theta,\phi) = \frac{jk}{2\pi} \int_A \frac{e^{-jkR}}{R} \left[(\hat{z} \times \mathbf{E}_a) \times \hat{R} \right] dS'$$

where $R = r - r'$.

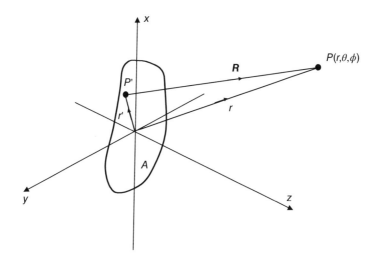

Figure P3.1 Arbitrary aperture in a ground plane

b. From Eq. P3.1 show that far from the aperture the non-zero electric field components are

$$E_\theta = \frac{jk}{2\pi} \frac{e^{-jkr}}{r} \left(N_x \cos \phi + N_y \sin \phi \right)$$

$$E_\phi = -\frac{jk}{2\pi} \frac{e^{-jkr}}{r} \cos \theta \left(N_x \sin \phi - N_y \cos \phi \right),$$

where

$$\mathbf{N} = \int_A \mathbf{E}_a \exp(jk\hat{r}\cdot\mathbf{r}) dS'.$$

P3.2 A coaxial transmission line with inner conductor radii a and b, respectively, is terminated in an infinite ground plane. Assume that the electric field in the aperture is

$$\mathbf{E}_a = -\hat{\rho} \frac{V}{\varepsilon_r \ln(b/a)} \frac{1}{\rho},$$

where V is the voltage between the conductors and ε_r is the dielectric constant of the material separating the conductors. Obtain the far-zone spherical components of the electric field radiated by the aperture.

P3.3 Determine the effect on the radiation pattern of a linear phase shift across a uniformly illuminated rectangular aperture with dimensions a × b. Assume the aperture field is given by

$$\mathbf{E}_a = \hat{x}E_o e^{-j\alpha x}; \quad |x| \le a; \quad |y| \le b.$$

P3.4 Repeat P3.3 to determine effect of a small quadratic phase error on the radiation of a uniformly illuminated aperture. Assume that

$$\mathbf{E}_a = \hat{x}E_o e^{-j\alpha x^2}; \quad |x| \le a; \quad |y| \le b$$

and $\alpha a^2 < 1$. Sketch the E-plane pattern when $\alpha a^2 = \pi/8$ and compare this with the pattern for the case $\alpha = 0$.

P3.5 Find the effect on the radiation pattern of a small random phase variation across a uniformly illuminated aperture. Assume that

$$\mathbf{E}_a = \hat{x}E_o e^{-j\alpha\theta(x)}; \quad |x| \le a; \quad |y| \le b$$

and $2\pi\alpha \ll 1$. The function θ is a uniformly distribution random process, where $0 \le \theta \le 2\pi$.

P3.6 From first principles show that the maximum gain of a uniformly illuminated circular aperture of diameter D is

$$G_{max} = \left(\frac{\pi D}{\lambda}\right)^2.$$

P3.7 Compare the far-zone fields radiated by a circular aperture containing a constant linearly polarized field when the aperture is
a. located in an infinite ground plane
b. located in free-space.
What is the major difference between the E-plane patterns and the H-plane patterns in each case?

P3.8 Compare the half-power beamwidth, the location of the first null and first sidelobe level of the radiation from a uniformly illuminated rectangular aperture, a cosine illuminated rectangular aperture and a uniformly illuminated circular aperture. Refer to Figure 3.4 for details of each type of illumination.

P3.9 An aperture antenna has been proposed for a microwave link application at 4 GHz where the distance between transmitter and receiver is 30 km. At a distance of 15 km there is a hill of height 30 m. The link consists of two identical antennas with diameter D that are mounted on towers 100 m high. Ignoring atmospheric effects, but including the curvature of the earth, estimate the minimum diameter of antennas required to ensure that the beam is unblocked out to the 6-dB point of the main beam. Assume a uniform aperture distribution and earth radius $R_e = 6371$ km.
Answer: $D \ge 148\lambda$.

P3.10 Obtain the radiated electric field in the Fresnel zone of a uniformly illuminated aperture of radius a and therefore the radiated power per unit solid angle. Show that the gain is $G = G_o[S(ka^2/4R)]^2$ where $G_o = 4\pi(\pi a^2)/\lambda$ is the far-field gain and R is the distance from the aperture origin to the observation point. Show that at $R = 2a^2/\lambda$, $G/G_o = 0.81$ and at the far-field distance $R = 8a^2/\lambda$, $G/G_o = 0.99$.

P3.11 Show that in the Fresnel zone of an aperture of radius a (see Figure P3.2), the phase from one annulus of width $\lambda/2$ to the next on the aperture changes sign so that the total contribution is almost zero. Assume that the aperture illumination is uniform. What happens to the field at P on the axis as R is increased?

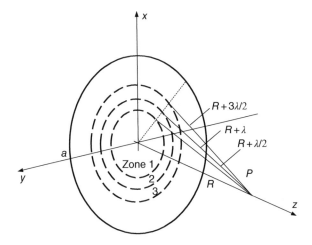

Figure P3.2 Fresnel zones on an aperture due to a source at P

P3.12 A circular aperture antenna has a far-field pattern function $A(\theta)$ in the E-plane and $B(\theta)$ in the H-plane. Based on this information, obtain expressions for the far-fields. From these show that the co-polar pattern in the $45°$-plane is given by $|A(\theta) + B(\theta)|/2$ and the cross-polar pattern in this plane is $|A(\theta) - B(\theta)|/2$.

P3.13 A commonly occurring integral in radiation problems is of the form

$$I(h) = \int_C f(z)e^{jh\Phi(z)}\,dz$$

where h is a large positive parameter, $\Phi(z) = -\cos(\theta - z)$, $f(z)$ is a complex illumination function and C is a contour in the z-plane.

a. Show that at the stationary point z_s an asymptotic expansion of $I(h)$ is

$$I(h) \sim \frac{e^{jh\Phi(z_s)}}{\sqrt{h}}\sum_{n=0}^{\infty}\frac{a_n}{h^n}$$

where a_n are expressed in terms of derivatives of $f(z)$ and $\Phi(z)$.

b. Verify the first coefficient of the expansion is $a_0 = \sqrt{2\pi}f(\theta)e^{j\pi/4}$.

P3.14 Obtain the radiation pattern of a uniformly illuminated circular aperture of diameter D, in the near-field region, the Fresnel zone and the far-field region. Plot and compare the patterns for an aperture of diameter $D = 50\lambda$ when the distance r is
(a) $r = 10\lambda$; (b) $r = 100\lambda$; and (c) $r = 5000\lambda$.

P3.15 Suppose an x-directed aperture distribution in a rectangular aperture with dimensions $a \times b$ is symmetric but triangular in the x-direction and uniform in the y-direction. Obtain the far-field radiation pattern of this distribution.

References

Abramowitz, M. and Stegun, I.A. (1965): 'Handbook of mathematical functions', Dover Inc., New York.

Bird, T.S. and Granet, C. (2013): 'Profiled horns and feeds' (Chapter 5) L. Shafai, S.K. Sharma & S. Rao (eds.) 'Vol. II: Feed systems' of 'Handbook of reflector antennas', Artech House, Chicago, IL, pp. 123–155.

Bodnar, D.G. (2007): 'Materials and design data' (Chapter 55). In: Volakis, J.L. 'Antenna engineering handbook', 4th ed., McGraw-Hill, New York.

Born, M. and Wolf, E. (1959): 'Principles of optics', Pergamon Press, London, UK.

Bouwkamp, C.J. and de Bruijn, N.G. (1946): 'The problem of optimum antenna current distribution', Philips Research Reports, Vol. 1, pp. 135–158.

Bracewell, R.N. (1961): 'Tolerance theory in large antennas', IRE Trans. Antennas Propagat., Vol. AP-9, pp. 49–58.

Brigham, E.O. (1974): 'The fast Fourier transform', Prentice-Hall Inc., Eaglewood Cliffs, New Jersey.

Felsen, L.B. and Marcuvitz, N. (1973): 'Radiation and scattering of waves', Prentice-Hall, Upper Saddle River, NJ.

Fröberg, C.-F. (1974): 'Introduction to numerical analysis', Addison-Wesley, London, UK.

Harrington, R.F. (1961): 'Time-harmonic electromagnetic fields', McGraw-Hill, New York.

Jackson, J.D. (1999): 'Classical electrodynamics', 3rd ed., John Wiley & Sons, Ltd, Chichester.

James, G.L. (1986): 'Geometrical theory of diffraction for electromagnetic waves', 3rd ed., Peter Peregrinus Ltd., London, U.K.

Jones, D.S. and Kline, M. (1958): 'Asymptotic expansion of multiple integrals and the method of stationary phase', J. Math. Phys., Vol. 37, pp. 1–28.

Kraus, J.D. and Carver, K.R. (1973): 'Electromagnetics', 2nd ed., McGraw-Hill, International Student Edition, Kagakuska Ltd, Tokyo, Japan.

Macdonald, H.M. (1902): 'Electric waves', Cambridge University Press, Cambridge, UK.

Masterman, P.H. (1973): 'A review of limited beam-steering microwave antennas for communications satellite earth stations', Ministry of Defence, London, UK. SRDE Research Report No. 73016.

Oppenheim, A.R. and Schafer, R.W. (1975): 'Digital signal processing', Prentice-Hall, Englewood Cliffs, NJ.

Pauli, W. (1938): 'On asymptotic series for functions in the theory of diffraction of light', Phys. Rev., Vol. 54, pp. 924–931.

Poulton, G.T., Lim, S.H. and Masterman, P.H. (1972): 'Calculation of input-voltage standing-wave ratio for a reflector antenna', Electron. Lett., Vol. 8, No. 25, pp. 610–611.

Riblet, H.J. (1948): 'Note on the maximum directivity of an antenna', Proc. IRE, Vol. 36, pp. 620–623.

Robieux, J. (1959): 'Lois génèrales de la liaison entre radiateurs d'ondes. Application aux ondes de surfaces et a la propagation', Ann. Radioelect., Vol. 14, pp. 187–229.

Silver, S. (1946): 'Microwave antenna theory and design', first published by McGraw-Hill, New York. Reprint published by Peter Peregrinus Ltd., London, UK, 1984.

van Kampen, N.G. (1949): 'An asymptotic treatment of diffraction problems: Pt. I', Physica, XIV, Jan., pp. 575–589. Pt. II published, Physica, XVI, Dec. 1950, pp. 817–821.

Volakis, J.L. (2007): 'Antenna engineering handbook', 4th ed., McGraw-Hill, New York.

Wood, P.J. (1980): 'Reflector antenna analysis and design', Peter Peregrinus Press, London, UK.

4

Waveguide and Horn Antennas

4.1 Introduction

Waveguide and horn antennas are based on the method of generating an electromagnetic wave from an exciter, or probe, at one end of a guiding structure with an open aperture. The wave travels to this aperture where it is mostly transmitted as radiation, and if the transition is well matched, only a small fraction of the wave is reflected back towards the source. The objective of the design is to obtain a smooth transition from the probe to free-space with as little reflection as possible and to produce a suitably directive beam. The first waveguide was first experimentally demonstrated by Oliver Lodge in 1894 and theoretically described by Rayleigh a little later (Sarkar et al., 2006). For experiments on polarization properties of crystals in 1895, Bose used a circular waveguide as a radiator as well as pyramidal horns in 1897 for further investigations on polarization as well as index of refraction. The set-up he used is shown in Figure 4.1 (Bose, 1927).

There appear to have been few developments on aperture antennas beyond this early work until the 1930s when given impetus for communications and radar. The Radiation Laboratory book by Silver (1946) provided many new horn designs and concepts for future work. Incremental progress continued until the 1960s when the theory of matched feeds for reflectors was developed (Minnett & Thomas, 1966; Rumsey, 1966). At that time it was realized that a corrugated waveguide or horn was a way of achieving conjugate matching to the focal field of a reflector. Further demanding requirements in communications (e.g. satellite) and radar resulted in further new horn designs, some of which will be described here. In this chapter, the basic properties of horns will be examined, and horns in common use will be described. As well, the material developed here will be used in following chapters as feeds for reflectors in Chapter 6, as elements of aperture arrays in Chapter 7 and in the design of applications outlined in Chapter 10.

Fundamentals of Aperture Antennas and Arrays: From Theory to Design, Fabrication and Testing,
First Edition. Trevor S. Bird.
© 2016 John Wiley & Sons, Ltd. Published 2016 by John Wiley & Sons, Ltd.
Companion website: www.wiley.com/go/bird448

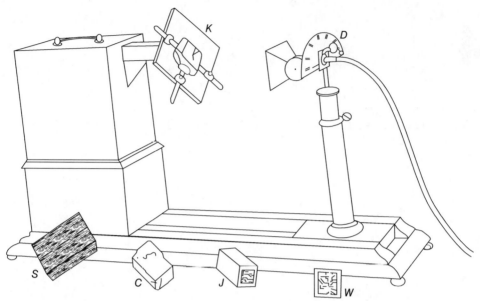

K, crystal-holder; S, a piece of stratified rock; C, a crystal; J, jute polariser; W, wire-grating polarsier; D, vertical graduated disc, by which the rotation is measured.

Figure 4.1 Apparatus used by Bose to measure polarization and double refraction at an evening lecture at the Royal Society in January 1897. *Source*: Reproduced from collected Physical Papers, Longmans, Green & Co. 1927 (Bose, 1927)

4.2   Radiation from Rectangular Waveguide

Suppose a rectangular waveguide of width a and height b (Figure 4.2) is excited only in its fundamental mode, the TE_{10} mode. The transverse fields of this mode are

$$E_x = E_o \cos\left(\frac{\pi y}{a}\right) e^{-j\beta z} \tag{4.1a}$$

$$H_y = Y_w E_x; \tag{4.1b}$$

where E_o is a constant, $\beta = \sqrt{k^2 - (\pi/a)^2}$ is the propagation constant of the TE_{10} mode in the z-direction and $Y_w = \beta/k\eta_o$ is the wave admittance of the mode. When the TE_{10} mode is incident on the open waveguide, some energy is reflected and some is stored in the aperture (as evanescent fields of higher-order modes). Typically, the reflection coefficient is less than -10 dB. The calculation of the self admittance of TE_{10} mode is detailed in Section 7.3.5.2. Reflection at the aperture tends to have only secondary effects on the radiation. Therefore, assume the fields in the aperture ($z = 0$) are approximately

$$\mathbf{E_a} = \hat{x} E_o \cos\left(\frac{\pi y'}{a}\right), \tag{4.2a}$$

$$\mathbf{H_a} = Y_w \hat{z} \times \mathbf{E_a}. \tag{4.2b}$$

The calculation of the radiated fields proceeds in the same way as for a uniformly illuminated rectangular aperture that was described in Section 3.4.1. Following from Eq. 4.2a, N_x is

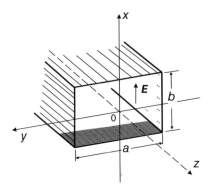

Figure 4.2 Geometry for radiating rectangular waveguide

the only non-zero component of Eq. 3.21, while the only non-zero component of Eq. 3.22 is given by

$$L_y = Y_w N_x. \tag{4.3}$$

The result differs from the uniform case, however, because the field in the y-direction is cosine distributed and the transform given as Eq. 3.28 is required. Therefore, for the TE_{10} mode only in the aperture, it follows that

$$N_x = E_o \frac{2ab}{\pi} S(\pi ub) C(\pi va), \tag{4.4}$$

where $u = \sin \theta \cos \phi / \lambda$ and $v = \sin \theta \sin \phi / \lambda$.

Using Eqs. 4.3 and 4.4 in the field expressions Eq. 3.20 gives the fields radiated by the waveguide as follows:

$$E_\theta(r,\theta,\phi) \approx \frac{jkabE_o}{2\pi^2} \frac{e^{-jkr}}{r} \left(1 + \frac{\beta}{k}\cos\theta\right) S(\pi ub) C(\pi va) \cos\phi \tag{4.5a}$$

$$E_\phi(r,\theta,\phi) \approx \frac{jkabE_o}{2\pi^2} \frac{e^{-jkr}}{r} \left(\frac{\beta}{k} + \cos\theta\right) S(\pi ub) C(\pi va) \sin\phi. \tag{4.5b}$$

Also the magnetic field components are as follows:

$$H_\theta(r,\theta,\phi) = -\frac{1}{\eta} E_\phi(r,\theta,\phi), \tag{4.5c}$$

$$H_\phi(r,\theta,\phi) = \frac{1}{\eta} E_\theta(r,\theta,\phi). \tag{4.5d}$$

The bracketed terms, $(1 + (\beta/k)\cos\theta)$ and $((\beta/k) + \cos\theta)$, are referred to as Huygens or obliquity factors and are typically slowly varying functions compared to the pattern functions

S and C. Note that for frequencies well above cut-off $\beta \approx k$, identical Huygens factors result for the two field components.

The principal plane radiation patterns occur in the x–z ($\phi = 0$) and the y–z planes ($\phi = \pm 90°$) and correspond to the E- and H-plane patterns, respectively.

E-plane ($\phi = 0$):

$$E_\theta = \frac{jkabE_o e^{-jkr}}{2\pi^2 r}\left(1 + \frac{\beta}{k}\cos\theta\right)S\left(\frac{kb}{2}\sin\theta\right). \tag{4.6}$$

The normalized power pattern is

$$P_E = \left[\left(1 + \frac{\beta}{k}\cos\theta\right)S\left(\frac{kb}{2}\sin\theta\right)\right]^2, \tag{4.7}$$

which is dominated by the S^2 function (see Figure 3.4). This dependence is expected because the E-plane aperture field is constant.

H-plane ($\phi = \pm 90°$):

$$E_\phi = -\frac{jkabE_o e^{-jkr}}{2\pi^2 r}\left(\frac{\beta}{k} + \cos\theta\right)C\left(\frac{ka}{2}\sin\theta\right). \tag{4.8}$$

The normalized power pattern is

$$P_H = \left[\left(\frac{\beta}{k} + \cos\theta\right)C\left(\frac{ka}{2}\sin\theta\right)\right]^2. \tag{4.9}$$

As the aperture field in the H-plane is cosine distributed, the H-plane radiation pattern is dominated by the C^2 function (refer to Figure 3.4).

The maximum gain of a TE_{10} mode excited rectangular waveguide is

$$G_{max} = \frac{32\,ab}{\pi\,\lambda^2}, \tag{4.10}$$

where it is assumed that $\beta \approx k$. By means of Eq. 3.49, the aperture efficiency of this aperture is $\eta_a = 8/\pi^2 \approx 0.811$. That is, the gain of a rectangular waveguide is 81.1% compared to the gain of a uniformly illuminated rectangular aperture with uniform phase.

4.3 Pyramidal Horn

Flaring the rectangular waveguide into a pyramidal horn, Figure 4.3a, provides not only a more directive radiation pattern but also a better transition from the feeding waveguide to free-space. The fields in the aperture of the horn may be found by treating the horn as a radial waveguide. As a first approximation, however, one may assume a TE_{10} mode is maintained in the flared section all the way to the aperture. This approximation works well providing the flare angle is

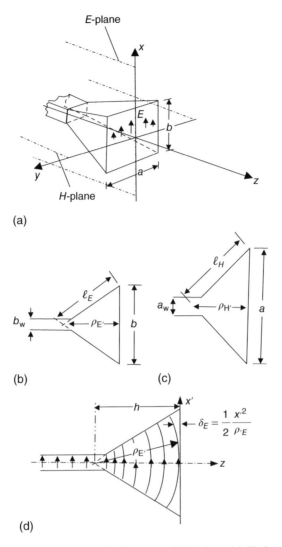

Figure 4.3 The pyramidal horn antenna. (a) Geometry; (b) E-plane; (c) H-plane; (d) quadratic phase factor in E-plane

not too great ($<10°$). For uniformity of notation, flare angle is defined as the angle between the centre-line and the linear taper. The angle between the two tapered sides will be called the full flare angle. An improved approximation is needed for greater flare angles because the wave phase-front at the aperture is no longer uniform. An accurate method for modeling flares and steps along the horn is described in a later section.

Consider the E-plane section of the horn, Figure 4.3d. When the TE_{10} mode in the waveguide reaches the flare, it expands outwards in order to satisfy the boundary conditions and forms a cylindrical wave. In what follows a TE_{10} mode is assumed as the basis of the representation of the field in the aperture. The TE_{10} mode is the fundamental mode of a rectangular waveguide with width a and height b such that $a > b$. When the waveguide is flared as shown in

Figure 4.3d, the phase of the field varies across the aperture. The difference is given by the extra distance from the wavefront to the aperture plane, δ_E. It can be shown that this distance is approximately given by

$$\delta_E = \frac{1}{2}\frac{x^2}{\rho'_E},$$

(4.11)

where ρ'_E is the radial distance from the apex to a point on the aperture, x, in the E-plane as shown in Figure 4.3d. Similarly, in the H-plane, the distance from the cylindrical wavefront of radius ρ'_H to the aperture is

$$\delta_H = \frac{1}{2}\frac{y^2}{\rho'_H}.$$

(4.12)

By the nature of Eqs. 4.11 and 4.12, the horn is said to have a quadratic phase dependence across its aperture. The radial distances are related to the basic geometry of the horn. Let the width and height of the input waveguide be a_w and b_w, respectively. If the length of the pyramidal section is h, by similar triangles, it follows that $\rho'_E = hb/(b-b_w)$ and $\rho'_H = ha/(a-a_w)$.

To account for the quadratic phase from the pyramidal section, an additional phase factor is applied to the TE_{10} mode transverse fields. At $z=0$, this results in approximate aperture fields of a pyramidal horn given by

$$\mathbf{E_a} = \hat{x}E_o \cos\left(\frac{\pi y}{a}\right)\exp\left[-\frac{jk}{2}\left(\frac{x^2}{\rho'_E}+\frac{y^2}{\rho'_H}\right)\right],$$

(4.13a)

$$\mathbf{H_a} = \frac{1}{\eta_o}\hat{z}\times\mathbf{E_a},$$

(4.13b)

where compared with Eqs. 4.5, the aperture is now assumed to be well above cut-off. Once again the radiated fields can be found from Eq. 3.20. The Fourier transforms of the aperture fields are found from Eq. 3.24:

$$N_x(u,v) = E_o \int_{-a/2}^{a/2} dy' \cos\left(\frac{\pi y'}{a}\right)\exp\left[j\left(2\pi vy'-\frac{ky'^2}{2\rho'_H}\right)\right]\times\int_{-b/2}^{b/2} dx' \exp\left[j\left(2\pi ux'-\frac{kx'^2}{2\rho'_E}\right)\right]$$

$$= E_o I_x(u)I_y(v)$$

(4.14a)

and

$$L_y = \frac{1}{\eta_o}N_x,$$

(4.14b)

where

$$I_x(u) = \int_{-b/2}^{b/2} dx' \exp\left[j\left(2\pi ux'-s_E^2 x'^2\right)\right]$$

(4.15)

and

$$I_y(v) = \int_{-a/2}^{a/2} dy' \cos\left(\frac{\pi y'}{a}\right) \exp\left[j\left(2\pi v y' - s_H^2 y'^2\right)\right] \tag{4.16}$$

where $s_E = \sqrt{k/2\rho_E'}$ and $s_H = \sqrt{k/2\rho_H'}$. The above integrals can be evaluated by several means such as by numerical integration (e.g. Simpson's rule) (Fröberg, 1974) or equivalently with a fast Fourier transform (FFT) algorithm (Oppenheim & Shafer, 1975). There is also a closed form solution in terms of cosine or sine integrals (Balanis, 1982, p. 583f) and another where the integrals are expressed in terms of complex Fresnel integrals. To obtain the latter, first consider I_x and complete the square in the exponential in the integrand as follows:

$$I_x(u) = \int_{-b/2}^{b/2} dx' \exp\left[j\left(2\pi u x' - s_E^2 x'^2\right)\right] = \int_{-b/2}^{b/2} dx' \exp\left[-j\left(\left(s_E x' - \frac{\pi u}{s_E}\right)^2 - \left(\frac{\pi u}{s_E}\right)^2\right)\right].$$

Substitute $\xi = (s_E x' - \pi u/s_E)$ so that I_x can be simplified as follows:

$$I_x(u) = \frac{\exp\left[j(\pi u/s_E)^2\right]}{s_E} \int_{-s_E b/2 - \pi u/s_E}^{s_E b/2 - \pi u/s_E} d\xi \exp\left(-j\xi^2\right)$$

$$= \frac{\exp\left[j(\pi u/s_E)^2\right]}{s_E} \left(\mathcal{K}\left(s_E \frac{b}{2} - \frac{\pi u}{s_E}\right) + \mathcal{K}\left(s_E \frac{b}{2} + \frac{\pi u}{s_E}\right)\right), \tag{4.17}$$

where $\mathcal{K}(z) = \int_0^z \exp\left(-j\xi^2\right) d\xi$ is the complex Fresnel integral (see Appendix E). The approach to evaluate I_y is similar except that initially the cosine in the integrand is expanded in its exponential components. Then the square is completed on both components. Thus,

$$I_y(v) = \int_{-a/2}^{a/2} dy' \cos\left(\frac{\pi y'}{a}\right) \exp\left[j\left(2\pi v y' - s_H^2 y'^2\right)\right]$$

$$= \frac{1}{2} \int_{-a/2}^{a/2} dy' \left\{ \exp\left[j\left(y'\left(2\pi v + \frac{\pi}{a}\right) - s_H y'^2\right)\right] + \exp\left[j\left(y'\left(2\pi v - \frac{\pi}{a}\right) - s_H y'^2\right)\right] \right\}$$

$$= \frac{1}{2} \int_{-a/2}^{a/2} dy' \left\{ \exp\left[-j\left(\left(s_H y' - \frac{\pi}{s_H}\left(v + \frac{1}{2a}\right)\right)^2\right)\right] \exp\left[j\left(\frac{\pi}{s_H}\left(v + \frac{1}{2a}\right)\right)^2\right] \right.$$

$$\left. + \exp\left[-j\left(\left(s_H y' - \frac{\pi}{s_H}\left(v - \frac{1}{2a}\right)\right)^2\right)\right] \exp\left[j\left(\frac{\pi}{s_H}\left(v - \frac{1}{2a}\right)\right)^2\right] \right\}.$$

Let $\alpha_\pm(v) = (\pi/s_H)(v \pm (1/2a))$. Then,

$$I_y(v) = \frac{1}{2s_H} \left\{ \begin{array}{l} \exp[j\alpha_+^2] \left[\mathcal{K}\left(s_H\frac{a}{2} - \alpha_+(v)\right) + \mathcal{K}\left(s_H\frac{a}{2} + \alpha_+(v)\right)\right] \\ + \exp[j\alpha_-^2] \left[\mathcal{K}\left(s_H\frac{a}{2} - \alpha_-(v)\right) + \mathcal{K}\left(s_H\frac{a}{2} + \alpha_-(v)\right)\right] \end{array} \right\}.$$

Therefore, Eq. 4.14a can be expressed as

$$\begin{aligned} N_x(u,v) = &\frac{E_o}{2s_E s_H} \exp\left[j\left(\frac{\pi u}{s_E}\right)^2\right] \left[\mathcal{K}\left(s_E\frac{b}{2} - \frac{\pi u}{s_E}\right) + \mathcal{K}\left(s_E\frac{b}{2} + \frac{\pi u}{s_E}\right)\right] \\ &\times \left\{ e^{j\alpha_+^2} \left[\mathcal{K}\left(s_H\frac{a}{2} - \alpha_+(v)\right) + \mathcal{K}\left(s_H\frac{a}{2} + \alpha_+(v)\right)\right] \right. \\ &\left. + e^{j\alpha_-^2} \left[\mathcal{K}\left(s_H\frac{a}{2} - \alpha_-(v)\right) + \mathcal{K}\left(s_H\frac{a}{2} + \alpha_-(v)\right)\right] \right\}. \end{aligned}$$

$$\text{(4.18)}$$

The electric field components in the far-field are then obtained from Eq. 3.20, giving

$$E_\theta(r,\theta,\phi) = \frac{jk}{4\pi}\frac{e^{-jkr}}{r} N_x(\theta,\phi)(1 + \cos\theta)\cos\phi \tag{4.19a}$$

$$E_\phi(r,\theta,\phi) = -\frac{jk}{4\pi}\frac{e^{-jkr}}{r} N_x(\theta,\phi)(1 + \cos\theta)\sin\phi \tag{4.19b}$$

where N_x is given by Eq. 4.18. The total radiated power from the TE_{10} mode excited pyramidal horn is

$$\begin{aligned} P_T &= \frac{|E_o|^2}{2\eta_o} \int_{-b/2}^{b/2} dx' \int_{-a/2}^{a/2} dy' \cos^2\left(\frac{\pi y'}{a}\right) \\ &= \frac{ab}{4\eta_o}|E_o|^2. \end{aligned}$$

$$\text{(4.20)}$$

Assuming the maximum gain occurs on axis, by Eqs. 3.48 and 4.20

$$\begin{aligned} G_{\max} &= \frac{8\pi r^2 \left(|E_\theta(0,0)|^2 + |E_\phi(0,0)|^2\right)}{ab|E_o|^2} \\ &= \frac{2k^2}{\pi ab}\left|\frac{N_x(0,0)}{E_o}\right|^2. \end{aligned}$$

$$\text{(4.21)}$$

When $u = 0 = v$ is set in Eq. 4.18, Eq. 4.21 becomes

$$G_{\max} = \frac{8k^2}{\pi ab}\left[\frac{\left|\mathcal{K}\left(s_E\frac{b}{2}\right)\right|\left|\mathcal{K}\left(s_H\frac{a}{2} - \frac{\pi}{2as_H}\right) + \mathcal{K}\left(s_H\frac{a}{2} + \frac{\pi}{2as_H}\right)\right|}{s_E s_H}\right]^2. \tag{4.22}$$

It may be shown that the maximum gain is a product of the gains of E- and H-plane sectoral horns times a geometric factor (see Problem P4.3).

A horn of given aperture and waveguide feed dimensions has an optimum length for maximum gain. The reason there is an optimum length is that as the horn length is increased, the gain increases initially until quadratic phase error dominates in the aperture field. Maximum gain occurs when the gain increases, due to an initial increase in length that is cancelled by quadratic phase error. After this maximum, gain falls with increasing length. A maximum gain pyramidal horn is often referred to as a standard gain horn (SGH) because accurate reproducible gain is achieved by accurately setting the horn dimensions. The SGH is widely used as a reference antenna for all types of measurements ranging from electromagnetic interference (EMI) tests to calibration of other antennas. Over the years, a considerable amount of work has gone into deriving accurate formulae for the gain. One of the reasons is that the gain of these horns is moderately high and predictable.

The SGH is sometimes referred to as an optimum gain horn. The usual definition of an optimum gain horn in relation to a pyramidal profile is a horn that has a maximum gain for a given length of horn. The greatest departure from uniform phase occurs at the edge of the aperture. For a horn with a linear profile, this occurs along the slope through any section of the horn be it rectangular, circular, or any other general cross section. In general, the phase error is in the form of Eq. 4.11, that is, $\alpha = ka^2/2L$, where L is the slant length from the apex.

A solution for the optimum geometry for maximum gain can be obtained as described by Bird and Love (2007). In this solution, it is assumed that the aperture dimensions are related to the long dimension of the flare in the E- and H-planes, ℓ_E and ℓ_H, respectively, through $a = \sqrt{\alpha_1 \ell_H}$ and $b = \sqrt{\beta_1 \ell_E}$, and the required gain is given by $G_r = g_1 ab/\lambda^2$ where the quantities α_1, β_1 and g_1 have been obtained from experience or through optimization. Typical values are $g_1 = 2\pi$, $\alpha_1 = 6\pi/k$ and $\beta_1 = 4\pi/k$. However, from optimization, improved design results are obtained if instead $g_1 = 1.992\pi$, $\alpha_1 = 6.10\lambda/k$, and $\alpha_1 = 4.14\lambda/k$ is chosen.

4.3.1   Design of a Standard Gain Pyramidal Horn

The steps for designing an optimum gain pyramidal horn are listed by Bird & Love (2007). The design commences from a specified gain G_r (initially specified in dBi, but converted to its dimensionless ratio for the calculations), and dimensions a_w and b_w of the input rectangular waveguide feed. The aim is to determine the remaining dimensions $(a, b, \rho'_E, \rho'_H, \ell_E, \ell_H$ and $h)$ that leads to the required gain. The reader should refer to the geometry in Figure 4.3.

1. The horn geometry needs to satisfy the following geometric constraint to be physically realizable:

$$\rho'_E \left(1 - \frac{b_w}{b} \right) = \rho'_H \left(1 - \frac{a_w}{a} \right), \tag{4.23}$$

where $\rho'_E = \sqrt{\ell_E^2 - (b/2)^2}$ and $\rho'_H = \sqrt{\ell_H^2 - (a/2)^2}$ are the distances to the vertex from the aperture in the E- and H-planes, respectively. Eq. 4.23 specifies that the length of the horn flared section should be the same in the two orthogonal planes. When all quantities are expressed in terms of one aperture dimension only (in this case, a). Eq. 4.23 results in a quartic polynomial for which in this case there is only one solution of interest (i.e. real solution and $0 \le a_w \le a$). This solution is expressed as follows:

$$a = \sqrt{A_1 + A_2} - \frac{b_w c_1 - (a_w^2/8)}{4A_1} + \frac{a_w}{4}, \tag{4.24}$$

where $c_1 = ((G_r \lambda^2)/g_1)(\alpha_1/\beta_1)$, $A_1 = \sqrt{A_2^2 + 3(U + (P/U))^2}$, $A_2 = (U - (P/U)) + (a_w^2/8)$ and $U = \left| Q + \sqrt{Q^2 + P^3} \right|^{1/3}$, $P = (c/12)(((G_r \lambda^2)/g_1) - (a_w b_w/4))$, $Q = (c_1^2/128)(a_w^2(\beta_1/\alpha_1) - b_w^2)$, $g_1 = 2\pi$, $\alpha_1 = 6\pi/k$ and $\beta_1 = 4\pi/k$.

2. Calculate the remaining horn parameters in Figure 4.3 from $b = (G_r \lambda^2)/g_1 a$, $\ell_H = a^2/\alpha_1$, $\ell_E = b^2/\beta_1$, $\rho'_E = \sqrt{\ell_E^2 - (b/2)^2}$, $\rho'_H = \sqrt{\ell_H^2 - (a/2)^2}$ and $h = \rho'_H(1 - a_w/a)$.

3. Refine the design as required with your favourite horn analysis software to take account of wall geometry effects over the required bandwidth to obtain the desired pattern and input match.

   Example: A standard gain X-band horn is to be designed using the procedure described above. A maximum gain of 22.6 dBi is required at 11 GHz. The input waveguide has dimensions $a_w = 2.286$ cm (0.9 inch) and $b_w = 1.016$ cm (0.4 inch). A solution can be found to Eq. 4.23 as $a = 16.563$ cm from which it follows that $b = 12.988$ cm, $\ell_E = 30.949$ cm, $\ell_H = 33.553$ cm, $\rho'_E = 30.260$ cm, $\rho'_H = 32.514$ cm and $h = 28.027$ cm. The computed E- and H-plane patterns of this antenna design are given in Figure 4.4. The gain computed from Eq. 4.22 is 22.62 dBi.

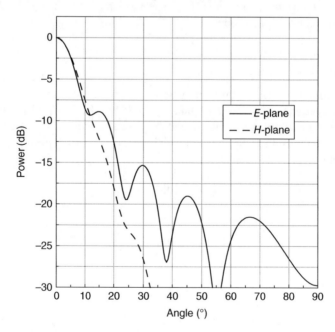

Figure 4.4 Principal plane patterns of a pyramidal horn ($a = 16.563$ cm, $b = 12.988$ cm, $h = 28.027$ cm, $a_w = 2.286$ cm, $b_w = 1.016$ cm) at 11 GHz

Although SGHs maximize gain for given aperture dimensions, there is no guarantee of the quality of the radiation pattern. High sidelobes may occur and, in addition, the peak field may not be on axis (i.e. $\theta = 0°$). Other horns such as smooth wall and corrugated conical horns to be discussed in the next section have similar maximum gain conditions. In some applications maximum gain is less important and emphasis of design is on pattern shape, sidelobes, etc. One example is when horns are used as feeds for reflectors. Then the aim is to illuminate the reflector efficiently and to minimize power loss due to spillover at the reflector edge (see Section 6.2.2). To achieve this, the pattern should have low sidelobes and the skirt of the pattern should decrease rapidly at the reflector edge. Efficient feeds for parabolic reflectors include rectangular, circular and corrugated waveguides as well as their more directive flared counterparts. Feed design for reflectors is discussed in Chapter 9.

4.3.2   Dielectric-Loaded Rectangular Horn

In some applications it is advantageous to have identical E- and H-plane patterns. This is not possible with a conventional pyramidal horn because of the difference in the aperture field distribution in the two principal directions, namely, uniform and cosine distributed. However, by placing a dielectric on the walls in the E-plane, the field can be made more uniform albeit over a limited frequency range (Tsandoulous & Fitzgerald, 1972). This type of horn is illustrated in Figure 4.5. The introduction of the dielectric leads to a set of hybrid modes in the horn where both longitudinal (in the direction of propagation) field components are present. In the dielectric-loaded waveguide, mode sets can be identified that can be either TE or TM to the surface of the dielectric that is, the x-direction in Figure 4.5. The component of field either E_x or H_x are zero. Hence the modes are referred to as TE_x or TM_x, respectively. An alternative terminology is to refer to these modes as either longitudinal section electric (LSE) or longitudinal section magnetic (LSM). The interested reader should consult the references for further details (Collin, 1960; Harrington, 1961).

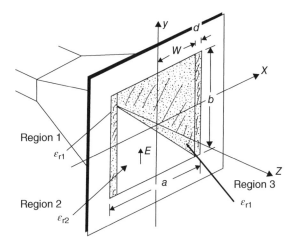

Figure 4.5   Dielectric slab-loaded pyramidal horn

The slabs in regions 1 and 3 have a thickness d and a dielectric constant ε_{r1}, while region 2, between the slabs, has width $2w$ and dielectric constant ε_{r2}. With the identification of the fundamental mode and its field distribution, an aperture field representation can be constructed in the same way as for unfilled pyramidal horn. The fundamental mode is the LSE_{10} mode. The transverse electric field of this mode is

$$E_y^{(1)} = E_o C_o \sin\left(k_1\left(\frac{a}{2}+x\right)\right); \quad -\frac{a}{2} \le x \le -w \tag{4.25a}$$

$$E_y^{(2)} = E_o \cos k_2 x; \quad -w \le x \le w \tag{4.25b}$$

$$E_y^{(3)} = E_o C_o \sin\left(k_1\left(\frac{a}{2}-x\right)\right); \quad w \le x \le \frac{a}{2}, \tag{4.25c}$$

where $C_o = \cos(k_2 w)/\sin k_1 d$ pertains to requiring continuity of the field at $|x| = (a/2-d) = w$. The wavenumbers in regions 1 and 2 are

$$k_1 = \sqrt{k^2 \varepsilon_{r1} - \gamma^2} \text{ and } k_2 = \sqrt{k^2 \varepsilon_{r2} - \gamma^2},$$

where γ is the propagation constant. The LSE_{10} mode propagation constant varies with frequency that is governed by the transcendental equation

$$k_2 \tan k_2 w = k_1 \cot k_1 d. \tag{4.26}$$

The roots may be found by means of conventional root finding techniques such as Newton–Raphson (Fröberg, 1974). A good approximation to the propagation constant for LSE_{mn} modes can be obtained from a perturbation formula derived from an approach provided by Gabriel and Bodwin (1965). This formula can been applied to many different types of dielectric-loaded waveguides and is expressed as

$$\left(\frac{\gamma}{k}\right)^2 = 1 - \left(\frac{k_{(o)}}{\omega/c}\right)^2 + \frac{\sum_{i=1}^{N} \iint_S dS \Phi_o (\varepsilon_{ri} - 1) \Phi_o}{\iint_S dS \Phi_o \Phi_o}, \tag{4.27}$$

where $\varepsilon_r a_i$ is the relative permittivity of region i ($i = 1, ..., N$), N is the number of dielectric regions, Φ_o is the first-order trial field solution and $k_{(o)}$ is the approximation to the wave number that occurs as in the solution to the wave equation $\nabla_t^2 \Phi_o + k_{(o)}^2 \Phi_o = 0$. Another possible approach is to use a variational expression as described by Berk (1956). In the present case of slab-loaded waveguide, to find the propagation constant of the LSE_{mn} mode select the trial functions $\Phi_o = E_y = E_o \sin(m\pi x/a)\cos(n\pi y/b)$ and $k_{(o)} = \sqrt{(m\pi/a)^2 + (n\pi/b)^2}$ as these correspond to related quantities of empty waveguide and can be expected to have the desired physical properties. After carrying out the required integrations in Eq. 4.26, an approximation to the propagation constant is

$$\gamma \approx \sqrt{k^2\left(1 + (\varepsilon_{r1} - 1)\frac{2d}{a}\left[1 - \frac{a}{2m\pi d}\sin\left(\frac{2m\pi d}{a}\right)\right]\right) - \left(\frac{m\pi}{a}\right)^2 - \left(\frac{n\pi}{n}\right)^2}, \tag{4.28}$$

where it has been assumed that $\varepsilon_{r2} = 1$. Eq. 4.28 can be used directly or as an estimate for Eq. 4.26 to obtain a more accurate answer. As an example, a rectangular waveguide with aperture dimensions 6.1 cm $\times$ 6.1 cm has dielectric slabs lining the E-plane walls (region 1 as in Figure 4.5). The slab thickness is $d = 4.1$ mm and dielectric constant is $\varepsilon_{r1} = 3.07$. At a frequency of 12.5 GHz, Eq. 4.28 predicts $\gamma = 2.5795$ rad/cm while the exact result obtained from Eq. 4.26 is $\gamma = 2.61302$ rad/cm. The error in this example is typical of what is achieved.

When the wavenumber in region 2 is zero, that is, $k_2 = 0$, the field in central region is uniform. From Eq. 4.25, this occurs when

$$d = \frac{\lambda}{4\sqrt{\varepsilon_{r1} - \varepsilon_{r2}}} \tag{4.29}$$

and where $\gamma = k\sqrt{\varepsilon_{r2}}$.

A quadratic phase function can be applied to the LSE$_{10}$ mode field as in Eq. 4.13a to model a flared horn except that this time the phase factor is $\exp\left(-j(\gamma/2)\left((x^2/\rho_H') + (y^2/\rho_E')\right)\right)$. In addition, on this occasion, it is assumed that the aperture is located in a large ground plane. The result, using the same notation as used for Eq. 4.5, is

$$E_\theta(r,\theta,\phi) = \frac{jk}{2\pi} \frac{e^{-jkr}}{r} N_y(u,v) \sin\phi \tag{4.30a}$$

$$E_\phi(r,\theta,\phi) = \frac{jk}{2\pi} \frac{e^{-jkr}}{r} N_y(u,v) \cos\theta \cos\phi, \tag{4.30b}$$

where on this occasion $N_y(u,v) = I_x(v)I_y(u)$. The function $I_y(u)$ is a result of the integration over the y-co-ordinate in the aperture, which transforms a uniform field and is identical to Eq. 4.17 except for some changes of notation. That is,

$$I_y(v) = \frac{\exp\left[j(\pi v/s_E)^2\right]}{s_E} \left(\mathcal{K}\left(s_E \frac{b}{2} - \frac{\pi v}{s_E}\right) + \mathcal{K}\left(s_E \frac{b}{2} + \frac{\pi v}{s_E}\right)\right), \tag{4.31}$$

where $s_E = \sqrt{\gamma/2\rho_E'}$. Similarly for the H-plane let $s_H = \sqrt{\gamma/2\rho_H'}$. The second integral for the function in u can be shown to give

$$
\begin{aligned}
I_x(u) = \frac{1}{2s_H} \Big[&F_+^{(2)}(u) + F_+^{(2)}(-u) + F_-^{(2)}(u) + F_-^{(2)}(-u) \\
&- jC_o\left(e^{j\pi ua}\left(F_+^{(1)}(-u) - F_-^{(1)}(-u)\right) + e^{-j\pi ua}\left(F_+^{(1)}(u) - F_-^{(1)}(u)\right)\right)\Big],
\end{aligned} \tag{4.32}
$$

where

$$
\begin{aligned}
F_\pm^{(1)}(u) = {}&\exp\left(js_H^2 \alpha_\pm^{(1)}(u)\left(a + \alpha_\pm^{(1)}(u)\right)\right) \\
&\times \left(\mathcal{K}\left(s_H\left(\frac{a}{2} + \alpha_\pm^{(1)}(u)\right)\right) + \mathcal{K}\left(s_H\left(-w - \alpha_\pm^{(1)}(u)\right)\right)\right)
\end{aligned}
$$

and

$$F_{\pm}^{(2)}(u) = \exp\left(j\left(s_H\alpha_{\pm}^{(2)}(u)\right)^2\right)\left(\mathcal{K}\left(s_H\left(\frac{a}{2}+\alpha_{\pm}^{(2)}(u)\right)\right)+\mathcal{K}\left(s_H\left(w-\alpha_{\pm}^{(2)}(u)\right)\right)\right),$$

where $\alpha_{\pm}^{(i)}(u) = (\pi u \pm k_i/2)/s_H^2$.

At the condition for uniform field in the aperture as specified by Eq. 4.29, the maximum gain of the slab-loaded horn is given by

$$G_{\max} = \frac{4b}{\pi(a-d)}\left[kw\left(1+\frac{2d}{w\pi}\right)\right]^2. \tag{4.33}$$

As an example of the radiation patterns obtained by slab loading a pyramidal horn, Figure 4.6 shows the results in the H-plane at 12.5 GHz of one of the horns pictured in Figure 1.1c. The horn has aperture dimensions $a = b = 61.0$ mm and a height of $h = 23$ cm. The input is WR-90 waveguide with dimensions 1.16 cm × 0.95 cm. Tapered dielectric slabs with $\varepsilon_r = 3.07$ and loss tangent $\tan \delta = 0.005$ were placed on the narrow (E-plane) walls where the thickness varies linearly from zero up to the required thickness of $d = 4.1$ mm over about 60% of the tapered

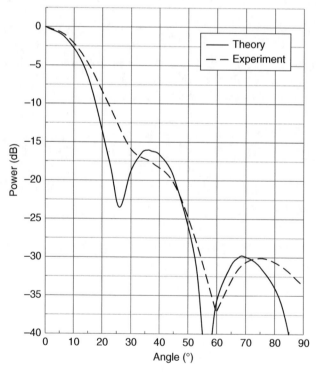

Figure 4.6 H-plane radiation pattern of a dielectric-loaded pyramidal horn at 12.5 GHz. Aperture dimensions $a = 6.1$ cm, $b = 6.1$ cm, $d = 0.41$ and $\varepsilon_{r1} = 3.07$

wall to give a uniform aperture field at a frequency of 12.7 GHz. The measured results at 12.5 GHz are in good agreement with the predictions given by Eq. 4.30 although the impact of the taper is seen by the slight discrepancy. The dielectric taper has to be sufficiently long to ensure that the sidelobes agree with the theory. The gain at this frequency was measured and found to be 17.89 dBi. This compares with a computed gain of 18.12 dBi obtained from Eqs. 4.30a, 4.30b, 4.31 and 4.32. The difference in theory and experiment is due mainly to dielectric losses, which were estimated to be 0.17 dB. This loss in the dielectric slabs as well as the adhesive used to fix them to the side walls can be a cause of significant loss, which may preclude the use of dielectric-loaded horns from some applications where gain is at a premium (e.g. satellite communications) unless a low loss dielectric with suitable properties can be found. The mismatch due to the dielectric loading is relative small for dielectric placed on the walls parallel to the E-plane as in this case (Bird & Hay, 1990).

4.4   Circular Waveguides and Horns

Antennas radiating from circular waveguides and horns find application in communications, radar and radio astronomy as feeds for reflectors, as reference antennas or for microwave links. Their geometrical symmetry and low cross-polarization are important factors in their widespread use. Examples of this type of antenna are illustrated in Figure 4.7.

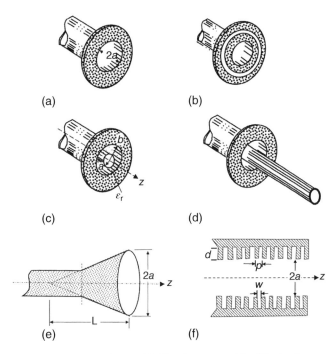

Figure 4.7   Circular waveguides and horns. (a) Circular waveguide; (b) circular waveguide with parasitic ring; (c) coaxial waveguide; (d) coaxial waveguide with extended central conductor; (e) conical horn and (f) corrugated waveguide

Circular waveguide (Figure 4.7a) is an efficient feed for moderately deep reflectors, and it has almost equal E- and H-plane patterns (pattern symmetry) for pipe diameters in the range 0.7–1.2 wavelengths depending on the size of the flange. Higher efficiencies are possible with the addition of parasitic rings as shown in Figure 4.7b. Coaxial waveguide, Figure 4.7c, has potentially greater flexibility in available radiation patterns due to the extra degree of freedom provided by the internal conductor. However, a large mismatch occurs at the aperture when the ratio of inner to outer conductor radii is greater than about 1/3. Figure 4.7d shows a self-supporting rear-radiating coaxial waveguide ('tomato can') feed for a reflector. Here the inner conductor that extends all the way to the vertex of the reflector allowing the feed to be driven from a transmission line in the centre conductor. Flaring a circular waveguide produces the conical circular horn shown in Figure 4.7e. Finally, the waveguide and horn side walls may be corrugated or covered with an anisotropic surface or material. This structure may be flared as in Figure 4.7e. With corrugations, when the depth of the corrugations, d, is about $\lambda/4$, almost pure polarized radiation patterns result. In this section, various circular aperture horns will be discussed and particularly those shown in Figure 4.7.

4.4.1   Circular Waveguide

The smooth wall circular waveguide Figure 4.7a antenna is usually excited in its fundamental TE_{11} mode. The transverse fields of this mode in cylindrical polar co-ordinates (ρ, ϕ, z) are

$$\mathbf{E}_t = E_o \left[\hat{\rho} \frac{J_1(k_c\rho)}{k_c\rho} \cos\phi - \hat{\phi} J_1'(k_c\rho) \sin\phi \right] e^{-j\beta z}, \tag{4.34a}$$

$$\mathbf{H}_t = Y_w \, \hat{z} \times \mathbf{E}_t, \tag{4.34b}$$

where $k_c a = 1.84118$ is the cut-off wavenumber of the TE_{11} mode in a circular waveguide of radius a, Y_w is the mode admittance and β is the propagation constant. The properties of the Bessel function, $J_1(x)$, and its first derivative, $J_1'(x)$, are summarized in Appendix B.

Assume that the TE_{11} mode is the only one produced in the aperture. Coupling at the aperture will generate other modes, but their effect on the radiated field is usually of second-order except close to cut-off. Expressing the field components in rectangular co-ordinates and making use of Bessel function recurrence relations for $J_p(z)/z$ and $J_p'(z)$ given in Appendix B, the aperture fields at $z=0$ are expressed as follows:

$$\mathbf{E}_a = \frac{E_o}{2} [\hat{x} J_0(k_c\rho) + J_2(k_c\rho) \cos 2\phi + \hat{y} J_2(k_c\rho) \sin 2\phi], \tag{4.35a}$$

$$\mathbf{H}_a = Y_w \, \hat{z} \times \mathbf{E}_a. \tag{4.35b}$$

Unlike the TE_{10} mode of rectangular waveguide, the TE_{11} mode has both x and y field components and, therefore, so has its vector transform. Eq. 4.35 and Figure 4.8a indicate that the principal aperture field polarization is x-polarized, the y-component has non-zero values away from the principal planes ($\phi=0, \pi$ and $\phi= \pm\pi/2$).

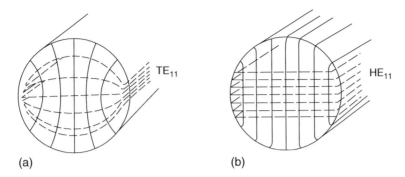

Figure 4.8 Electric and magnetic fields in (a) smooth wall circular waveguide and (b) corrugated waveguide. Solid curve: electric; dashed curve: magnetic

To determine the radiated field for circular waveguide using Eq. 3.30, the transforms **N**, Eq. 3.24a, are required. To evaluate these integrals that involve Bessel functions use is made of Eqs. B.3 and B.5. From these equations,

$$N_x = \frac{\pi a E_o}{k_c^2 - w^2} \left[\begin{array}{c} k_c J_1(k_c a) J_0(wa) - w J_0(k_c a) J_1(wa) \\ - \cos 2\phi (w J_2(k_c a) J_1(wa) - k_c J_1(k_c a) J_2(wa)) \end{array} \right] \tag{4.36a}$$

$$N_y = \frac{\pi a E_o}{k_c^2 - w^2} \sin 2\phi [w J_2(k_c a) J_1(wa) - k_c J_1(k_c a) J_2(wa)], \tag{4.36b}$$

where $w = k \sin \theta$. Substitute Eq. 4.36 into Eq. 3.20 and, by means of Bessel function recurrence relations, the far-zone fields are found to be given by

$$E_\theta(r,\theta,\phi) = \frac{jkaE_o e^{-jkr}}{2} \frac{}{r} \cos \phi \left(1 + \frac{\beta}{k} \cos \theta \right) \frac{J_1(k_c a)}{k_c} \frac{J_1(wa)}{wa}, \tag{4.37a}$$

$$E_\phi(r,\theta,\phi) = -\frac{jkaE_o e^{-jkr}}{2} \frac{}{r} \sin \phi \left(\frac{\beta}{k} + \cos \theta \right) J_1(k_c a) \frac{k_c J_1'(wa)}{k_c^2 - w^2}. \tag{4.37b}$$

Comparing Eq. 4.37 with Eq. 4.5 it is seen that $J_1(wa)/wa$ and $J_1'(k_c a)/(k_c^2 - w^2)$ perform corresponding functions as $S(kb \sin \theta/2)$ and $C(ka \sin \theta/2)$ do in the principal planes of rectangular waveguide. Plots of the E- and H-patterns are shown in Figure 4.9 for several different waveguide radii.

The power radiated by the TE$_{11}$ mode is

$$P_T = \frac{E_o^2 \pi a^2}{4\eta_o} \frac{\beta}{k} J_1^2(k_c a) \left(1 - \frac{1}{(k_c a)^2} \right). \tag{4.38}$$

From Eq. 3.48 the maximum gain is

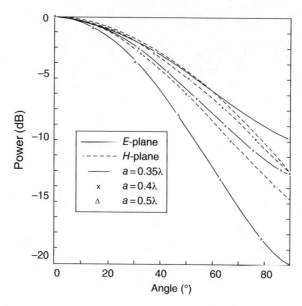

Figure 4.9 Radiation patterns of circular waveguide (E–H field model). The radius of the waveguide is a parameter

$$G_{max} = \frac{(1/2)(ka)^2 (1 + (\beta/k))^2 (k/\beta)}{\left[(k_c a)^2 - 1 \right]}.$$ (4.39)

At frequencies well above cut-off, $\beta \approx k$ and Eq. 4.39 simplifies to

$$G_{max} = 0.837 \, (ka)^2,$$

which corresponds to an aperture efficiency of 83.7%.

Equations 4.37 approximate the field components of radiated by a thin wall waveguide that has no currents flowing on the outside wall. This is because they were derived for an aperture containing both electric and magnetic currents, which in turn are governed by both the electric and magnetic fields. As a result, Eqs. 4.37 are referred to as the E–H field model. In practice the wall currents have a significant effect on the radiation pattern. This model contrasts with an exact solution (Weinstein, 1969) which is obtained with the Wiener–Hopf method that has external currents on an infinitely thin waveguide wall. To demonstrate differences due to changed aperture conditions, consider a circular waveguide that is terminated in an infinite ground plane. The radiated field in this case is obtained from Eq. 3.26, and where N_x and N_y are given by Eq. 4.36. Thus,

$$E_\theta(r,\theta,\phi) = jkE_o \frac{e^{-jkr}}{r} \cos \phi \frac{J_1(k_c a) J_1(wa)}{k_c \quad wa},$$ (4.40a)

$$E_\phi(r,\theta,\phi) = -jkE_o \frac{e^{-jkr}}{r} \sin \phi \cos \theta J_1(k_c a) \frac{k_c J_1'(wa)}{k_c^2 - w^2}.$$ (4.40b)

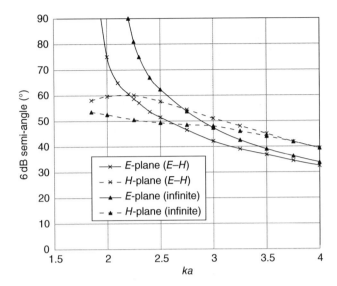

Figure 4.10 Six decibel half-beamwidth of circular waveguide

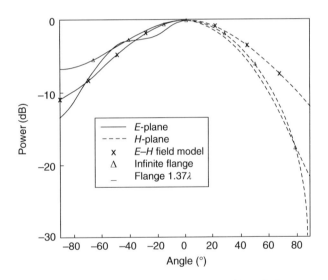

Figure 4.11 Effect of flange on circular waveguide radiation patterns

Because Eq. 4.40 was determined from magnetic currents, and hence electric field in the aperture, they are sometimes referred to as the *E*-field model.

Comparing Eqs. 4.37 and 4.40, the effect of the ground plane is to only change the Huygens' factors. The 6 dB half-beamwidth (half the angle between the 6 dB points of the pattern) of the *E*- and *H*-plane patterns is given in Figure 4.10 for both models. In Figure 4.11, the *E*- and *H*-plane patterns are plotted for a waveguide of radius $a = 0.37\lambda$. Also shown is the computed pattern for the same pipe but this time terminated in a finite flange (as shown in Figure 4.1a) of

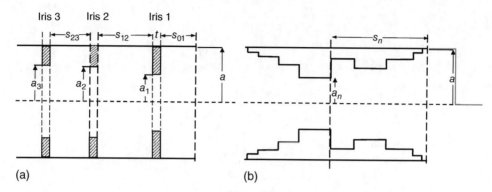

Figure 4.12   Matching techniques for circular apertures. (a) Irises and (b) stepped sections

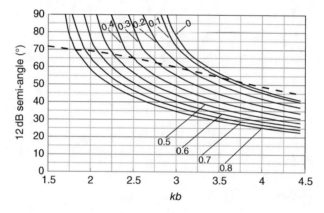

Figure 4.13   Radiation from a coaxial aperture in a ground plane. The 12 dB half-power beamwidth is plotted versus normalized frequency given as a function of a/b. Solid line: E-plane; dashed line: H-plane

radius 1.37λ. These last results agree quite well with measured results. They were obtained using an E-field model in conjunction with the geometrical theory of diffraction (James, 1986) to take into account the effect of the finite flange. In Figure 4.12, it is observed that the E–H field model gives a reasonable approximation to the E-plane pattern but not so in the H-plane. On the other hand, the E-field model approximates the E- and H-plane patterns quite well. These conclusions are fairly typical of the methods described here.

4.4.1.1   Matching at a Circular Aperture

Well above cut-off, circular waveguides and horns have an inherently low reflection coefficient, which may be sufficient for many applications. Better matching may be required in more demanding applications such as for low noise and broadband operation. Two basic methods are employed for improving the match at a circular aperture, and these are illustrated in Figure 4.12. Most techniques are axially symmetric to ensure the matching does not increase cross-polarization. The first approach is to use several circular irises of varying depth from the inner wall into the centre of the circular pipe, annular windows or annular rings. The iris provides a mainly capacitive reactance and for small hole radii, r, and thickness $t \ll \lambda$, relative to the guide wavelength. The susceptance for when the TE_{11} mode only propagates, that is, $1.841 < ka < 2.404$, is approximately given by (Marcuvitz, 1986)

$$B \approx \frac{Y_w}{\beta a} \left(\frac{1.995}{(r/a)^3} - 3.666 \right),\tag{4.41}$$

where β is the propagation constant of the TE_{11} mode and Y_w is the wave admittance. This expression is only approximate and has best accuracy for frequencies close to TE_{11} cut-off and at higher frequencies when $r/a < 0.5$. For more accurate design, computer programs such as mode matching should be used.

A second approach is to introduce a series of steps of different length and height as shown in Figure 4.12b. This approach enables matching over a broad frequency band. The availability of computer packages allows accurate analysis of the structures chosen for matching. For broadbanding a design, a computer simulator is required, preferably one with an optimizer. One approach that has proved especially effective in waveguide is the mode-matching method. This is described in a following section.

The traditional approach was to interpolate the susceptance from curves generated from measurement (e.g. Marcuvitz, 1986) or computed results (e.g. James, 1987; Sharstein & Adams, 1988). These data provide an estimate of the available susceptances for matching. The same can be done with Eq. 4.41. An approximate formula for the admittance of a circular aperture is given by Eq. 7.88. To proceed with matching using a Smith chart, locate the aperture admittance for the waveguide dimension and operating frequency on the chart. Move back from the aperture towards the source to cancel out, or partly cancel, the susceptance shown with an iris of selected dimensions. The first iris aims to allow the transfer by rotation of this cancelled admittance to a location where a second iris is able to further transfer the resulting admittance close to the centre of the chart to achieve a good match. By judicious choice of the iris susceptances, it is possible to obtain a trajectory on the Smith chart that is insensitive to frequency so as to achieve a moderate bandwidth. This design should be checked by trial and error.

4.4.2 Coaxial Waveguide

The coaxial waveguide illustrated in Figure 4.7c is particularly useful as a feed for a reflector with a short focal length, and it potentially has the additional flexibility of a central conductor which can be used for self-support. For convenience, the inner conductor radius is defined here as a and the outer conductor radius is b. Shorted coaxial apertures are also used in the flange of a circular waveguide to improve the pattern symmetry and reduce cross-polarization (see Figure 4.8b). An advantage of coaxial waveguide antennas that operate predominantly in the TE_{11} mode is that the beamwidth is broader than the equivalent open-ended circular waveguide radiation. This characteristic can be achieved also with excellent pattern symmetry and low cross-polarization, all of which are important for many applications. For example, TE_{11}-mode coaxial waveguide antennas can be used as a single feed or array element for reflectors with short focal lengths and also as an element in closely spaced directly radiating arrays for radio astronomy (see Section 10.3.3). In multibeam feed applications, the TE_{11}-mode coaxial waveguide antenna has the advantage that the outer diameter is smaller than equivalent circular feeds that use external choke rings or set-back flanges to achieve wide beamwidth and pattern symmetry.

There are some significant differences with circular waveguide, however, caused principally by the central conductor. The first difference is that the fundamental mode, the TEM mode, does not radiate very well but is strongly reflected from the open end. The mode that is used for antenna applications is the TE_{11} mode, and the cut-off and field distribution is very similar to the TE_{11} mode of circular waveguide, which is the limiting mode when the radius of the central conductor becomes small. The field distribution of the TE_{11} mode of coaxial waveguide is similar to the one for circular waveguide except that now the central field lines curve to accommodate the additional boundary condition on the central conductor. As a result, the properties of the radiation pattern are different as shown in Figure 4.13. The width of the H-plane pattern tends to narrow as the radius of the centre conductor increases. This means that there is thus an optimum range of conductor radii in which the E- and H-plane patterns are similar.

The radiation pattern of the TE_{11} mode of coaxial waveguide can be obtained in the same way as for circular waveguide. The electric field of the TE_{11} mode in coaxial waveguide at $z=0$ is expressed as

$$\mathbf{E}_t(\rho,\phi) = A_{11}\left[\hat{\rho}\frac{Z_1(k_c\rho,k_ca)}{k_c\rho}\cos(\phi-\psi) - \hat{\phi}Z_1'(k_c\rho,k_ca)\sin(\phi-\psi)\right], \qquad (4.42)$$

where $\mathbf{H}_t = Y_w\hat{z}\times\mathbf{E}_t$, ψ is the reference phase angle relative to the initial line (x-axis) and the function $Z_p(x,y)$ is a compound Bessel function that is defined in Appendix B. A prime on this function indicates the first derivative with respect to the first argument. It also has the property that as the second argument approaches zero, then $Z_p(x,y)\big|_{y\to 0} = J_p(x)$. Thus, Eq. 4.42 reduces to Eq. 4.34a in the limit of zero centre conductor radius. The cut-off wavenumber k_c of the TE_{1n} modes is given by $Z_1'(k_cb,k_ca)=0$. A useful approximation to the cut-off for small inner conductor radii is

$$k_cb \approx \frac{2}{1+(a/b)}. \qquad (4.43)$$

In rectangular components, the aperture field of a coaxial waveguide is

$$\mathbf{E}_a(\rho,\phi) = \frac{A_{11}}{2}\left[\begin{array}{l}\hat{x}(Z_0(k_c\rho,k_ca) + Z_2(k_c\rho,k_ca))\cos(2\phi-\psi)\\ +\hat{y}Z_2(k_c\rho,k_ca)\sin(2\phi-\psi)\end{array}\right]. \qquad (4.44)$$

In many applications, the coaxial waveguide will have a flange or thick wall. Therefore, in what follows it will be assumed here that the coaxial aperture terminates in an infinite ground plane. By means of Eqs. 3.26, it can be shown that the far-zone fields of coaxial waveguide are

$$E_\theta(r,\theta,\phi) = jkbA_{11}\frac{e^{-jkr}}{r}\frac{Z_1(k_cb,k_ca)}{k_c}\frac{J_1(wb)}{wb}\cos(\phi-\psi), \qquad (4.45a)$$

$$E_\phi(r,\theta,\phi) = -jkbA_{11}\frac{e^{-jkr}}{r}Z_1(k_cb,k_ca)\frac{k_cJ_1'(wb)}{k_c^2-w^2}\cos\theta\sin(\phi-\psi). \tag{4.45b}$$

These equations are similar to Eq. 4.40 and are identical in the limit as $ka \to 0$. However, for the coaxial case there are significant differences as will be described in the following.

The radiation properties of the TE_{11} mode from coaxial waveguide in a large ground plane are summarized in Figures 4.13 and 4.14. The former shows the half-angle between the beams at the 12 dB level as a function of the normalized frequency kb with the inner to outer conductor ratio a/b as a parameter, while the latter shows the maximum cross-polar level in the 45°-plane. The 12-dB semi-angle in the H-plane is approximately $\theta_{H12dB} = -10(kb)+92$ degrees. For a given outer conductor radius a there is always a frequency at which the 12-dB semi-angle is identical in the E- and H-planes. At frequencies below the optimum kb for pattern symmetry, the E-plane 12-dB semi-angle is greater than θ_{H12dB}, and when the frequency is above the optimum kb, the E-plane 12-dB semi-angle is smaller than θ_{H12dB}. At the frequency for pattern symmetry, the E- and H-plane patterns are almost identical over the main beam and, therefore, low cross-polarization is obtained. The frequency where the minimum of the cross-polar pattern occurs is $kb \approx 1.25(a/b)^2 - 3.3(a/b) + 3.6$. Figure 4.15 also shows that the level of the cross-polar maxima increases with increasing a/b.

As a first example, suppose an application at 1.5 GHz requires a 12 dB beamwidth of 140°. From Figure 4.13, it is estimated that this beamwidth could be achieved with $kb = 2.05$ and $a/b = 0.6$. For this choice of parameters, the peak cross-polarization level from Figure 4.14 will be approximately -30 dB relative to the peak co-polar level. Thus, at the specified centre frequency, the outer conductor radius should be 65.2 mm, while the inner conductor radius should be 39.1 mm.

As a second example, consider the radiation patterns in the 45° plane of a coaxial waveguide with a large metallic flange that are shown in Figure 4.15. The antenna is required to operate at a frequency of 4.5 GHz (i.e. $ka = 2.54$), and for this the outer conductor has a radius $b = 2.7$ cm and a centre conductor radius of $a = 1.44$ cm. Figure 4.15 shows that at this frequency and for

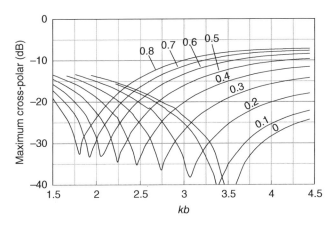

Figure 4.14 Maximum cross-polarization level radiated by a coaxial aperture in the 45° plane versus normalized frequency as a function of a/b

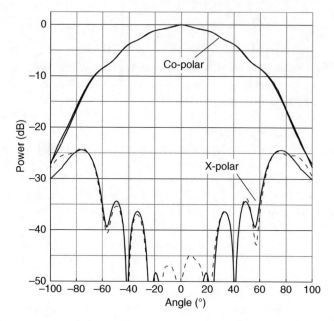

Figure 4.15 Principal plane radiation patterns in the 45° plane of a TE$_{11}$ mode coaxial waveguide at 4.5 GHz. The aperture dimensions are $b = 27$ mm and $a = 14.4$ mm. Solid line: theory; dashed line: experiment

the selected waveguide dimensions the peak cross-polar level is about -22 dB, which is close to the level predicted in Figure 4.14.

It is important also to understand some of the basic radiation properties of the fundamental TEM mode. This mode radiates poorly as the aperture is not well matched to free-space and the radiation efficiency is low. The transverse electric field distribution of this mode is

$$\mathbf{E}_a(\rho) = \frac{E_{TEM}}{\rho} [\hat{x} \cos(\phi - \psi) + \hat{y} \sin(\phi - \psi)]. \tag{4.46}$$

The reflection coefficient of this mode at the end of the waveguide is

$$\Gamma \approx \frac{1 - y_{TEM}}{1 + y_{TEM}},$$

where

$$y_{TEM} = \frac{1}{\ln(b/a)} \int_0^\infty \frac{dw}{w\sqrt{1 - w^2}} (J_0(kbw) - J_0(kaw))^2 \tag{4.47}$$

for the aperture radiating into free-space. When the conductor radii are small in terms of the wavelength, that is, $a, b/\lambda \ll 1$, then the Bessel functions can be replaced by series which can then be integrated term by term. The resulting normalized admittance is (Galejs, 1969)

$$y_{\text{TEM}} \approx Y_c \left[\frac{(kD)^4}{360} \left(1 - \frac{(kD)^2}{5} \right) + j \frac{kD}{60\pi} \left(\ln \left(\frac{8D}{(b-a)} \right) - \frac{1}{2} + \frac{2}{3}(kD)^2 \right) \right],$$ (4.48)

where $Y_c = 2\pi[\eta_o \ln(b/a)]^{-1}$ is the characteristic admittance of the coaxial line and $D = (b+a)/2$. Typically, the value of y_{TEM} is such that the reflection coefficient is very close to unity as the admittance value is quite small. This is due mainly to the abrupt termination of the current on the inner conductor at the aperture.

The far-field radiation pattern due to TEM mode in infinite ground plane is

$$E_\theta = kE_{\text{TEM}} \frac{\exp(-jkr)}{r} \left[\frac{J_o(kb \sin \theta) - J_o(ka \sin \theta)}{k \sin \theta} \right]$$ (4.49)

$$E_\phi = 0.$$

The pattern given by Eq. 4.49 is omnidirectional and has a null on axis at $\theta = 0$. This can be easily seen from the Bessel function series for small arguments. As $\sin \theta \to 0$, the factor in the square brackets approaches $(-b/4)\left(1 - (a/b)^4\right)(kb \sin \theta)^3$.

4.4.2.1 Matching of a Coaxial Aperture

The coaxial waveguide modes have a significant mismatch at the aperture. As described above, the fundamental TEM mode is the worst affected, while the other modes are only well matched over a relatively narrow band of frequencies. Extending the conductor into free-space improves the match somewhat but this tends to narrow the beamwidth in the H-plane. Several common methods are normally used for improving the input match of TE$_{11}$-mode coaxial waveguide antennas, such as tuning screws and irises. The latter approach is especially effective as broadband matching is possible through a combination of 'inductive' and 'capacitive' irises (Bird et al., 1986) as shown in Figure 4.16. An inductive iris is achieved in coaxial waveguide

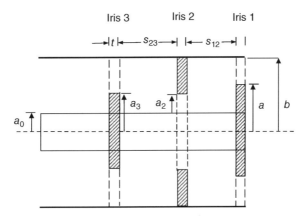

Figure 4.16 Matching the aperture of a TE$_{11}$ mode coaxial waveguide antenna

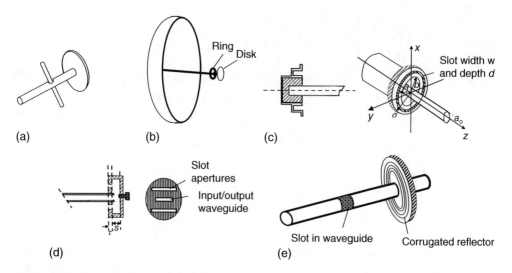

Figure 4.17   Rear-radiating feeds. (a) dipole and disk fed by coaxial cable (after Silver, 1946); (b) the same with ring backed by a small reflector; (c) waveguide excited cup feed with ring-slot (i.e. annular ring) flange (Poulton & Bird, 1986); (d) waveguide with radiating slots (Cutler, 1947) and (e) hat feed (Kildal, 1987)

by an iris that extends down from the outer conductor producing a gap between the inner conductor and the inner diameter of the iris (Iris 2 in Figure 4.17). Conversely, a capacitive iris (Iris 3 in Figure 4.16) is produced by extending the iris upwards from the inner conductor leaving a gap between the outer conductor and the outer diameter of the iris. This simple matching method using two irises can achieve a 20 dB return-loss bandwidth of about 30%. It has been found that the method is most effective when $a/b < 0.35$. The method to be described is based on two simple concepts: (i) alternate capacitive and inductive irises extending back from the aperture and (ii) step the inner conductor diameter at the aperture down to moderate value so that (i) can affect a broadband match. The matching method is illustrated in Figure 4.16. The thickness of the iris is assumed to be $t \ll b$.

Assuming a coaxial aperture with $a/b < 0.3$, the approach using two irises is easily explained with the aid of the Smith chart. The admittance presented by the iris-aperture combination follows a clockwise trajectory on the Smith chart with increasing frequency. The first length of waveguide from the load to the source, s_{12}, should transfer the aperture admittance so that it lies inside the unit conductance circle with the real part in the vicinity of 1.5–2 and the imaginary part (y_1) is positive. The inductive iris (Iris 2) is chosen to approximately cancel the imaginary part of the admittance (y_1) at the centre frequency. The second length of waveguide, s_{23}, transfers the admittance to the unit conductance circle with a negative susceptance (y_2). The final capacitive iris (Iris 3) is chosen to cancel the negative susceptance (y_2) and to bring the locus near to the centre of the Smith chart (i.e. $y_3 \approx 1 + j0$).

For larger inner conductors, a third iris (Iris 1) is placed at the aperture to improve the match but this is usually possible only over a narrower frequency band. The approach described above for $a/b < 0.3$ is then used to match the iris at the aperture as illustrated in Figure 4.16.

It has been found that Iris 3 influences the low end of the frequency band. If the susceptance of Iris 3 is increased, the bottom end of the 20 dB band moves lower in frequency, while if the susceptance is decreased, the bottom end of the band moves higher. The radius a_3 has a similar

Table 4.1 Two iris matching of TE_{11} mode coaxial waveguide

a/b	0.3	0.4	0.5	0.6	0.7
Centre frequency (kb)	2.73	2.46	2.24	2.05	1.92
s_{12}/b	0.585	0.526	0.457	0.398	0.300
a_2/b	0.495	0.463	0.430	0.390	0.356
s_{23}/b	0.300	0.199	0.184	0.182	0.205
a_3/b	0.419	0.486	0.572	0.662	0.729
t/b	0.07	0.07	0.07	0.07	0.07
a_o/b	0.3	0.3	0.3	0.3	0.3
% bandwidth	25.6	19.5	17.4	12.6	9.0

effect. If the length s_{12} is reduced, the upper-band edge increases, while if s_{12} is increased, this frequency reduces. The spacing s_{23} has an effect on the level of the mid-band reflection coefficient. If s_{23} is reduced, the mid-band level increases, and when s_{23} is increased, the mid-band level reduces. A slight displacement of the band occurs also with a shift towards lower frequencies as s_{23} is reduced. Table 4.1 lists the combination of lengths and iris dimensions to achieve a broadband match at coaxial apertures with various ratios b/a. The percentage bandwidth is also shown in each case.

4.4.2.2 Coaxial Apertures with an Extended Central Conductor

A coaxial aperture from which the inner conductor extends outwards far beyond it (Figure 4.7d) has several important applications. It can be used as a prime-focus feed that is self-supporting from the vertex of the reflector, or it can be used as part of a probe in medical applications. In the first example, strut blockage can be eliminated while in the second the centre conductor may support another instrument such as a hypodermic needle. Several types of rear-radiating feed are shown in Figure 4.17. One configuration uses a cup at the end of the central circular conductor. Resonant slots, dipoles or a waveguide in the cup are used to excite an annular aperture principally in the TE_{11} mode. The design of a rear-radiating feed may be considered in two independent stages, namely, (i) the transition from the central transmission line support to the radiating element and (ii) radiation in the presence of the central conductor. It is usually the first stage that sets the various feed types in Figure 4.17 apart and provides the name. For instance, one of the first feeds of this type is called the Cutler feed. This usually consists of several radiating slots that are excited from the transmission line by a resonant cavity, which was invented by C.C. Cutler (1947). A tuning screw can be used to improve the match to the input. This feed is a very narrowband design. As its appearance can look like a 'tomato can', it sometimes goes by that name as well. A modification to this design uses coaxial waveguide sections at the back of the cup to improve the match. These sections can be easily designed to operate over a modest bandwidth (typically 5%). Wider bandwidth can sometimes be achieved with a TEM transmission line in the centre conductor that is connected to dipoles in the cup. Another form of a cup feed (Poulton & Bird, 1986) uses waveguide transitions to achieve matching from a TE_{11} circular waveguide input to the coaxial aperture. A simplified picture of the operation of the transition is shown in Figure 4.18. The transition is potentially capable of bandwidths in excess of 10% with suitable optimization. Importantly, the symmetry of the transition allows the polarization properties to be conserved. In this

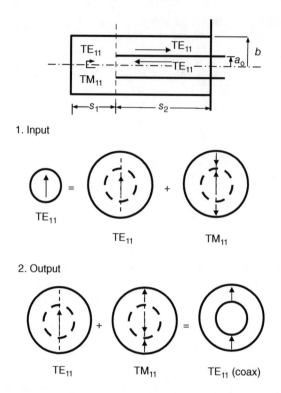

Figure 4.18 Two mode representation of the cup feed waveguide transition. *Source*: Reproduced by permission of The Institution of Engineers, Australia

transition, a TE_{11} mode incident in the central waveguide will excite TE_{1n} and TM_{1n} modes in the cup region, although the cup diameter is chosen so that only the first TE_{11} and TM_{11} modes propagate. By making $a_o/b \approx 0.5$ and choosing the lengths s_1 and s_2 appropriately the sum of the TE_{11} and TM_{11} modes in the cup waveguide can be made to approximate the TE_{11} mode of coaxial waveguide at the output leading to the aperture. Wideband operation is possible because the phase velocity of the TE_{11} and TM_{11} modes are approximately the same in the cup.

Another successful design is the Kildal hat feed (Kildal, 1987). This uses a circumferential aperture instead of transverse apertures as for the Cutler feed. The brim of the hat is a smooth or a corrugated flange, and further improvement is achieved using a number of axial slots placed symmetrically around the waveguide between the aperture and the flange (refer to Figure 4.17d). The bandwidth of operation depends on achieving wideband coupling from the central waveguide to the aperture and various methods are used in practice.

The predominant radiating mechanism of most cup feeds is due to the TE_{11} coaxial waveguide mode that is created in the cup. This mode radiates through slots or from the waveguide aperture itself in the presence of the supporting central conductor. The radiation from an annular aperture is shown in Figure 4.19. It shows a coaxial waveguide with an infinite flange and a central conductor of infinite extent. In the region $z > 0$, the radiated field is equivalent to that produced by an annulus of magnetic current transverse to the axis of the central conductor. The electric field radiated from the coaxial waveguide aperture in the presence of the conductor is

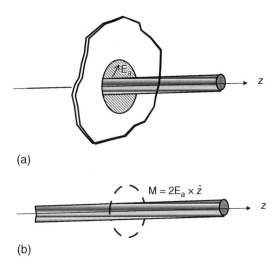

(a)

(b)

Figure 4.19 Radiation of a coaxial waveguide aperture terminated in an infinite flange with central conductor of infinite extent. (a) Original problem and (b) equivalent representation. *Source*: Reproduced by permission of The Institution of Engineers, Australia

$$\mathbf{E}(r,\theta,\phi) = \iint_D \underline{\underline{\mathbf{\Gamma}}}^{(e)} \cdot \mathbf{M}\, dS, \tag{4.50}$$

where it is assumed the aperture is located in an infinite conducting plane and $\mathbf{M} = 2\mathbf{E}_a \times \hat{z}$ is the magnetic current on the aperture. The quantity $\underline{\underline{\mathbf{\Gamma}}}^{(e)}$ is the Greens dyadic for the electric field, and this corresponds to a solution for an infinitesimal magnetic source in the vicinity of an infinite cylinder, the central conductor. Sums of TE and TM coaxial modes can be used to approximate aperture fields in a annular aperture. Rectangular slot apertures can be treated the same way with rectangular mode functions. For coaxial modes, the orthogonality of the sinusoidal azimuthal functions ensures that a mode with an azimuthal period p couples only to modes with the same period. Thus, TE_{pm} and TM_{pm} modes incident on the aperture excite only TE_{pn} and TM_{pn} modes at the aperture, where integers m and n can be same or different.

A far-field approximation to Eq. 4.50 can be obtained for the TE and TM modes for the geometry shown in Figure 4.20. For the TE_{pn} mode, which is the m-th modal contribution of the radiated field (e.g. $m = 1$ corresponds to TE_{11}), the radiated fields are given by Bird (1987)

$$E_\theta(r,\theta,\phi) \sim C_m k j^p p \frac{\exp(-jkr)}{r} \cos\left(p\phi - \phi_o\right) \frac{M_m(k\sin\theta, b, a)}{k\sin\theta} \tag{4.51a}$$

$$E_\phi(r,\theta,\phi) \sim -C_m k j^p \frac{\exp(-jkr)}{r} \sin\left(p\phi - \phi_o\right) \cos\theta \frac{k_m^2 L_m(k\sin\theta, b, a)}{k_m^2 - (k\sin\theta)^2}, \tag{4.51b}$$

where C_m is a constant, ϕ_o is the polarization angle relative to $\phi = 0$, k_m is the cut-off wavenumber of coaxial mode m (given approximately by Eq. 4.27). The remaining two functions M_m and L_m are defined by

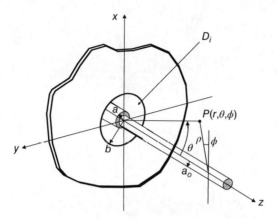

Figure 4.20 Radiation from a coaxial aperture with an extended centre conductor

$$M_m(\xi,x,y) = \frac{1}{k_m}\left[Z_p(k_mx,k_ma)U_p(\xi x,\xi a_o) - Z_p(k_my,k_ma)U_p(\xi y,\xi a_o)\right]$$

$$L_m(\xi,x,y) = \frac{1}{k_m}\left[xZ_p(k_mx,k_ma)V_p'(\xi x,\xi a_o) - yZ_p(k_my,k_ma)V_p'(\xi y,\xi a_o)\right],$$

where a_o is the radius of the central conductor, which, in general, is different from the inner conductor radius of the aperture a. The functions $U_n(x, y)$ and $V_n(x, y)$ are compound Bessel functions and are defined in Appendix B. A prime on these functions indicates the first derivative with respect to the first argument. When the external conductor becomes small, that is, $a_o \rightarrow 0$, $U_n(x,y) \rightarrow J_n(x)$ and similarly $V_n(x,y) \rightarrow J_n(x)$. Thus, Eqs. 4.51 are generalizations of Eqs. 4.45 for the TE$_{11}$ coaxial waveguide ($p = 1 = m$). The functions M, L, U and V all include factors that take into account the external conductor. Furthermore, as $a_o \rightarrow 0$ the expressions reduce to the those for an empty circular waveguide.

Compared with the radiation from circular or coaxial waveguide apertures, the presence of the external central conductor increases the beamwidth in the E-plane and at the same time reduces the H-plane beamwidth. The phase centres are widely separated in the principal planes, and cross-polarization is quite high. For cross-polar levels less than -20 dB, the central conductor radius $a=0=a_o$ should be kept relatively small, typically $kb>2.5$ and $a/b<0.3$. A cup feed that is excited by circular waveguide requires $ka\sqrt{\varepsilon_r} < 1.85$, where ε_r is the dielectric constant of the material in the centre of the waveguide, which is required to ensure propagation occurs. To use the stepped waveguide transition, the inner–outer conductor diameter should be about half the outer conductors', that is, $a/b \approx 0.5$ and correspondingly $kb > 3.9/\sqrt{\varepsilon_r}$. For this range of parameters, the H-plane beamwidth is relatively narrow. This means the best gain with a reflector is achieved with $f/D \sim 0.45$. When the surrounding flange is finite, diffraction from the rim can be used to improve the radiation performance. The dependence of the beamwidth and peak cross-polar level on the flange diameter can be predicted by methods such as GTD as shown in Figure 4.21. These are compared with some measured results. The addition of a coaxial ring slot in the flange, which is excited by parasitic coupling from the main aperture, improves the radiation performance over a narrowband (typically 8%).

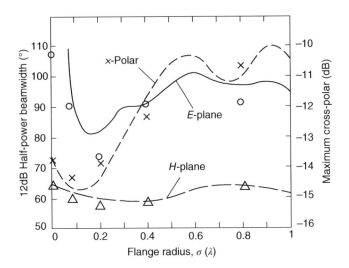

Figure 4.21 Computed beamwidth in the principal planes and maximum cross-polar in the 45° plane of a cup feed versus flange radius (σ) for a coaxial waveguide with extended centre conductor ($b = 0.36\lambda$, $a_o/b = 0.533$) shown in Figure 4.14c. The symbols o, x and Δ correspond to measured values

As an example of the improvements that are possible, Figure 4.22 shows the properties of a cup feed (see Figure 4.17c) with aperture dimensions given by $b = 27$ mm, $a/b = 0.533$ with a ring slot as in Figure 4.17c that is located a distance $\tau = 17.6$ mm away with a width $w = 7.5$ mm and depth $d = 15$ mm. The reflection coefficient of the model (infinite length centre conductor) obtained from theory is shown in Figure 4.22a along with experimental results for a conductor length of $h = 990$ mm. The radiation patterns of a C-band cup feed in the 45° plane at 4 GHz are shown in Figure 4.22b with and without a ring-slot in the flange. A 7 dB reduction in the peak cross-polar level is achieved by having a ring-slot in the flange. Another benefit is that the ring-slot flange brings the phase centres in the principal planes closer together, which in a feed application improves antenna efficiency and reduces aberrations. Adding a second ring slot was shown to make little further improvement to the radiation performance, although it considerably reduces the operating bandwidth.

4.4.3   Conical Horn

If a circular aperture is flared into a conical horn as in Figure 4.7e, the spherical wavefront produced in the aperture can be included in the aperture field by the method described for the pyramidal horn in Section 4.3. For a conical horn, the distance from the wavefront to the aperture plane is

$$\delta = \frac{1}{2}\frac{\rho'^2}{L},\qquad(4.52)$$

$\rho' = \sqrt{x'^2 + y'^2}$ is the radial distance to the source point and L is the distance from the horn apex to the aperture (see Figure 4.7e). In this formulation it is assumed that the flared aperture terminates in an infinite conducting plane. Multiplying Eq. 4.34a by Eq. 4.52 and carrying out the transformation required in polar co-ordinates that is necessary to obtain the far-field

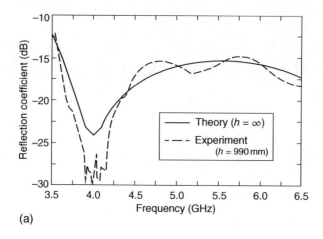

(a)

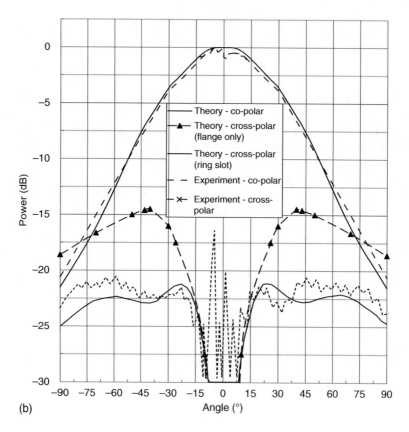

(b)

Figure 4.22 Characteristics of the cup feed shown in Figure 4.17c with $b = 27$ mm, $a/b = 0.533$, $\tau = 17.6$ mm, $w = 7.5$ mm and $d = 15$ mm. (a) Input reflection coefficient – full line: theory with infinite length centre conductor; dashed line: experiment with 990 mm long conductor (b) Co- and cross-polar radiation patterns and at 4 GHz – full line, theory with ring-slot flange; short dash x, experiment with ring-slot and flange ($\sigma = 80$ mm); dash & triangle, theory with flange only ($\sigma = 33$ mm)

components from Eq. 3.26 in the following approximate far-zone electric fields of a conical horn:

$$\begin{Bmatrix} E_\theta \\ E_\phi \end{Bmatrix}(r,\theta,\phi) = \pm \frac{jkE_\text{o}e^{-jkr}}{2}\frac{}{r}\left(Q_0(k_\text{c},k\sin\theta,k/2L,a)\mp Q_2(k_\text{c},k\sin\theta,k/2L,a)\right)\begin{Bmatrix} \cos\phi \\ \cos\theta\sin\phi \end{Bmatrix},$$

(4.53)

where

$$Q_m(\alpha,\beta,\gamma,a) = \int_0^a J_m(\alpha\rho')J_m(\beta\rho'\sin\theta)\exp\left(-j\gamma\rho'^2\right)\rho'd\rho'$$ (4.54)

and $k_\text{c}a = 1.841184$. Eq. 4.53 is usually quite accurate providing the flare angle, θ_o, is less than about 30°. Eq. 4.54 can be expressed in closed form as shown in Eq. B.6. The conical horn can be modeled more accurately for general flare angles through the mode matching method as described in Section 4.5.2. The maximum gain of a conical horn can be obtained from Eqs. 3.48 and 4.38 and is

$$G_\text{max} = \frac{2|(k/a)Q_0(k_\text{c},0,k/2L,a)|^2}{J_1(k_\text{c}a)^2\sqrt{1-(k_\text{c}/k)^2}\left[1-\left(1/\left((k_\text{c}a)^2\right)\right)\right]}.$$

As an illustration of the results given by Eqs. 4.53, the E- and H-plane patterns were calculated for a horn with $a = 1.7\lambda$ and $L = 3.5\lambda$. The results are shown in Figure 4.23 along with

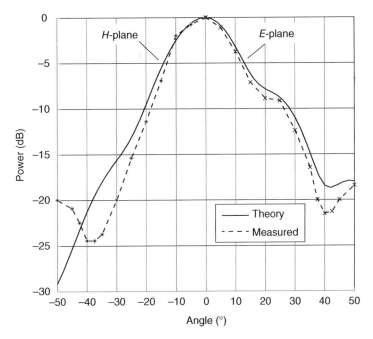

Figure 4.23 Theoretical (Eq. 4.53) and measured (King, 1950) radiation patterns of a conical horn of aperture radius $a = 1.7\lambda$ and length $L = 3.5\lambda$

some classic measured results (King, 1950). The computed E-plane pattern agrees reasonably well with the measured data, but the H-plane pattern is slightly narrower at wider angles possibly due to how the aperture is terminated. The computed gain is 17.53 dBi which agrees well with the 17.7 dBi quoted by King.

The phase centre of the conical horn is different in the two principal planes due mainly to the differences in how the aperture fields terminate at the wall. These phase centres has been computed from Eqs. 4.53 and 3.46 as a function of the cone angle, θ_o. As the phase is available as a function the phase centre calculation is easily modified to replace the summations with integrals, so that by Eq. 3.46 the phase centre in the azimuth plane $\phi = \phi_p$ becomes

$$kz_o\left(\phi_p\right) = \frac{b_1 - a_1 b_0}{a_2 - a_1^2},$$

where $z_o(\phi_p)$ is the phase centre location on the z-axis relative to the aperture in the plane,

$a_q = \left(\int_{-\theta_L/2}^{\theta_L/2} \cos^q\theta\, d\theta\right) \Big/ \theta_L,\; b_q = \left(\int_{-\theta_L/2}^{\theta_L/2} \Phi(\theta, \phi_p) \cos^q\theta\, d\theta\right) \Big/ \theta_L,\; \Phi(\theta, \phi_p)$ is the phase

function of the pattern and θ_L is a symmetrical angular range. In this case, the angular range chosen is the 12 dB beamwidth, and the phase function is computed from the field given in Eq. 4.53. The phase centres shown in Figure 4.24 are for a conical horn of radius $a = 5\lambda$ and have been normalized to the height of the cone. Note that when $1 + z_o/L = 0$, $z_o = -L$, the phase centre is at the apex of the cone inside the horn. Furthermore, when $\Delta > 0.6$, it is seen that phase centre can be just in front of or behind the apex.

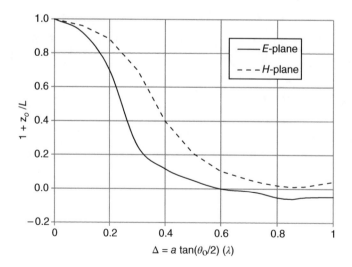

Figure 4.24   Phase centre of a conical of aperture radius $a = 5\lambda$ versus the cone angle θ_o. The height of the cone is $L = a/\tan(\theta_o)$

4.4.4 Corrugated Radiators

As has been observed previously, smooth-wall circular waveguides and horns have, in general, different E- and H-plane patterns. It is desirable for some application to have very similar patterns. Radiators with equal principal plane patterns are said to be axisymmetric. The reason they are not axisymmetric for smooth waveguides is because the electric and magnetic field components in the pipe satisfy different boundary conditions on the wall. The tangential electric field experiences a short circuit, while the magnetic field contends with an open circuit. One result of this is that the electric field lines are curved as in Figure 4.8a. By introducing corrugations or metamaterials on the inside wall, as, for example, in Figure 4.7f, the impedance experienced by the tangential field components may be modified. The principal surface impedance and admittance are given by

$$X_\phi = -j\frac{1}{\eta_o}\frac{E_\phi}{H_z}\bigg|_{\rho=a} \quad \text{and} \quad Y_z = j\eta_o\frac{H_\phi}{E_z}\bigg|_{\rho=a}, \tag{4.55}$$

where the Eqs. 4.55 are determined by the width (w), depth (d) and pitch of the slots (p). When there are many corrugations per guide wavelength (in practice >5 per guide wavelength is usually sufficient) and the corrugation depth is approximately $\lambda/4$ (a quarter free-space wavelength) (i.e. $d \approx 0.5\lambda$), the parallel plate transmission lines formed by the slots in the wall transform the apparent short circuit on the surface into an open circuit, that is, $Y_z \approx 0$. This allows a non-zero axial electric field on the corrugated surface, $\rho = a$, while still giving a zero circumferential component, that is, $X_\phi \approx 0$. Consequently, the modes of corrugated waveguide are no longer TE or TM to the longitudinal (propagation) direction as in dielectric-lined waveguides. As both axial field components are present (i.e. non-zero), the modes are called hybrid modes. Starting from the axial field components

$$E_z(\rho,\phi) = \eta_o a_1^e J_1(k_1\rho)\cos\phi \tag{4.56a}$$

and

$$H_z(\rho,\phi) = a_1^h J_1(k_1\rho)\cos\phi, \tag{4.56b}$$

where $k_1 = \sqrt{k^2\varepsilon_r - \beta^2}$ is the transverse wavenumber in the region at centre of the waveguide with dielectric constant ε_r, the remaining field components can be obtained from Maxwell's equations. Satisfaction of the impedance wall conditions results in

$$a_1^h = a_1^e\left(\frac{k_1}{\beta}\right)((k_1a)Y_z + L_1(k_1a)),$$

where

$$\left(\frac{\beta}{k}\right)^2 = ((k_1a)X_\phi + L_1(k_1a))((k_1a)Y_z + L_1(k_1a))$$

is the characteristic equation in which $L_1(x) = -xJ_1'(x)/J_1(x)$. This equation provides the transverse wavenumber and thereby the propagation constant as a function of frequency. For a corrugated surface with $Y_z = 0$ and $X_\phi = 0$

$$a_1^h = a_1^e \left(\frac{k_1^2 a}{\beta}\right) L_1(k_1 a) \tag{4.57a}$$

and

$$\left(\frac{\beta}{k}\right)^2 = (L_1(k_1 a))^2. \tag{4.57b}$$

There are two roots to Eq. 4.57b. Of interest here is the positive root which gives the principal mode of corrugated waveguide that is used for aperture antennas. This is the HE_{11} mode, which has the field distribution illustrated in Figure 4.8b. The root corresponding to this mode is

$$\bar{\beta} = \left(\frac{\beta}{k}\right) = L_1(k_1 a). \tag{4.58}$$

Equation 4.58 has to be solved to find $k_1 a$. The rectangular components of the transverse electric field of HE_{11} mode are approximately given by (Clarricoats & Olver, 1984)

$$E_x(\rho,\phi) = \left(\frac{E_o}{2}\right)[(\bar{\beta}+\Lambda)J_0(k_1\rho) + (\bar{\beta}-\Lambda)J_2(k_1\rho)\cos 2\phi] \tag{4.59a}$$

$$E_y(\rho,\phi) = -\left(\frac{E_o}{2}\right)[(\bar{\beta}-\Lambda)J_2(k_1\rho)\sin 2\phi], \tag{4.59b}$$

where $k_1 = \sqrt{k^2 \varepsilon_r - \beta^2}$ and $E_o = -ja_1^e(k/k_1)$. The function $\Lambda = L_1(k_1 a)/\bar{\beta}$ is related to the ratio of longitudinal electric and magnetic field components and is the inverse of the so-called hybrid factor. The magnetic field can be obtained from Maxwell's equations.

In antenna feed applications, optimum operation occurs when a circularly symmetric radiation pattern is achieved. Equation 4.59b shows that this occurs when $\Lambda = 1/\bar{\beta} \approx 1$, which is known as the balanced hybrid condition, where the y-component of the aperture field vanishes and there is zero cross-polarization. At the balanced hybrid condition, $\bar{\beta} \approx k$ and $L_1(x_{01}) = 1$, where $k_1 a = x_{01} = 2.40482$. An approximation to Eq. 4.58 based on the Taylor series in the vicinity of x_{01} is

$$L_1(x) \approx \left[1 + x_{01}(x - x_{01}) + \left(\frac{x_{01}}{2}\right)(x - x_{01})^2\right]. \tag{4.60}$$

This approximation has an error less than 10% for most values of x of practical interest. At the balanced hybrid condition, Eq. 4.56 shows that apart from the free-space wave impedance factor, the amplitude of the longitudinal field components is identical. Also, from Eq. 4.60, an approximation to the operating condition is

$$\frac{\beta}{k} = \sqrt{1 - \frac{k_1^2}{k}} \approx \left[1 + x_{01}((k_1 a) - x_{01}) + \left(\frac{x_{01}}{2}\right)((k_1 a) - x_{01})^2\right],$$

which gives an estimate that is accurate within 1% or better for $ka > x_{01}$ and has the advantage of not requiring Bessel function evaluations. As an example, consider a corrugated waveguide with a 2 cm inner diameter ($a = 1$ cm), operating at a frequency of 30 GHz giving $ka = 6.287$. The approximation above estimates $k_1 a = 2.374$, while Eq. 4.58 gives $k_1 a = 2.373$. The smallest real root of the approximate equation for $ka > 3$ is accurately given by

$$k_1 a \approx x_{01}(1 - \Delta),$$

$$\text{where } \Delta = \frac{[(1 + 1/K^2) - \sqrt{1 + (2 - x_{01} - x_{01}^2)/K^2}]}{(x_{01} + x_{01}^2 + 1/K^2)}$$

where $K = ka$ and $\Delta \rightarrow 0$ as $K \rightarrow \infty$ in accordance with the exact solution.

An improved approximation to the transverse fields in the vicinity of the balanced hybrid condition of the HE_{11} mode that takes into account the bell-shape amplitude, which tapers from the axis to vanish at the corrugations, is

$$\mathbf{E}_t = \hat{x} E_1 J_0(k_\rho \rho), \tag{4.61a}$$

$$\mathbf{H}_t = \frac{1}{\eta_o} \hat{z} \times \mathbf{E}_t, \tag{4.61b}$$

where $k_\rho = x_{01}/a$. As seen in Figure 4.8b the electric (and magnetic) field lines are parallel virtually everywhere in the transverse plane. It is left as an exercise to the reader to show from Eqs. 4.61a and 3.20 the far-zone electric fields radiated by the HE_{11} mode at the balanced hybrid condition are

$$\left\{ \begin{matrix} E_\theta \\ E_\phi \end{matrix} \right\} (r, \theta, \phi) = \pm \frac{jk}{4\pi} \frac{e^{-jkr}}{r} (1 + \cos\theta) N_x \left\{ \begin{matrix} \cos \\ \sin \end{matrix} \right\} \phi, \tag{4.62}$$

where $N_x = \dfrac{2\pi E_1}{k_\rho^2 - w^2} (k_\rho a) J_1(k_\rho a) J_0(wa)$ and $w = k \sin\theta$. As noted earlier, the radiation patterns in the E- and H-planes are identical. These patterns are governed by the Bessel function of zero order, and several examples are shown in Figure 4.25. It is observed that a simple approximation to the Bessel function in Eq. 4.62 over most of the main beam is $J_0(z) \sim \exp\left[-(z/2)^2\right]$ and this can be used to study possible designs and radiation patterns.

The total power radiated by the HE_{11} mode at the balanced hybrid condition is

$$P_T = \frac{E_1^2 \pi a^2}{2\eta_o} J_1^2(k_\rho a),$$

from which the maximum gain of corrugated waveguide can be shown to be

$$G_{max} = 0.692 \, (ka)^2.$$

This gives a maximum aperture efficiency η_a of 69.2%.

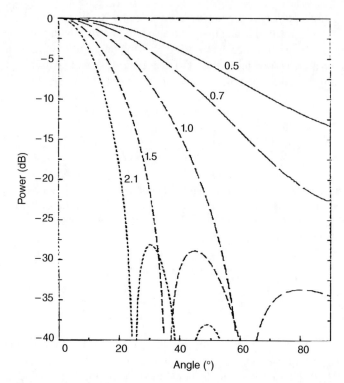

Figure 4.25 Radiation pattern of circular corrugated waveguide. The parameter is the waveguide radius in wavelengths

As an alternative approach, consider a corrugated waveguide terminated in a large ground plane. To determine the radiated fields from Eq. 3.26, substitute Eqs. 4.59a and 4.59b into the transform vector **N**. Making use of the Bessel integral identities in Appendix B, the result is

$$E_\theta(r,\theta,\phi) = \frac{jkE_o}{2}\frac{e^{-jkr}}{r}\left((\bar{\beta}+\Lambda)N_1\cos\phi + (\bar{\beta}-\Lambda)N_2\sin\phi\right) \qquad (4.63a)$$

$$E_\phi(r,\theta,\phi) = -\frac{jkE_o}{2}\frac{e^{-jkr}}{r}\cos\theta\left(-(\bar{\beta}+\Lambda)N_1\sin\phi + (\bar{\beta}-\Lambda)N_2\cos\phi\right), \qquad (4.63b)$$

where

$$N_1(\theta,\phi) = \left[\Omega_0(k_1a, wa) + \Omega_2(k_1a, wa)\cos 2\phi\right]$$

$$N_2(\theta,\phi) = -\Omega_2(k_1a, wa)\sin 2\phi$$

$$\Omega_0(\alpha,\beta) = \frac{a^2}{\alpha^2-\beta^2}\left(\alpha J_1(\alpha)J_0(\beta) - \beta J_0(\alpha)J_1(\beta)\right)$$

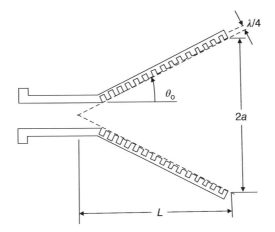

Figure 4.26 Conical corrugated horn. The nominal corrugation depth is $\lambda/4$ but near the throat region the depth is reduced gradually from $\lambda/2$ to $\lambda/4$ for good matching to the smooth wall input waveguide

$$\Omega_2(\alpha,\beta) = \frac{a^2}{\alpha^2-\beta^2}\left(\beta J_2(\alpha)J_1(\beta)-\alpha J_1(\alpha)J_2(\beta)\right)$$

and $w = ka\sin\theta$.

Terminating the periodic corrugated waveguide at an aperture results in reflections and the presence of higher order modes. This transition to free-space can obviously be improved by either extending the corrugations into the aperture plane with a smooth curve (see Thomas, 1978) or by means of a taper (see Figure 4.26). Conical corrugated horns may be treated in the same way as smooth wall conical horns. Approximate aperture fields are obtained by multiplying Eq. 4.61a with the quadratic phase factor $\exp(-jk\delta)$, where δ is given by Eq. 4.52. At the balanced hybrid condition, the radiated fields of a conical corrugated horn with a linear taper of length L and aperture radius a are approximately given by

$$\begin{Bmatrix} E_\theta \\ E_\phi \end{Bmatrix}(r,\theta,\phi) = \pm\frac{jkE_1}{2}\frac{e^{-jkr}}{r}(1+\cos\theta)Q_0\left(k_\rho,k\sin\theta,k/2L,a\right)\begin{Bmatrix} \cos \\ \sin \end{Bmatrix}\phi, \qquad (4.64)$$

where Q_0 is given by Eqs. 4.54 and B.6. Assuming negligible reflection at the aperture, the maximum gain that is predicted by Eq. 4.64 is

$$G_{max} = \left(\frac{2k}{a_I}\right)^2\left|\frac{Q_0\left(k_\rho,0,k/2L,a\right)}{J_1(k_c a_I)}\right|^2,$$

where a_I is the radius of the input waveguide. Eqs. 4.59 and 4.64 are helpful for the design of horns with a moderate flare angle ($<30°$) and linear profile. For example, some results are shown in Figure 4.27 for a horn with $a = 2.1\lambda$, $L = 20\lambda$ and $\theta_o = 6°$ obtained from Eq. 4.64 where there is excellent in excellent agreement with experiment (Loefer et al., 1976). Also, there is similarly good agreement with experiments at 12.5 GHz when Eq. 4.64 is applied to the

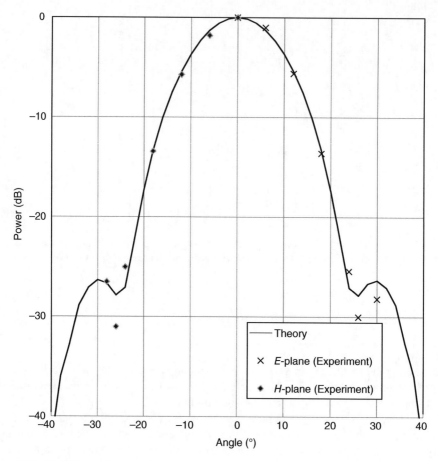

Figure 4.27 Radiation pattern of corrugated horn $a = 2.1\ \lambda$, $L = 20\ \lambda$, $\theta_0 = 6°$ slot depth 0.228λ, theory (Eq. 4.64) versus experiment (Loefer et al., 1976)

corrugated feed horn shown in Figure 1.1d, which has a wider flare angle $\theta_0 = 11.8°$, $L = 65.33$ cm and diameter $2a = 27.3$ cm. In both designs 4–5 corrugations per wavelength were used along the length of the horn, which is usually sufficient to represent a periodic surface. Matching of corrugated horns and wideband design of corrugated horns is described in the references (Thomas, 1978; Thomas et al., 1986; Olver et al., 1994). An approach that provides a good match over moderate bandwidth is to have a uniform waveguide input section where the corrugations commence from a depth with a depth of $\lambda/2$ in order to simulate an electric wall. As the horn is flared, the depth of the corrugations are gradually reduced until a depth of $\lambda/4$ is reached, whereupon they are continued at this depth until the aperture is reached as shown in Figure 4.26.

4.4.5   Cross-Polarization

Cross-polarization became important in antenna design with the introduction of terrestrial and satellite radio systems using two orthogonal polarizations. Signals may be transmitted either linearly polarized in vertical and horizontal components or in circular polarization where the signals can rotate in a right- or left-hand sense. Frequency 're-use' by dual polarization effectively doubles the system bandwidth. Circular waveguides and horns are desirable for dual polarized applications because of their geometrical symmetry. However, as has been seen, this does not ensure low cross-polarization and hence low interference. For example, in Figure 4.8a, the TE_{11} mode in smooth wall waveguide has a radial-oriented electric field at the wall, and therefore 'cross-polarization', occurs to satisfy the boundary conditions. Cross-polarization is transferred to the far-fields, but this can be reduced by the excitation of other suitable modes or with parasitically excited slots in the flange.

Most of the circular waveguides and horns described in Section 4.4 radiate a total electric field of the general form

$$\mathbf{E}(r,\theta,\phi) = c\frac{e^{-jkr}}{r}\left[\hat{\theta}A(\theta)\cos\phi + \hat{\phi}B(\theta)\sin\phi\right], \tag{4.65}$$

where $A(\theta)$ and $B(\theta)$ are the E- and H-plane pattern functions, and c is a constant. In some circumstances, only the E- and H-plane patterns may be known and Eq. 4.65 is a reasonable starting point for design. The form of Eq. 4.65 is a consequence of geometrical symmetry and the aperture fields having only a single period in azimuth (i.e. $\cos\phi$ or $\sin\phi$ dependence).

Eq. 4.65 can be resolved into components parallel and perpendicular to a reference field as described in Section 3.5.4. The parallel component ($\hat{p}$) is the co-polar field, and the cross-polar field is the orthogonal component ($\hat{q}$). The reference polarization is the one that gives an electric field parallel to the x–z (E-) plane for all angles θ and also perpendicular to the y–z (H-) plane. That is, the reference field is

$$\hat{p} = \hat{\theta}\cos\phi + \hat{\phi}\sin\phi. \tag{4.66a}$$

The orthogonal cross-polar field vector is

$$\hat{q} = \hat{r}\times\hat{p} = -\hat{\theta}\sin\phi + \hat{\phi}\cos\phi. \tag{4.66b}$$

Resolving Eq. 4.65 into co-polar and cross-field components results in

$$E_p = \mathbf{E}\cdot\hat{p} = c\frac{e^{-jkr}}{r}\left[A(\theta)\cos^2\phi + B(\theta)\sin^2\phi\right] \tag{4.67a}$$

and

$$E_q = \mathbf{E}\cdot\hat{q} = -c\frac{e^{-jkr}}{r}\sin 2\phi\left[\frac{A(\theta)-B(\theta)}{2}\right]. \tag{4.67b}$$

When $A(\theta) = B(\theta)$, Eq. 4.67a indicates that the co-polar component is independent of ϕ, that is, the radiation pattern is axisymmetric. An antenna with this property is called a Huygens source. Also, the cross-polar component, given by Eq. 4.67b, is zero. Some antennas, such as corrugated waveguides and horns operate close to these conditions. Generally for smooth wall circular radiators $A(\theta) \neq B(\theta)$ although the phase of the pattern functions is approximately

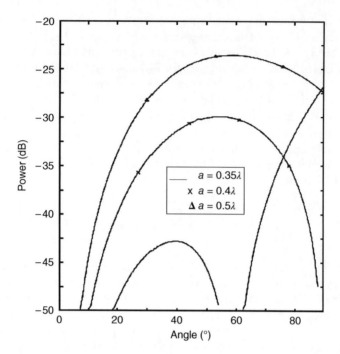

Figure 4.28 Circular waveguide cross-polar patterns in 45° plane (*E–H* field model)

the same. Eq. 4.67b shows that peak cross-polarization occurs between the principal planes at $\theta = \pm 45°$ and equals $|A(\theta) - B(\theta)|/2$. Therefore, cross-polarization due to antennas of the type described by Eq. 4.65 depends on the difference of the *E*- and *H*-plane pattern functions. Further, as the phase of these functions is approximately the same, the peak cross-polar level is approximately given by the difference in the *E*- and *H*-plane patterns. On the other hand, the co-polar pattern in the inter-cardinal planes is $|A(\theta) + B(\theta)|/2$, that is, the average of the *E*- and *H*-plane patterns.

Cross-polar patterns have a characteristic null on axis. Some examples are shown in Figure 4.28 for circular waveguides with *E*- and *H*-plane patterns given in Figure 4.10. The patterns that are computed with the simple *E–H* and *E*-field models are sometimes too inaccurate for design because of the importance of wall currents in cross-polarization. More sophisticated models that include the effects of the flange can predict cross-polarization quite accurately, as illustrated in Figure 4.29.

Now consider the case of corrugated waveguide once again in a little more detail. The field components in the *p* and *q* directions as derived from Eq. 4.63 are in the following form:

$$E_p = \mathbf{E} \cdot \hat{p}$$

$$\approx c \frac{e^{-jkr}}{r} \left[(\bar{\beta} + \Lambda)(\Omega_0 + \Omega_2 \cos 2\phi) \cos^2 \phi - (\bar{\beta} - \Lambda)\Omega_2 \sin 2\phi \sin \phi \cos \phi \right.$$

$$\left. - \left[(\bar{\beta} + \Lambda)(\Omega_0 + \Omega_2 \cos 2\phi)_1 \sin^2 \phi + (\bar{\beta} - \Lambda)\Omega_2 \sin 2\phi \cos \phi \sin \phi \right] \right].$$

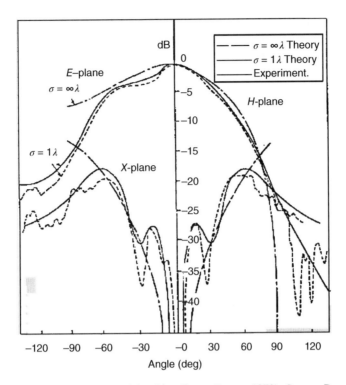

Figure 4.29 Patterns of circular waveguide with a flange (James, 1979). *Source*: Reproduced by permission of The Institution of Engineers, Australia

That is,

$$E_p \approx c \frac{e^{-jkr}}{r} \left((\bar{\beta} + \Lambda)(\Omega_0 + \Omega_2 \cos 2\phi) - (\bar{\beta} - \Lambda)\Omega_2 \sin^2 2\phi \right).$$

Also,

$$E_q = \mathbf{E} \cdot \hat{q} \approx c \frac{e^{-jkr}}{r} \left((\bar{\beta} + \Lambda)(\Omega_0 + \Omega_2 \cos 2\phi) \cos \phi \sin \phi - (\bar{\beta} - \Lambda)\Omega_2 \sin 2\phi \sin^2 \phi \right)$$
$$- \left[(\bar{\beta} + \Lambda)(\Omega_0 + \Omega_2 \cos 2\phi)_1 \cos \phi \sin \phi + (\bar{\beta} - \Lambda)\Omega_2 \sin 2\phi \cos^2 \phi \right].$$

That is,

$$E_q = \mathbf{E} \cdot \hat{q} \approx -c \frac{e^{-jkr}}{r} \sin 2\phi (\bar{\beta} - \Lambda)\Omega_2.$$

Clearly the peak cross-polarization occurs in the ±45° planes. Of interest is this value relative to the peak co-polar level. Thus, for the corrugated waveguide,

$$\left. \frac{E_q(\text{peak})}{E_p} \right|_{\phi = \pm 45°} \approx \frac{(\bar{\beta} - \Lambda)\Omega_2}{(\bar{\beta} + \Lambda)(\Omega_0 + \Omega_2)}. \tag{4.68}$$

This ratio is zero at the balanced hybrid condition. At frequencies close to the balanced hybrid condition, the approximation Eq. 4.60 applies and also $\beta \approx k$. Hence

$$\frac{E_q(\text{peak})}{E_p}\bigg|_{\phi=\pm 45°} \approx \left(x_{01}(x-x_{01}) + \left(\frac{x_{01}}{2}\right)(x-x_{01})^2\right)\frac{\Omega_2}{2(\Omega_0 + \Omega_2)}.$$

Equality of the E- and H-plane patterns may be difficult or impossible to achieve in practice. However, by ensuring the E- and H-plane patterns cross-over at around the 8–13 dB level reasonably low cross-polarization may be realized for feed applications. Parasitic rings (Figure 4.7b) are also useful for tailoring the radiation pattern to minimize cross-polarization (James, 1979).

4.5 Advanced Horn Analysis Topics*

Among the possible further topics to discuss in relation to waveguide and horn antennas, three important aspects have been selected for further study because of their important consequences for practical design. These topics are flange effects, modelling of stepped horns through mode matching and the design of horns with a general profile in order to achieve specific performance requirements. These advanced topics will be considered in the following sections.

4.5.1 Flange Effects*

As has been seen from Figure 4.29 and elsewhere, the flange surrounding the aperture of a waveguide or horn can have a significant effect on the principal radiation patterns as well as cross-polarization. The flange tends to have a second-order effect on the input reflection coefficient although it can be significant for small aperture horns. When a horn radiates, the field is scattered by any obstacle nearby in the 360° space surrounding it. Because the field is

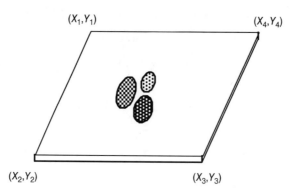

Figure 4.30 Layout of a finite flange and the vertices defining its shape

usually small at the rear of the horn, the amount scattered directly back to the aperture is usually small, except when there are other external objects, and therefore its influence on other parts of the pattern is secondary. A flange or aperture rim can occur very near the actual source of radiation as they are often in the same plane (i.e. at 90°). Currents are set up on these nearby objects, and they radiate in all directions including back towards the aperture which in turn influences the radiation from the horn. Usually, it is a good approximation to assume the edges and rims are secondary radiators that are superimposed on the direct radiation from the aperture.

There are several approximate methods by which to calculate corrections for the flange. The methods are geometrical theory of diffraction (GTD) (James, 1986) and the physical theory of diffraction (PTD) (Hay et al., 1996) which can be employed to correct the radiation patterns for the effect of diffraction from a finite flange (such as in Figure 4.30). While GTD is based on the laws of geometric optics, PTD uses the assumptions of physical optics in conjunction with the field equations. Although GTD is normally computationally faster than PTD, PTD tends to be more accurate and the pattern obtained is a continuous function.[1] It is recommended that PTD be used if the edge of the flange is close to an aperture.

In both methods the flange can be assumed to be circular, polygonal or a general shape. Within the accuracy of the methods used a rectangular co-ordinate system such as in Figure 4.30 is convenient for describing most flanges. The edge of the flange is treated as a vertex of two planes (see Figure 4.31) with an internal angle β. To treat a flange with finite thickness it is best to choose $\beta = 90°$ (Figure 4.31a). A thin flange is treated by setting $\beta \approx 0°$. This means the edge will have a profile like that in Figure 4.31b. Note that a wedge angle of $\beta = 90°$ models the rear of the antenna as an infinite conductor extending behind the aperture, and the fields normal and tangential to the conductor will satisfy the usual boundary conditions. The fields are unspecified inside the conductor and, therefore, to estimate the horn's front-to-back ratio it is preferable in that case to use a thin flange approximation. Although these models are approximations, they often give satisfactory corrections to the radiation pattern, providing the flange is not too near the aperture ($> 0.1\lambda$). The alignment of the field with the edge of the flange is an important factor in assessing the likely effect of the flange on the radiation pattern. For example, in the E-plane, the electric field is normal to the edge and this results in a larger scattered field due to a larger diffraction coefficient than in the H-plane. Therefore, the E-plane pattern is more greatly impacted by a finite flange. The reader should consult the references for further details (Balanis, 1982; James, 1986; Hay et al., 1996).

4.5.2   Mode Matching in Horns*

The representation of the field in a tapered, stepped or a more general profiled structure can be handled in several ways. One of these is the method of mode matching, which is quite accurate and has an appealing physical description that is given in this section. A general model of a simple electromagnetic horn, shown in Figure 4.32, radiates into a multi-dielectric semi-infinite region. Transition sections may be employed in the horn for matching purposes along with choke rings in

---

[1] Discontinuities cause problems when the data are used to compute secondary radiation from reflectors.

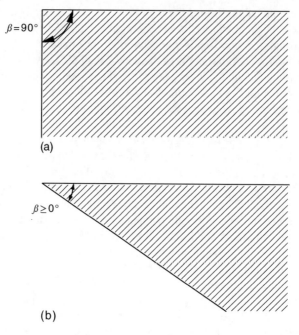

Figure 4.31   Modelling the edge of a finite flange (a) 90° corner; and (b) wedge of angle β

order to control the radiation. There are essentially two types of approach needed to handle the transverse discontinuities in the closed waveguide and semi-infinite regions, and each approach requires knowledge of the modes in the waveguide region. A prototype section for mode matching in a closed waveguide is shown in Figure 4.33. Modal field solutions may be obtained by analytical methods for waveguides with circular, elliptical and rectangular cross section, while structures with more general cross sections can be analysed with a numerical method, such as the finite elements. Whatever technique is used to obtain the transverse fields, it is assumed that a set of known modes with transverse fields is available in the form $(\mathbf{e}_{pi}, \mathbf{h}_{pi})$, where the subscript p refers to the mode and i is the section number. These forward and reverse travelling waves have coefficients $\bar{a}_{pi}$ and $\bar{b}_{pi}$ resulting in column vectors $\bar{\mathbf{a}}_1, \bar{\mathbf{b}}_1, \bar{\mathbf{a}}_2$ and $\bar{\mathbf{b}}_2$. Note that in conventional scattering wave notation, on the input port $\mathbf{a}_1 = \bar{\mathbf{a}}_1$ and $\mathbf{b}_1 = \bar{\mathbf{b}}_1$, and on the output port $\mathbf{a}_2 = \bar{\mathbf{b}}_2$ and $\mathbf{b}_2 = \bar{\mathbf{a}}_2$.

In section i of the horn, the total transverse field $\left(\mathbf{E}_t^{(i)}, \mathbf{H}_t^{(i)}\right)$ is approximated as a finite sum of $M(i)$ modes as follows:

$$\mathbf{E}_t^{(i)} = \sum_{p=1}^{M(i)} \left(\bar{a}_{pi}e^{-j\gamma_{pi}z} + \bar{b}_{pi}e^{+j\gamma_{pi}z}\right)\mathbf{e}_{pi}(x,y)Y_{pi}^{-1/2} \tag{4.69a}$$

$$\mathbf{H}_t^{(i)} = \sum_{p=1}^{M(i)} \left(\bar{a}_{pi}e^{-j\gamma_{pi}z} - \bar{b}_{pi}e^{+j\gamma_{pi}z}\right)\mathbf{h}_{pi}(x,y)Y_{pi}^{+1/2} \tag{4.69b}$$

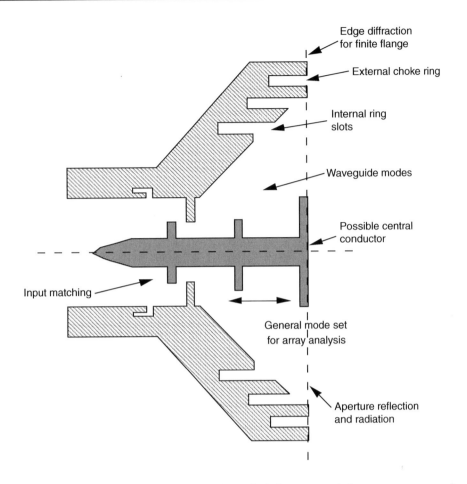

Figure 4.32   Geometry of a typical electromagnetic horn

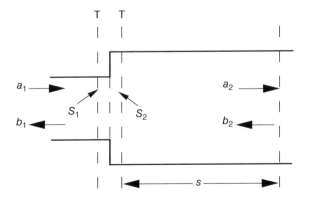

Figure 4.33   Prototype section for mode matching in waveguides

where $\gamma_{pi} = \beta_{pi} - j\alpha_{pi} =$ propagation constant of mode p in region i

$$\mathbf{h}_{pi} = \hat{z} \times \mathbf{e}_{pi}$$

and

$$\iint_{S_i} dS \mathbf{e}_{pi} \times \mathbf{h}_{qi} \cdot \hat{z} = 2\delta_{pq}. \tag{4.70}$$

Y_{pi} is the wave admittance of mode p and δ_{pq} is the Kronecker delta. By enforcing continuity of the fields at $z = 0$ and by vector post-multiplying $\mathbf{E}_t^{(1)} = \mathbf{E}_t^{(2)}$ by $\mathbf{h}_{q1}$ and pre-multiplying $\mathbf{H}_t^{(1)} = \mathbf{H}_t^{(2)}$ by $\mathbf{e}_{q2}$ and integrating across the junction taking into account orthogonality Eq. 4.70 results in the following mode-matching equations:

$$\begin{aligned}
\mathbf{D}_1^{-1}(\bar{\mathbf{a}}_1 + \bar{\mathbf{b}}_1) &= \mathbf{C}\mathbf{D}_2^{-1}(\bar{\mathbf{a}}_2 + \bar{\mathbf{b}}_2) \\
\mathbf{C}^{\mathbf{T}}\mathbf{D}_1(\bar{\mathbf{a}}_1 - \bar{\mathbf{b}}_1) &= \mathbf{D}_2(\bar{\mathbf{a}}_2 - \bar{\mathbf{b}}_2)
\end{aligned} \tag{4.71}$$

After rearranging

$$\begin{bmatrix} \bar{\mathbf{b}}_1 \\ \bar{\mathbf{a}}_2 \end{bmatrix} = \begin{bmatrix} \mathbf{S}_{11} & \mathbf{S}_{12} \\ \mathbf{S}_{21} & \mathbf{S}_{22} \end{bmatrix} \begin{bmatrix} \bar{\mathbf{a}}_1 \\ \bar{\mathbf{b}}_2 \end{bmatrix}$$

where $\mathbf{S}_{11}, \mathbf{S}_{12}$ and so on are scattering parameters of the junction given by

$$\mathbf{S}_{11} = -(\mathbf{I} + \mathbf{XY})^{-1}(\mathbf{I} - \mathbf{XY}),$$

$$\mathbf{S}_{12} = 2(\mathbf{I} + \mathbf{XY})^{-1}\mathbf{X},$$

$$\mathbf{S}_{21} = 2(\mathbf{I} + \mathbf{YX})^{-1}\mathbf{Y}, \text{ and}$$

$$\mathbf{S}_{22} = (\mathbf{I} + \mathbf{YX})^{-1}(\mathbf{I} - \mathbf{YX}).$$

In addition, $\mathbf{I}$ is the unit matrix, $\mathbf{X} = \mathbf{D}_1 \mathbf{C}^{\mathbf{T}} \mathbf{D}_2^{-1}$ and $\mathbf{Y} = \mathbf{D}_2^{-1} \mathbf{C} \mathbf{D}_1 = \mathbf{X}^{\mathbf{T}}$, where

$$\mathbf{D}_{1,2} = \begin{bmatrix} Y_{11,12}^{1/2} & 0 & 0 & \cdots \\ 0 & Y_{21,22}^{1/2} & 0 & \cdots \\ 0 & 0 & Y_{31,32}^{1/2} & \cdots \\ \cdots & \cdots & \cdots & \cdots \end{bmatrix}$$

The elements of the matrix $\mathbf{C}$ are given by

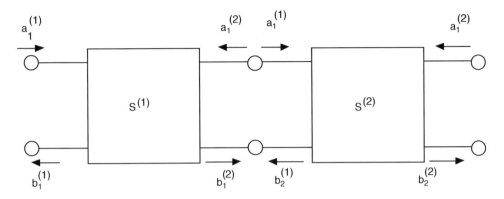

Figure 4.34 Concatenation of two scattering matrices

$$C_{pq} = \frac{1}{2} \iint_{S_i} dS \, \mathbf{e}_{p1} \times \mathbf{h}_{q2} \cdot \hat{z}. \tag{4.72}$$

Two further fundamental operations are needed to analyse more than a single step in cross section. Uniform waveguide interconnections are treated by a reference plane extension, and the other is the combining scattering matrices of many sections in cascade by concatenating the scattering matrices of each section.

The cascading is illustrated in Figure 4.34 for two sections in scattering matrix notation.

Let $M(i)$ be the number of modes present at port i. To find the equivalent scattering matrix for a cascade of two two-port networks that have known scattering matrices $S^{(1)}$ and $S^{(2)}$, let the scattering matrix for network i be partitioned as follows:

$$\begin{bmatrix} \mathbf{b}_1^{(i)} \\ \mathbf{b}_2^{(i)} \end{bmatrix} = \begin{bmatrix} \mathbf{S}_{11}^{(i)} & \mathbf{S}_{12}^{(i)} \\ \mathbf{S}_{21}^{(i)} & \mathbf{S}_{22}^{(i)} \end{bmatrix} \begin{bmatrix} \mathbf{a}_1^{(i)} \\ \mathbf{a}_2^{(i)} \end{bmatrix}. \tag{4.73}$$

Continuity is enforced between the networks when

$$\mathbf{a}_2^{(1)} = \mathbf{b}_1^{(2)} \quad \text{and} \quad \mathbf{a}_1^{(2)} = \mathbf{b}_2^{(1)}.$$

The reflected wave amplitude at the input of network 2 is

$$\begin{aligned} \mathbf{b}_1^{(2)} &= \mathbf{S}_{11}^{(2)} \mathbf{a}_1^{(2)} + \mathbf{S}_{12}^{(2)} \mathbf{a}_2^{(2)} \\ &= \mathbf{S}_{11}^{(2)} \mathbf{b}_2^{(1)} + \mathbf{S}_{12}^{(2)} \mathbf{a}_2^{(2)} \\ &= \mathbf{S}_{11}^{(2)} \left(\mathbf{S}_{21}^{(1)} \mathbf{a}_1^{(1)} + \mathbf{S}_{22}^{(1)} \mathbf{b}_1^{(2)} \right) + \mathbf{S}_{12}^{(2)} \mathbf{a}_2^{(2)}. \end{aligned}$$

That is,

$$\mathbf{b}_1^{(2)} = \mathbf{\Delta} \mathbf{S}_{11}^{(2)} \mathbf{S}_{21}^{(1)} \mathbf{a}_1^{(1)} + \mathbf{\Delta} \mathbf{S}_{12}^{(2)} \mathbf{a}_2^{(2)},$$

where $\Delta = \left(I - S_{11}^{(2)} S_{22}^{(1)} \right)^{-1}$. This last equation is now substituted into $b_1^{(1)}$ and $b_2^{(2)}$ to produce the scattering matrix equations for the combined network

$$b_1^{(1)} = \left(S_{11}^{(1)} + S_{12}^{(1)} \Delta S_{11}^{(2)} S_{21}^{(1)} \right) a_1^{(1)} + S_{12}^{(1)} \Delta S_{12}^{(2)} a_2^{(2)}$$

$$b_2^{(2)} = S_{21}^{(2)} \left(S_{21}^{(1)} + S_{22}^{(1)} \Delta S_{11}^{(2)} S_{21}^{(1)} \right) a_1^{(1)} + \left(S_{22}^{(2)} + S_{21}^{(2)} S_{22}^{(1)} \Delta S_{12}^{(2)} \right) a_2^{(2)}. \tag{4.74}$$

That is,

$$S = \begin{bmatrix} S_{11}^{(1)} + S_{12}^{(1)} \Delta S_{11}^{(2)} S_{21}^{(1)} & S_{12}^{(1)} \Delta S_{12}^{(2)} \\ S_{21}^{(2)} \left(I + S_{22}^{(1)} \Delta S_{11}^{(2)} \right) S_{21}^{(1)} & S_{22}^{(2)} + S_{21}^{(2)} S_{22}^{(1)} \Delta S_{12}^{(2)} \end{bmatrix}. \tag{4.75}$$

Eq. 4.75 is the scattering matrix of concatenated uniform sections 1 and 2.

As an illustration of the results obtained from the mode-matching method, consider transverse steps in rectangular waveguides in the H- and E-planes, respectively. The results will be used to demonstrate the convergence of the mode-matching solution with an increasing numbers of modes. The question of convergence of the mode-matching method has been debated in the literature (Lee et al., 1971; Masterman & Clarricoats, 1971). In the examples to follow, consider a square waveguide input with side length 0.7λ and output waveguide with dimensions of 0.7λ by 1λ. Two cases arise corresponding to this step occurring in the E- or the H-plane. A TE_{10} mode is incident at the input and the reflection coefficient in the square waveguide is plotted in Figs. 4.36 and 4.37 as a function of the number of modes in the output waveguide. In the H-plane case, TE_{m0} ($m = 1, 3, 5, \ldots$) modes are only excited, while in the E-plane case, TE_{1n} and TM_{1n} ($n = 0, 2, 4, \ldots$) modes are excited in the output. Results are shown for each case in Figures 4.35 and 4.36 under two situations: (a) $M(1) = M(2)$ i.e. mode number is the same in each region, and (b) $M(1)/M(2) = $ (input dimension)/(output dimension). The latter (b) has the potential advantage of requiring fewer modes, which offers substantial reduction in computation time for structures composed of many steps. Most importantly, it ensures the edge condition is satisfied at the step as $M(2)$ increases exponentially (Lee et al., 1971; Hockham, 1975). Figures 4.35 and 4.36 show that there is little advantage in accuracy of (b) over (a) for single steps in waveguide cross section. However, if a thin iris were to separate the two waveguides, (b) would ensure convergence to the correct solution.

A mode-matching method is now described for modelling the radiation from a horn. In the exterior region, the fields may be represented as integrals of a suitable Green's function and the equivalent electric and magnetic currents on the aperture surface. This will be described in Section 7.3 for apertures located in a large ground plane that is parallel to $z = 0$. Continuity of the transverse fields at the aperture and then an application of Galerkin's method results in

$$b^{(0)} = S^{(0)} a^{(0)}, \tag{4.76}$$

where $S^{(0)}$ is the mode scattering matrix at the aperture for modes transitioning to free-space, $a^{(0)}$ is the vector of incident mode amplitudes and $b^{(0)}$ is the amplitude of the reflected mode amplitudes at the apertures.

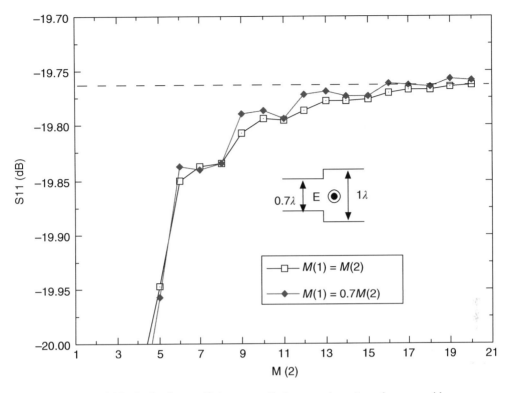

Figure 4.35   Reflection coefficient at an *H*-plane step in rectangular waveguide

The number of modes required for an accurate representation of the aperture field depends on the operating frequency. If the waveguide or horn operates in the fundamental mode and all other modes are well below cut-off, a good estimate of reflection is obtained from a few modes only. Use of several high-order modes is recommended, however, for accurate predictions. Satisfaction of the edge condition is not critical except when there is a thin iris at the aperture, although Hockham (1975) showed that for rectangular waveguide, inclusion of TE_{m0} ($m = 3, 5, \ldots$) and TE_{0n} ($n = 2, 4, \ldots$) modes improve solution convergence.

The final part of the analysis combines the mode scattering matrices of the horn transitions with the mode scattering matrix of the aperture. A network model for an array of horns that combines radiation and horn steps is shown in Figures 7.12 and 4.37 where for simplicity only horn (1) is shown. Assume modes of amplitude a_I are input to all horns which is represented by a combined mode scattering matrix S. Separately, at the input of element p of the array there are forward and backward waves with amplitudes $a_I^{(p)}$ and $b_I^{(p)}$. These amplitudes are related to the forward and backward waves at the aperture through

$$\begin{bmatrix} b_I \\ a_O \end{bmatrix} = \begin{bmatrix} S_{11} & S_{12} \\ S_{21} & S_{22} \end{bmatrix} \begin{bmatrix} a_I \\ b_O \end{bmatrix}, \tag{4.77}$$

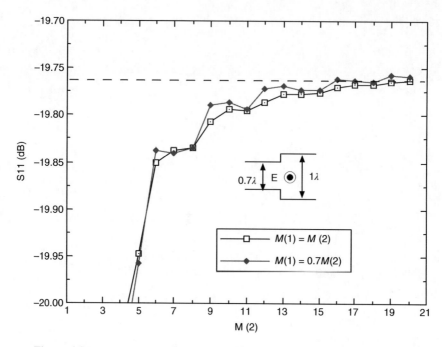

Figure 4.36   Reflection coefficient at an *E*-plane step in rectangular waveguide

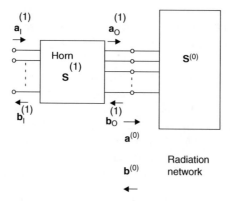

Figure 4.37   Network representation of a horn

where $\mathbf{a}_O^{(p)}$ and $\mathbf{b}_O^{(p)}$ are the amplitudes of incident and reflected modes at the aperture and where the sub-matrices of the scattering matrix are partitioned as follows:

$$
\mathbf{S_{ij}} =
\begin{bmatrix}
S_{ij}^{(1)} & 0 & 0 & \cdots \\
0 & S_{ij}^{(2)} & 0 & \cdots \\
0 & 0 & S_{ij}^{(3)} & \cdots \\
\cdots & \cdots & \cdots & \ddots
\end{bmatrix}.
$$

where $S_{ij}^{(n)}$ refers to the scattering matrix for horn n. By means of Eqs. 4.76 and 4.77, a relationship is obtained between the reflected mode amplitudes and the incident mode amplitudes, as follows,

$$\mathbf{b_I} = \left[\mathbf{S_{11}} + \mathbf{S_{12}} \mathbf{S}^{(0)} \left(\mathbf{I} - \mathbf{S_{22}} \mathbf{S}^{(0)} \right)^{-1} \mathbf{S_{21}} \right] \mathbf{a_I}$$

$$\mathbf{a}^{(0)} = \left[\left(\mathbf{I} - \mathbf{S_{22}} \mathbf{S}^{(0)} \right)^{-1} \mathbf{S_{21}} \right] \mathbf{a_I} \qquad\qquad (4.78)$$

$$\mathbf{b}^{(0)} = \left[\mathbf{S}^{(0)} \left(\mathbf{I} - \mathbf{S_{22}} \mathbf{S}^{(0)} \right)^{-1} \mathbf{S_{21}} \right] \mathbf{a_I}$$

From Eq. 4.78 it is seen that all coefficients depend on the matrix $\left(\mathbf{I} - \mathbf{S_{22}} \mathbf{S}^{(0)} \right)$, which is almost unity for large apertures as they are reasonably well matched to the external aperture and the remainder of the horn. This occurs for horns with gentle linear tapers. In that instance, the reflection coefficient is determined by the mismatch at the horn input and the mode distribution in the aperture is determined by the transmission within the horn itself.

4.5.3   *Profiled Horns**

The design of a horn is usually a compromise of several competing performance options such as efficiency, low sidelobes and cross-polar levels as well as achievement of a minimum reflection coefficient. To do this in a systematic manner, it is advisable to use a structured approach such as a numerical optimization method and to profile the horn accordingly. In the method to be outlined here, an initial horn profile is represented by a cubic spline passing through a series of node-points (Bird & Granet, 2013). Thus approximated, the horn can then be modelled by means of several techniques for the purpose of analysis. If the mode-matching technique is adopted, short uniform waveguide sections are selected between the node points.

The traditional approach is to select a horn geometry and a representation of it and to model its performance compared with a reference structure. However, the availability of accurate and fast analysis methods, coupled with optimization methods, has made automatic geometry determination possible with fast computers. The geometry can be changed in a systematic way to improve the horn's performance. These changes, and the design approach itself, should have physical basis such as limits on the maximum gain and sidelobe level for a given aperture taper.

An alternative design approach is to fix the basic structure such as the geometry, any substrate thicknesses and dielectric constants. Other constraints can be implemented in the method by setting limits on the input reflection coefficient, minimum or maximum gain, efficiency, peak cross-polarization, sidelobe levels, radiation pattern symmetry, half-power beamwidth, and minimum or maximum dimensions such as the wall thickness and the total length. In feed applications, the reflector edge illumination may need to have limits.

A simple way to implement these constraints is through a penalty function and optimizer, which is a well-established optimization strategy. The process then modifies/changes the geometry in three-dimensional space while at the same time checks to determine how these changes

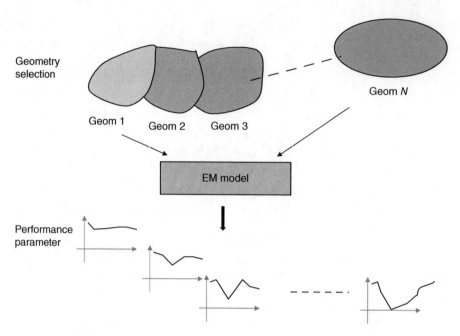

Figure 4.38 Diagrammatic representation of geometry optimization

alter the performance. This process is shown diagrammatically in Figure 4.38. The process amounts to modifying the geometry in one space while applying constraints in other spaces.

The actual implementation of a penalty function and numerical optimization is described in the function Section 6.9.2. The techniques adopted here are very similar to the optimization of the excitation coefficients for arrays and the approach has been adapted for horn profiling. In the remainder of this section, aspects of the optimization related to profiling horns are discussed and some results are described.

A penalty function approach commences with creating a performance index. This index is constructed usually from a sum of constraint functions. Let L stand for function to be minimized. Initially L is set equal to zero. At each frequency f_i a contribution to $L(i)$ is obtained at all NF frequency points ($i = 1, \ldots,$ NF) in the band. For example, let constraints be applied to $RL(i) =$ return loss, $XP(i) =$ peak cross-polar, and $CP(i) =$ co-polar level at frequency i. Suppose the target limits of each of these is designated by a 'T'. Thus, $RLT(i)$, $XPT(i)$ and so on are the targets. Let NRL, NXP etc. be integer powers on each constraint and wRL, wXP etc. be weighting functions on them. Therefore, at frequency i, $L(i)$ is formed as follows:

Let $L(i) = 0$, then
If $RL(i) < RLT(i)$, let $L(i) = wRL(RLT(i) - RL(i))^{NRL}$.
If $XP(i) > XPT(i)$, let $L(i) = L(i) + wXP(XP(i) - XPT(i))^{NXP}$.
If $CP(i) < CPUB(i)$, let $L(i) = L(i) + wCP1(CP(i) - CPUB(i))^{NCP}$.
If $CP(i) > CPLB(i)$, let $L(i) = L(i) + wCP2(CPLB(i) - CP(i))^{NCP}$ and so on as required.

Note that if all constraints are satisfied, then the index $L(i)$ is zero.

Once a penalty function has been created, it can be minimized or maximized using standard techniques as described in the next sub-section. In profiling, it can be beneficial to employ a

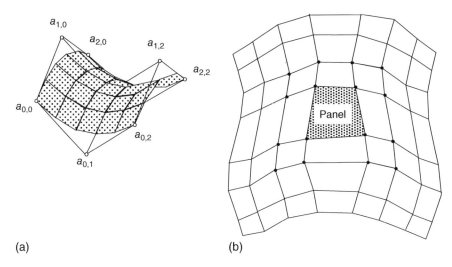

Figure 4.39   Spline representation of (a) Bezier surfaces with control polygons $a_{i,j}$ and (b) general two dimensional surface made up of panels

combination of optimization methods. For example, particle swarm optimization (PSO) could be used to get into the vicinity of a global optimum, and this could be followed by a gradient search technique for faster convergence to the minimum.

When using optimizers, usually a first trial design is made with the constraints deliberately relaxed to ensure convergence. Then the requirements are tightened until converge fails. During this process, care must be taken as the result may not be optimum in any sense. If the problem is well formulated, good results are nearly always obtained. Any uncertainty may correspond to the problem of local versus global extrema in optimization.

Representation of the surface depends on whether it is to be rotationally symmetric or not. For rotational symmetric surfaces such as horns or dielectric rods, a cubic spline with knot points uniformly distributed along the z-axis can be used to represent the profile. This ensures a smooth and continuous profile while minimizing the number of parameters to optimize. In this case, $\mathbf{p}(u) = \sum_{i=0}^{m} N_i(u)\,\mathbf{p}_i$, where $N_i(u)$ is the spline function (de Boor, 1978). For general surfaces, the surface is discretized into panels as shown in Figure 4.39. The shape function on the panel is specified by a B-spline surface, and the coefficients of these shape functions are optimized. Such a shape function is

$$\mathbf{p}(u,v) = \sum_{i=0}^{m}\sum_{j=0}^{n} N_{ip}(u)N_{jq}(v)\mathbf{p}_{ij}, \tag{4.79}$$

where the $\mathbf{p}_{ij}$ are optimized (de Boor, 1978). The design objectives for profiled horns often involves minimizing the overall length for a given diameter horn. This has benefits in applications with weight and space limitations. In some reflector systems, the space available is very limited and a shorter horn would be convenient and possibly less expensive as long as the performance was not compromised. For an axisymmetric horn, the profile is initially represented by a series of points. Cubic splines are then fitted to these points. The coefficients of the spline

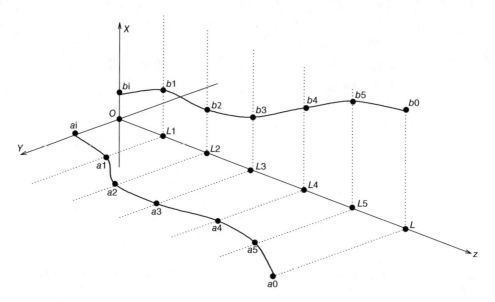

Figure 4.40   Horn profile representation in two planes. In a circular symmetric horn $b_i = a_i$

functions are then optimized. This usually means a smaller number of optimizer variables compared with a point-wise representation. The basis of the approach is optimization based on minimizing a penalty function usually over a band of frequencies. Further details of the format of a penalty function are provided in Section 6.9.2.

Most horns in common use can be represented by one or two profiles as illustrated in Figure 4.40. The single profile corresponds to the axisymmetric horn. The two profiles correspond to a rectangular (Bird & Granet, 2007) or elliptical horn. In each plane there are design parameters on which the optimizer is required to satisfy where possible.

4.5.3.1   Optimization*

The coefficients of the spline functions are adjusted to satisfy the required constraints. For example, a gradient search or a genetic algorithm could be used to minimize the performance index over a band of frequencies. A flow chart of the optimization approach is shown in Figure 4.41. The length of the horn could be part of the numerical optimization, but often it is simpler instead to do this manually and run the software several times to adjust it. The variables involved in a typical single horn profile optimization are listed in Table 4.2.

4.5.3.2   Parametric Profiles*

It is great assistance for convergence to choose an initial profile at the start of the optimization that is a good approximation to the final result. Some useful starting profiles are defined below for a small number of parameters including the input radius a_i, aperture radius a_o length L of the profiled section and a shape function (Figure 4.42).

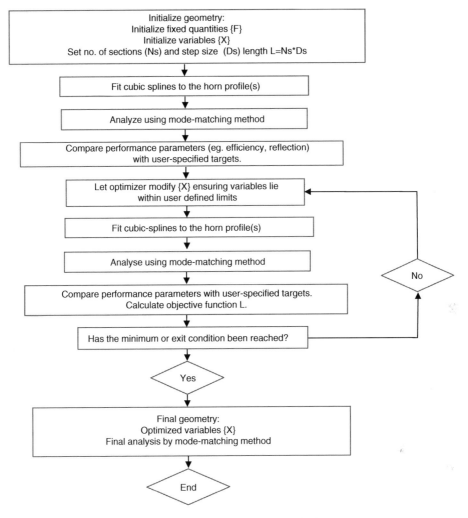

Figure 4.41   Flow chart of optimization

Table 4.2   Typical variables involved in a horn profile optimization

a_i	Input width/height – the first extreme node in each plane (fixed values)
a_o	Output width/height – the second extreme node in each plane
d_o	Allowed displacement of the output width/height (usually set as a % of a_o) since the required beamwidth of the radiation pattern is known beforehand and a_o can be estimated
a_1, a_2, a_3, a_4, a_5	Width/height of the five inner nodes
d_1, d_2, d_3, d_4, d_5	Allowed displacements of each of the inner nodes (making a_1, a_2, a_3, a_4, a_5 constrained variables)
L_1, L_2, L_3, L_4, L_5	Positions of each of the inner nodes as % of the horn's length

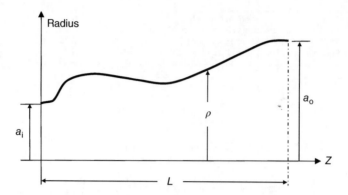

Figure 4.42   Geometry for parametric profiles

Linear:

$$\rho(z) = a_i + (a_o - a_i)\frac{z}{L} \tag{4.80a}$$

Gaussian (or hyperbolic):

$$\rho(z) = a_i \sqrt{1 + \left(\frac{z}{L}\right)^2 \left(\left(\frac{a_o}{a_i}\right)^2 - 1\right)} \tag{4.80b}$$

Sine to the power p:

$$\rho(z) = a_i + (a - a_i)\sin^p\left(\frac{\pi z}{2L}\right). \tag{4.80c}$$

The parametric profiles given by Eq. 4.80 can provide suitable results by themselves. For example, the corrugated horn shown in Figure 4.43 was designed with a sine raised to the power p profile by adjusting the parameter p in Eq. 4.80c in order to give low sidelobes which was required in order to avoid interference with neighbouring satellites in a space-borne application (Granet et al., 2000). The radix $p = 0.8$ gave the best predicted results. Further details of this horn are given in Section 10.3.1.

As an example of a horn with a profile designed by optimization is the two wavelength diameter circularly symmetric horn shown in Figure 4.44. The horn profile was designed for maximum gain and a peak cross-polarization less than −25 dB in the frequency band 11.7–12.2 GHz. This horn was designed by the techniques described previously and was fabricated as shown in Figure 4.44. The profile is plotted in Figure 4.45. The computed and measured results across the band are given in Table 4.3. Two different computer simulations were used to analyse the horn. One method was mode matching (MM) and the other was with CST Microwave Studio (MS). There is generally good agreement between the two methods and also with experiment (Expt) as is seen in Table 4.3. Measured and computed patterns at 11.95 GHz are shown in Figure 4.46. The reflection coefficient was less than −25 dB at frequencies above 11.7 GHz.

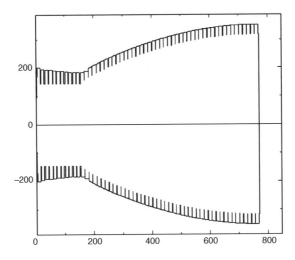

Figure 4.43 Profiled corrugated circular horn with sine-to-power $p = 0.8$

Figure 4.44 Smooth wall circular horn with aperture diameter 2λ designed for maximum gain and low cross-polarization

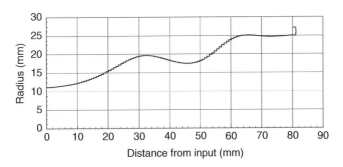

Figure 4.45 Profile of smooth wall circular horn shown in Figure 4.44

Table 4.3 Measured and computed gain of the profiled circular horn shown in Figures 4.44 and 4.45

Frequency	MM			MS			Expt		
	Reflection coefficient (dB)	Computed gain (dBi)	Computed efficiency (%)	Reflection coefficient (dB)	Computed gain (dBi)	Computed efficiency (%)	Reflection coefficient (dB)	Gain (dBi)	Efficiency (%)
11.7	−25.5	15.51	93.96	−23.5	15.55	94.83	−26.7	15.5 ± 0.3	92.9
11.95	−26.4	15.72	94.53	−24.2	15.74	94.97	−28.2	15.6 ± 0.3	91.1
12.2	−27.1	15.95	95.63	−25.4	15.96	95.85	−28.7	15.8 ± 0.3	92.0

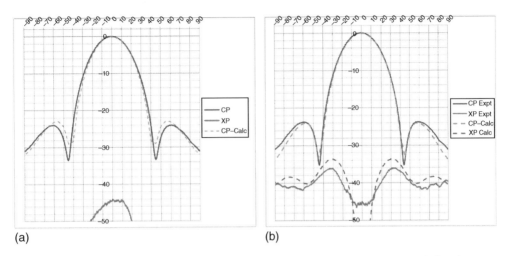

(a)   (b)

Figure 4.46 Radiation patterns at 11.95 GHz of a profiled 2λ diameter circular horn designed for maximum gain. (a) E-plane and (b) 45-degree plane patterns. Solid curve is measured and dashed curve is computed using the mode-matching method

The predicted maximum aperture efficiency was about 95%, while in practice the highest efficiency realised was about 92%. This compares with 84% for an ideal circular horn with a uniform aperture field. More will be said in Section 10.3.1 about horn profile optimization for maximum efficiency.

4.6 Problems

P4.1 Follow the procedure given in Section 4.3.1 to design a standard gain pyramidal horn with a gain of 24 dBi at 13.6 GHz. The feed waveguide input dimensions are $a_w = 1.905$ cm and $b_w = 0.9525$ cm.
(Answer: $a = 15.657$ cm, $b = 12.408$ cm, $\ell_E = 34.92$ cm, $\rho_E' = 34.365$ cm, $\ell_H = 37.068$ cm, $\rho_H' = 36.232$ cm, and $h = 31.823$ cm.) Note: A commercially available standard gain horn from the Scientific Atlanta Company with a measured gain of 24 dB at 13.6 GHz has aperture dimensions $a = 15.189$ cm and $b = 12.470$ cm.

P4.2 Show that Eq. 4.23 for a pyramidal horn leads to a quartic polynomial in either aperture dimensions a or b. To do this, express all geometric quantities in terms of the required dimension, for example, $\rho_H' = a^2/\alpha_1$. From this polynomial, verify that only one solution is a valid solution for this application.

P4.3 Use the results of Section 4.3 to obtain far-field expressions for sectoral horns that are flared in either the E- or H-planes but are uniform in the orthogonal direction. Use these results to obtain the maximum gain for both horns. Verify the product of these expressions results in Eq. 4.21 times a geometric factor $(2/\pi)^3 k^2 (a_w b_w)$ where a_w and b_w are width and height of the input waveguide.

P4.4 Follow the steps described in Section 4.4.3 of the text to obtain the far-fields for the smooth wall conical horn (given in Eq. 4.53). At the aperture assume that the propagation constant of the TE_{11} mode is $\beta \approx k$.

P4.5   Repeat P4.4 for a corrugated horn operating in the HE_{11} mode at the balanced hybrid condition. Show that the far-field radiation pattern is given by Eq. 4.64.

P4.6   For a Ku-band conical corrugated horn operating from 11.7 to 12.5 GHz, determine
       a. the corrugation depths at the lowest and upper frequencies and also choose a suitable slot depth for this band with reasons; and
       b. estimate the required width of the corrugations to approximate a long periodic surface.

P4.7   Assume the transverse electric field of the LSE_{10} or the TEx_{10} mode in a slab-loaded dielectric waveguide is given by Eq. 4.25. Use Maxwell's equations and the boundary conditions to obtain the remaining field components.

P4.8   Starting from the perturbation expression Eq. 4.27, show that for an air-filled rectangular waveguide ($\varepsilon_{r2} = 1$), width a and height b, centrally loaded with a slab of dielectric of width c and dielectric constant ε_{r1} and with the interfaces parallel to the narrow wall, the propagation constant of the LSE_{mn} modes in this type of dielectric loaded waveguide is approximately given by

$$\gamma = \sqrt{k^2 \left(1 + (\varepsilon_{r1}-1)\frac{c}{a}\left\{1 + \frac{a}{m\pi c}\sin\left[m\pi\left(1-\frac{c}{a}\right)\right]\right\}\right) - \left(\frac{m\pi}{a}\right)^2 - \left(\frac{n\pi}{b}\right)^2}$$

P4.9   Show that for a large aperture, the thickness of dielectric required to make the propagation constant of the LSE_{10} mode in slab-loaded dielectric waveguide in Figure 4.5 equal to the free-space wavenumber, that is, $\gamma = k$, is

$$d \approx \frac{a}{2}\left[\frac{6}{(ka)^2(\varepsilon_r-1)}\right]^{1/3},$$

       where a is the width of the rectangular waveguide and ε_r is the dielectric constant of the material.

P4.10 Obtain the approximate phase centre of an antenna with the complex far-field E- and H-plane patterns given by $P_E = A(\theta)\cos\phi$ and $P_H = B(\theta)\sin\phi$, respectively, when
       a. $A(\theta) = C = B(\theta)$, and
       b. $A(\theta) = C\exp(-ja\theta)$ and $B(\theta) = C\exp(-j\beta\theta)$,
       where C, α, and β are constants.

P4.11 Use the mode-matching approach to obtain the scattering parameters for a symmetrical E-plane step in rectangular waveguide. Assume a single mode approximation on both sides of the junction.

P4.12 The susceptance of a thin iris in circular waveguide is given by Eq. 4.41. Use this expression to select parameters to obtain a low reflection coefficient at the aperture of a circular waveguide of one wavelength in diameter. The TE_{11} mode aperture admittance normalized to the free-space wave admittance is $1.0105 - j0.0189$ (calculated from Eq. 7.88).

P4.13 Design a conical horn to produce a gain of 25 dBi at a frequency of 10 GHz.

P4.14 The peak gain of a conical horn is a compromise between the aperture diameter $2a$ and its length L (King, 1950). Use the equation for the maximum gain of a conical horn to estimate the diameter that achieves the peak gain for a horn of length (a) $L = 3.5\lambda$ and (b) $L = 5\lambda$.

Answer: (a) $a = 1.65\lambda$ and (b) $a = 1.972\lambda$.

P4.15 Design a linearly tapered circular horn with the same aperture dimensions as the horn shown in Figures 4.44, 4.45 and 4.46 (2λ at 11.95 GHz) to achieve a minimum first sidelobe in the E-plane. How long is this horn, what is the minimum first sidelobe level and what is the peak gain?

P4.16 A rectangular horn with dimensions $a = 2.7$ cm and $b = 3$ cm is to be used as the basis of a slab-loaded horn for operation at 12 GHz. A material with a dielectric constant of 3.6 is to be placed on the sidewalls. What thickness of slab is required to obtain a uniform field in both the E- and H-planes? Plot the radiation pattern in the H-plane.

P4.17 Design a linearly tapered corrugated horn by means of Eq. 4.64 for operation at 30 GHz to produce a half-power beamwidth of $40°$.

P4.18 Design a coaxial horn with minimum cross-polarization at 7.5 GHz and a 12 dB half-beamwidth in the H-plane of $120°$ and an inner conductor that is large enough to include a circular waveguide feed for 30 GHz.

P4.19 The axial field components of the fundamental HE_{11} mode of a dielectric rod of radius a with relative permittivity ε_r are given by

$$E_z(\rho,\varphi,z) = \sin \varphi e^{-j\gamma z} \begin{cases} B_1 J_1(k_1\rho); & \rho \le a \\ B_2 K_1(h_2\rho): & \rho \ge a \end{cases}$$

$$H_z(\rho,\varphi,z) = \cos \varphi e^{-j\gamma z} \begin{cases} C_1 J_1(k_1\rho); & \rho \le a \\ C_2 K_1(h_2\rho): & \rho \ge a \end{cases}$$

where $k_1 = \sqrt{k^2\varepsilon_r - \gamma^2}$, $k_2 = -jh_2 = -j\sqrt{\gamma^2 - k^2}$, $J_1(x)$ is the ordinary Bessel function and $K_1(x)$ is the modified Bessel function of the second kind of order 1, B_1, C_1 and so on are constants that may be obtained from the boundary conditions at $\rho = a$, $\gamma = \beta - j\alpha$ is the propagation constant and (ρ, ϕ) is cylindrical polar-co-ordinate. Describe how the far-field radiation patterns may be calculated when the rod is terminated.

P4.20 Obtain an expression for the maximum gain of a coaxial waveguide aperture.

P4.21 A rectangular pyramidal horn has its aperture covered by a radome of uniform thickness $d \ll \lambda$. Assume the radome is illuminated by a uniform spherical wave radiating from the phase centre of the horn. Use geometric optics to obtain the aperture field on the outside surface of the radome and the equivalent currents.

a. Use the equivalent currents to calculate the radiated far-fields.

b. Determine the effect on the radiation patterns and gain of the horn shown in Figure 4.4 when radome material with dielectric constant of 1.4 and $d = 0.5$ mm is placed over the aperture.

References

Balanis, C.A. (1982): 'Antenna theory: analysis and design', Harper and Rowe, New York.

Berk, A.D. (1956): 'Variational principles for electromagnetic resonators and waveguides', IRE Trans. Antenna Propagat., Vol. **AP-4**, No. 2, pp. 104–111.

Bird, T.S., James, G.L. and Skinner, S.J. (1986): 'Input mismatch of TE11 mode coaxial waveguide feeds', IEEE Trans. Antennas Propag., Vol. **AP-34**, pp. 1030–1033.

Bird, T.S. (1987): 'TE11 mode excitation of flanged circular coaxial waveguides with an extended centre conductor', IEEE Trans. Antennas Propag., Vol. **AP-35**, pp. 1358–1366.

Bird, T.S. and Hay, S.G. (1990): 'Mismatch in a dielectric-loaded rectangular waveguide antenna', Electron. Lett., Vol. **26**, pp. 59–61.

Bird, T.S. and Granet, C. (2007): 'Optimization of profile of rectangular horns for high efficiency', IEEE Trans. Antennas Propag., Vol. **AP-55**, pp. 2480–2488.

Bird, T.S. and Love, A.W. (2007): 'Horn antennas', Chapter 14. Volakis, J.L. 'Antenna engineering handbook', 4th ed., McGraw-Hill, New York.

Bird, T.S. & Granet, C. (2013): 'Profiled horns and feeds', L. Shafai, S.K. Sharma & S. Rao (eds.) 'Vol. II: Feed systems' of 'Handbook of reflector antennas', Artech House, Chicago, IL, pp. 123–155.

Bose, J.C. (1927): 'Collected physical papers', Longmans, Green & Co., London, UK.

Clarricoats, P.J.B. and Olver, A.D. (1984): 'Corrugated horns for microwave antennas', Peter Peregrinus Ltd., London, UK.

Collin, R.E. (1960): 'Field theory of guided waves', McGraw-Hill, New York.

Cutler, C.C. (1947): 'Directional microwave antenna', US Patent No. 2,422,184, June 17.

de Boor, C. (1978): 'A practical guide to splines', Springer-Verlag, New York.

Fröberg, C.-F. (1974): 'Introduction to numerical analysis', Addison-Wesley, London, UK.

Gabriel, G.J. and Bodwin, M.E. (1965): 'The solution of guided waves in inhomogeneous anisotropic media by perturbation and variational methods', IEEE Trans. Microw. Tech., Vol. **MTT-13**, pp.364–370.

Galejs, J. (1969): 'Antennas in inhomogeneous media', Oxford, UK, Pergamon Press.

Granet, C., Bird, T.S. and James, G.J. (2000): 'Compact multimode horn with low sidelobes for global earth coverage', IEEE Trans. Antennas Propag., Vol. **AP-48**, No. 7, pp. 1125–1133.

Harrington, R.F. (1961): 'Time-harmonic electromagnetic fields', McGraw-Hill, New York.

Hay, S.G., Cooray, F.R. and Bird, T.S. (1996): 'Accurate modelling of edge diffraction in arrays of circular and rectangular horns', Journees Internationales de Nice sur les Antennes – JINA **96**, 12–14 November, pp. 645–648.

Hockham, G.A. (1975): 'Use of the 'edge condition' in the numerical solution of waveguide antenna problems', Electron. Lett., **11**, pp.418–419.

James, G.L. (1979): 'Cross-polarization performance of flanged cylindrical and conical waveguides', Proc. IREE (Aust.), Vol. **40**, pp. 180–184.

James, G.L. (1986): 'Geometrical theory of diffraction for electromagnetic waves', 3rd Edition, Peter Peregrinus Ltd., London, UK.

James, G.L. (1987): 'Admittance of irises in coaxial and circular waveguides for TE11-mode excitation', IEEE Trans. Microw. Theory Tech., Vol. **MTT-35**, No. 4, pp. 430–434.

King, A.P. (1950): 'The radiation characteristics of conical horn antennas', Proc. IRE, Vol. **38**, pp. 249–251.

Kildal, P.-S. (1987): 'A hat-feed: a dual-mode rear-radiating waveguide antenna having low cross polarization', IEEE Trans. Antennas Propag., Vol. **AP-35**, No.9, pp. 1010–1016.

Lee, S.W., Jones, W.R. and Campbell, J.J. (1971): 'Convergence of numerical solutions of iris-type discontinuity problems', IEEE Trans. Microw. Theory Tech., Vol. **MTT-19**, pp. 528–536.

Loefer, G.R., Newton, J.M., Schuchardt, J.M. and Dees, J.W. (1976): 'Computer analysis speeds corrugated horn design', Microwaves, pp. 58–65.

Marcuvitz, N. (1986): 'Waveguide handbook', Peter Peregrinus Ltd., London, UK.

Masterman, P.H. and Clarricoats, P.J.B. (1971): 'Computer field-matching solution of waveguide transverse discontinuities', Proc. IEE, Vol. **118**, pp. 51–63.

Minnett, H.C. and Thomas, B.M.A. (1966): 'Fields in the image space of symmetrical focussing reflectors', IEEE Trans. Antennas Propag., Vol. **AP-14**, pp. 654–656.

Olver, A.D., Clarricoats, P.J.B., Kiskh, A.A. and Shafai, L. (1994): 'Microwave horns and feeds', IEEE Press, New York.

Oppenheim, A.R. and Shafer, R.W. (1975): 'Digital signal processing', Prentice-Hall, Englewood Cliffs, NJ.

Poulton, G.T. and Bird, T.S. (1986): 'Improved rear-radiating waveguide cup feeds', IEEE Antennas & Propagation Symposium, Philadelphia, PA, 8–13 June, pp. 79–82.

Rumsey, V.H. (1966): 'Horn antennas with uniform power patterns around their axes', IEEE Trans. Antennas Propag., Vol. **AP-14**, pp. 656–658.

Sarkar, T.K., Mailloux, R.J., Oliner, A.A., Salazar-Palma, M. and Sengupta, D.L. (2006); 'History of wireless', John Wiley & Sons, Hoboken, NJ.

Sharstein, R.W. and Adams, A.T. (1988): 'Thick circular iris in a TE11 mode circular waveguide', IEEE Trans. Microw. Theory Tech., Vol. **MTT-36**, No. 11, pp. 1529–1531.

Silver, S. (1946): 'Microwave antenna theory and design', McGraw-Hill Book Co., New York. Reprint published by Peter Peregrinus Ltd., London, UK, 1984.

Thomas, B. MacA. (1978): 'Design of corrugated conical horns', IEEE Trans. Antennas Propag., Vol. **AP-26**, pp.367–372.

Thomas, B. MacA., James, G.L. and Greene, K.J. (1986): 'Design of wide-band corrugated conical horns for Cassegrain antennas', IEEE Trans. Antennas Propag., Vol. **AP-34**, pp. 750–757.

Tsandoulous, G.N. and Fitzgerald, W.D. (1972): 'Aperture efficiency enhancement in dielectric loaded horns', IEEE Trans. Antennas Propag., Vol. **AP-20**, No. 1, pp.69–74.

Weinstein, L.A. (1969): 'The theory of diffraction and the factorization method', The Golem Press, Boulder, CO.

5

Microstrip Patch Antenna

5.1 Introduction

The microstrip patch and other printed antennas shown in Figure 5.1 are now commonplace as they are readily combined with electronic components and integrated circuits. The feed line can be incorporated into existing microstrip line circuitry branching to amplifiers, mixers, down-converters and semiconductor sources (Gupta et al., 1979). Another major advantage of the microstrip patch is that it can be flush-mounted planar or conformal with other surfaces, such as an aerofoil, with only a minimum of space required for the feed line. The patch may be fed by a transmission line also etched on the dielectric sheet (substrate) as shown in Figure 5.1 or by a probe through the back of the ground plane. The shape of the patch varies significantly although the basic radiation mechanism is similar. In the examples shown in Figure 5.1, one or more resonances can be established in part of the geometry that is coupled to the input. From these resonances, radiation can be created with different radiation characteristics depending on the geometry, and this has resulted in a variety of useful designs (Sainati, 1996). Printed antennas were discovered in the 1950s (Deschamps & Sichak, 1953) and their success in portable equipment generated much research interest and improvements (Munson, 1974). These advances led to a greater understanding of the radiation mechanism. The fundamental mode of the microstrip line is quasi-TEM, and this is the one assumed in the antennas shown in Figure 5.1. Other higher modes are excited to a limited extent although these make a small contribution to the overall radiation. In other microstrip structures they can become more important and some can radiate as leaky modes. In this chapter, a simple model of radiation is developed and some basic properties of a rectangular patch are described with the added incentive of describing another aperture antenna.

Fundamentals of Aperture Antennas and Arrays: From Theory to Design, Fabrication and Testing,
First Edition. Trevor S. Bird.
© 2016 John Wiley & Sons, Ltd. Published 2016 by John Wiley & Sons, Ltd.
Companion website: www.wiley.com/go/bird448

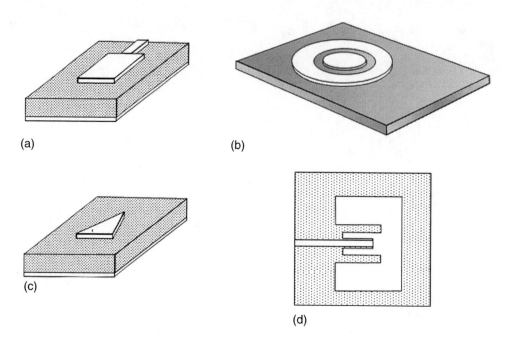

Figure 5.1 Typical microstrip patch antennas. (a) Rectangular. (b) Annular ring. (c) Triangular. (d) E-shaped

5.2 Microstrip Patch Aperture Model

The rectangular microstrip shown in Figure 5.2 consists of a thin metallic patch separated by a dielectric substrate of thickness h and dielectric constant ε_r from a ground plane. The patch has width w and length ℓ when referred to the connecting input line. A signal assumed initially travelling down the microstrip line encounters a change of width initially at $z = -\ell/2$ and characteristic impedance where the patch appears as a low impedance parallel plate transmission line with characteristic impedance $Z_{0p} \approx \eta h / w \sqrt{\varepsilon_r}$. After some reflection, the signal continues until it encounters the open circuit at $z = \ell/2$. At this discontinuity, the signal undergoes a large reflection and travels back along the length of the patch to the transition where it also radiates. As described, two slot radiators are formed at each end of the patch with the ground plane as illustrated in Figure 5.2 with a separation ℓ. As a result, they can be interpreted as a two-element aperture array. The slots are separated by a length ℓ. In addition, if the patch height is small, the field in each slot will be approximately uniform as shown in Figure 5.2a. Further, each slot has an image as shown. Therefore, the aperture model has a height $2h$. Consider first a single slot in the x–y plane as shown in Figure 5.3.

The aperture field is approximated by

$$\mathbf{E}_a = \begin{cases} \hat{x} E_o; & -h \leq x \leq h; \quad -w/2 \leq y \leq w/2 \\ 0; & \text{elsewhere} \end{cases}, \tag{5.1a}$$

and

$$\mathbf{H}_a = \frac{\sqrt{\varepsilon_r}}{\eta} \hat{z} \times \mathbf{E}_a. \tag{5.1b}$$

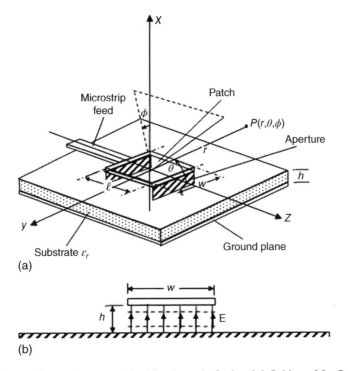

Figure 5.2 Rectangular patch antenna (a) with microstrip feed and definition of far-field point P and (b) approximate electric field in parallel plate region under the patch

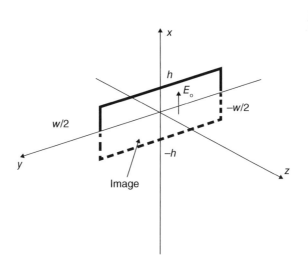

Figure 5.3 Model of microstrip patch radiation

Since the electric field is assumed to be zero everywhere except over the slot, this situation is similar to the problem of a uniformly excited aperture in an infinite ground plane. There is, however, a small field in the region $z < 0$. Therefore by the methods of Section 2.2, a perfect electric conductor can be introduced behind the aperture. The equivalent sources on the aperture in Figure 5.3 are in the form

$$\mathbf{M}_s = -2\hat{z}_a \times \mathbf{E}_a \tag{5.2a}$$

$$\mathbf{J}_s = 0. \tag{5.2b}$$

The far-fields radiated by these sources are obtained directly from Eqs. 3.26 as

$$E_\theta = \frac{jk}{2\pi} \frac{e^{-jkr}}{r} N_x \cos\phi \tag{5.3a}$$

$$E_\phi = \frac{jk}{2\pi} \frac{e^{-jkr}}{r} \cos\theta N_x \sin\phi, \tag{5.3b}$$

where

$$N_x(u,v) = 2hwE_oS(2\pi uh)S(\pi vw); \tag{5.4}$$

is the only non-zero component and as usual $u = 1/\lambda(\sin\theta\cos\phi)$ and $v = 1/\lambda(\sin\theta\sin\phi)$. If, as in many applications, $h \ll \lambda$ then

$$N_x(u,v) \approx 2hwE_oS(\pi vw). \tag{5.5}$$

Returning to the original geometry, the field radiated from the microstrip patch is the super-position of two radiators given by Eqs. 5.3 and phased according to the length ℓ. Assuming $\ell \ll r$, the distance to the far-zone region, the combined radiation may be found by the principle of superposition. Combining the radiation from the apertures at the ends 1 and 2 of the patch

$$E_\theta = E_{\theta 1} + E_{\theta 2}$$
$$\approx \frac{k}{\pi} hwE_oS(\pi vw)\cos\phi\left(\frac{e^{-jkR_1}}{R_1} + \frac{e^{-jkR_2}}{R_2}\right)$$

and

$$E_\phi = E_{\phi 1} + E_{\phi 2}$$
$$\approx -j\frac{k}{\pi} hwE_oS(\pi vw)\cos\theta\sin\phi\left(\frac{e^{-jkR_1}}{R_1} + \frac{e^{-jkR_2}}{R_2}\right)$$

where R_1 and R_2 are the distances from each end to the far-field. At large distances from the apertures, $r \gg \approx \ell/2$ the angles are $\theta_1 \approx \theta \approx \theta_2$ and similarly $\phi_1 \approx \phi \approx \phi_2$. Therefore, $R_1 \approx r - (\ell/2)\cos\theta$ and $R_2 \approx r + (\ell/2)\cos\theta$. By making these approximations into the phase functions and also approximating $1/R_1 \approx 1/r \approx 1/R_2$ in the amplitudes of the fields, the electric field components become

$$E_\theta(r,\theta,\phi) \approx \frac{2k}{\pi} hwE_oS\left(\frac{kw}{2}\sin\theta\sin\phi\right)\cos\left(\frac{k\ell}{2}\cos\theta\right)\cos\phi\frac{e^{-jkr}}{r} \tag{5.6a}$$

and

$$E_\phi(r,\theta,\phi) \approx -j\frac{2k}{\pi} hwE_oS\left(\frac{kw}{2}\sin\theta\sin\phi\right)\cos\left(\frac{k\ell}{2}\cos\theta\right)\cos\theta\sin\phi\frac{e^{-jkr}}{r}. \tag{5.6b}$$

This result can be visualized as the product of the element pattern due to a single aperture multiplied by an array factor that is due to free-space phasing and the phase of the radiation to the observation point.

As an example, consider a patch with length $\ell = \lambda/2\sqrt{\varepsilon_r} = \lambda_g/2$ so that $k\sqrt{\varepsilon_r}\ell = \pi$. In the principal planes, the patterns are as follows:

E-plane ($\phi = 0$)

$$E_\theta(r,\theta,\phi) = \frac{2jkhwE_o}{\pi}\frac{e^{-jkr}}{r}\cos\left(\frac{\pi}{2}\cos\theta\right) \tag{5.7a}$$

H-plane ($\theta = \pm\pi/2$)

$$E_\theta(r,\theta,\phi) = \frac{2jkhwE_o}{\pi}\frac{e^{-jkr}}{r}S\left(\frac{\pi w}{\lambda}\sin\phi\right)\cos\phi. \tag{5.7b}$$

The E- and H-plane patterns of a microstrip patch of width $w = 0.25\lambda$ on a substrate of thickness $h = 0.1\lambda$ and dielectric constant $\varepsilon_r = 2.54$ are shown in Figure 5.4.

The maximum gain from a microstrip fed patch can be estimated using Eq. 3.48. To do this, the power input is assumed to come from a uniform TEM field in the line of width w_I as given by Eq. 3.36. Thus,

$$P_T \approx \frac{\sqrt{\varepsilon_r}}{2\eta_o} w_I h |E_I|^2, \tag{5.8}$$

where in Eq. 5.8 E_I is the peak of the uniform electric field at the input. This peak field is related to the peak field in the patch E_o by the input reflection coefficient, in the usual way by

$$\Gamma_i = \frac{Z_{0p} - Z_{0I}}{Z_{0p} + Z_{0I}}. \tag{5.9}$$

Z_{0p} is the characteristic impedance of the parallel plate transmission line of the patch, which is given by $Z_{0p} \approx \eta_o h/w\sqrt{\varepsilon_r}$, and Z_{0I} is the characteristic impedance of the input microstrip, which is approximately (Gupta et al., 1979; Sainati, 1996)

$$Z_{0I} = \frac{\eta_o}{\sqrt{\varepsilon_r}(\tau + 1.393 + 0.667\ln(\tau + 1.444))}, \tag{5.10}$$

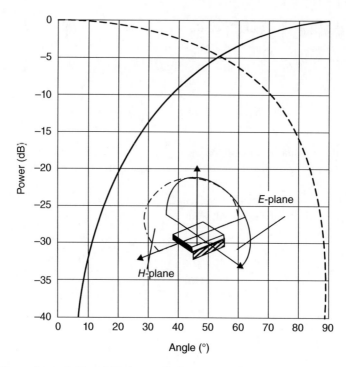

Figure 5.4 Microstrip patch E- and H-plane radiation patterns for a patch with dimension $w = 0.25\lambda$, $h = 0.1\lambda$; $\ell = 0.8\lambda$, $\varepsilon_r = 2.54$. Solid: E-plane; dashed: H-plane

where

$$\tau = \frac{w_I}{h} + 0.398\frac{T}{h}\left(1 + \ln\left(\frac{2h}{T}\right)\right),$$

and T is the thickness of the microstrip. The characteristic impedance given by Eq. 5.10 is accurate to within 1% for $w_I/h > 1$. Combining Eqs. 5.8 and 5.9 with Eqs. 5.7 in Eq. 3.48 results in a maximum gain of

$$G_{max} \approx \frac{16\pi}{\sqrt{\varepsilon_r}}\left(\frac{wh}{\lambda^2}\right)\left(\frac{w}{w_I}\right)\left(1 - |\Gamma_I|^2\right). \tag{5.11}$$

The final term in the square brackets of Eq. 5.11 is the power transmission coefficient. Typically, $|\Gamma_I| < 0.25$ (i.e. -6 dB) and $w > 2w_I$. Suppose this is the case in the previous example for $w = 0.25\lambda$, $h = 0.1\lambda$ and $\varepsilon_r = 2.54$. For this microstrip patch, the maximum gain estimate given by Eq. 5.11 is $G_{max} > 1.7$ dBi.

An equivalent circuit for the patch antenna has been developed based on a transmission line model as shown in Figure 5.5 (Sainati, 1996). According to Figure 5.2a, the transmission line has a characteristic impedance Z_{op}, is of length ℓ and has a slot radiator at each end. The slots are modelled by a conductance G_r and a capacitance C_e in parallel to represent, respectively, radiation and energy storage. The input impedance depends on where the patch is fed such as at one end

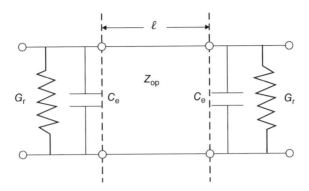

Figure 5.5 Circuit model of patch antenna

with a microstrip line input or an asymmetrically located probe. An approximate value for the conductance is found from $G_r = P_r/|E_o h|^2$ where P_r is the total radiated power by a single slot, which can be found from Eq. 5.3, and $E_o h$ is the peak voltage across the slot. The result is

$$G_r = \frac{2\sqrt{\varepsilon_r}}{\eta_o}\left(\frac{w}{\lambda}\right)^2 \int_0^\pi d\theta \sin\theta \left[M_0(\theta) - \sin^2\theta \, M_2(\theta) \right] \qquad (5.12)$$

where

$$M_p(\theta) = \int_0^{2\pi} d\phi \sin^p\phi \, S^2\left(\frac{\pi w}{\lambda}\sin\theta\sin\phi\right) S^2\left(\frac{\pi 2h}{\lambda}\sin\theta\cos\phi\right).$$

The edge capacitance at a given frequency is given by $C_e = \tan(\beta\ell_{eff})/(\omega Z_{op})$, where ℓ_{eff} is the effective increase in length of the parallel plate region due to fringing; typically, $\ell_{eff} \sim 0.1 - 0.15\lambda$.

Equation 5.11 in particular emphasizes four important aspects of patch design. These are the physical extent of the patch, the matching at the input from the feed line to the patch, the efficient transition of power to the patch and, of course, these often have to be achieved over a reasonable bandwidth. Other more sophisticated models of the microstrip patch can be devised and are required for accurate design including the effects of mutual coupling in an array environment (e.g. James et al., 1982; Sainati, 1996; Jackson, 2007). For details of other patch geometries, the reader should also consult the references. The topic of mutual coupling in patch arrays is left until Chapter 7.

5.3 Microstrip Patch on a Cylinder

In some applications, it is desirable to locate the patch on a curved surface. The simple model described in the previous section can be used in a non-planar geometry as will be described here. Other models could be adapted to a curved surface in the same way by following the same approach. Suppose the microstrip patch is to be mounted on a large conducting cylinder as illustrated in Figure 5.6. The radius of the cylinder R_o is assumed very much greater than the

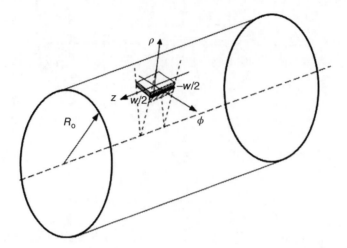

Figure 5.6   Radiation model of a microstrip antenna mounted on a large cylinder

microstrip dimensions, that is, $R_o \gg w, h, \ell$. For simplicity, it will be assumed that the axis of the patch is in the circumferential direction as shown in Figure 5.6. Once again two radiating apertures are assumed as in the planar case. From the figure, the electric field is polarized in the radial direction and the normal to the aperture is $\hat{\phi}$. Therefore, this time the magnetic current is given by $\mathbf{M}_s = -\hat{\phi} \times (\hat{\rho} E_o)$ at each aperture. The substrate thickness will be assumed to be small and, therefore, the apertures appear as two uniform magnetic line sources of length w which are parallel to the axis of the cylinder and separated by a distance ℓ. For simplicity, these sources are assumed on the cylinder and to be separated by the angle $\Delta\phi = \ell/R_o$.

As the method of images cannot be used for a source on a cylinder, the sources must be considered to radiate in the presence of a large cylinder. This is done by adopting a high frequency approximation for the elemental field radiated by a line source on the axis of a cylinder as described in Section 8.2. This field is given by Eq. 8.1 (Wait, 1959) as

$$dE_\phi(t,\theta) \sim -dM_z \frac{1}{(2\pi)^2 R_o} \frac{\exp(-jkr)}{r} \exp(jkz' \cos\theta) \sum_{n=-\infty}^{\infty} j^n \frac{\exp(jn(\phi-\phi'))}{H_n^{(2)'}(\gamma)} \qquad (5.13a)$$

$$dE_\theta(t,\theta) = 0 = dE_r(t,\theta), \qquad (5.13b)$$

where $H_n^{(2)'}(\gamma)$ is the derivative of the Hankel function of the second kind order n with argument $\gamma = kR_o \sin\theta$ and the primed co-ordinates are the source co-ordinates.

Initially consider the field radiated by the uniform line source at $\phi' = +\Delta\phi/2$. Eq. 5.13a gives

$$E_\phi^{(1)}(r,\theta,\phi) \sim -\frac{hE_o}{2\pi^2 R_o} \frac{\exp(-jkr)}{r} \int_{-w/2}^{w/2} \exp(jkz' \cos\theta) dz' \times \sum_{n=-\infty}^{\infty} \frac{j^n \exp(jn(\phi-\Delta\phi/2))}{H_n^{(2)'}(\gamma)}$$

$$\approx -\frac{hE_o w}{2\pi^2 R_o} \frac{\exp(-jkr)}{r} S\left(\frac{kw}{2} \cos\theta\right) \sum_{n=-\infty}^{\infty} \frac{j^n \exp(jn(\phi-\Delta\phi/2))}{H_n^{(2)'}(\gamma)}.$$

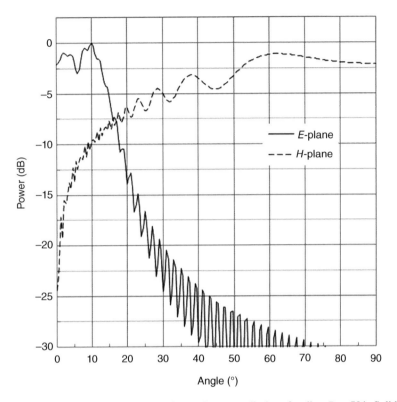

Figure 5.7 Radiation patterns of a microstrip patch on a cylinder of radius $R_o = 50\lambda$. Solid: E-plane $\theta = 90°$ plane; dashed: H-plane $\phi = 0°$ plane. Microstrip parameters $w = \lambda/4$, $\ell = \lambda_g/2$, $h = 0.005\lambda$ and $\varepsilon_r = 2.54$

The field due to the line source at $\phi' = -\Delta\phi/2$ is similar. Combining the contributions of both uniform line sources results in

$$E_\phi(r,\theta,\phi) \approx -\frac{hE_o hw \exp(-jkr)}{2\pi^2 R_o} \frac{}{r} S\left(\frac{kw}{2}\cos\theta\right) \sum_{n=-\infty}^{\infty} \frac{j^n \exp(jn\phi)}{H_n^{(2)'}(\gamma)} \times \left(\exp\left(\frac{jn\Delta\phi}{2}\right) + \exp\left(\frac{-jn\Delta\phi}{2}\right)\right)$$

$$= -\frac{hE_o w \exp(-jkr)}{\pi^2 R_o} \frac{}{r} S\left(\frac{kw}{2}\cos\theta\right) \sum_{n=-\infty}^{\infty} \frac{j^n \exp(jn\phi)}{H_n^{(2)'}(\gamma)} \cos\left(\frac{n\sqrt{\varepsilon_r}\ell}{2R_o}\right)$$

$$= -\frac{hE_o w \exp(-jkr)}{\pi^2 R_o} \frac{}{r} S\left(\frac{kw}{2}\cos\theta\right) \left(\frac{1}{H_0^{(2)'}(\gamma)} + 2\sum_{n=1}^{\infty} \frac{j^n}{H_n^{(2)'}(\gamma)} \cos\left(\frac{n\sqrt{\varepsilon_r}\ell}{2R_o}\right) \cos(n\phi)\right).$$

$$(5.14)$$

The maximum of the field occurs normal to the patch. As the electric field is circumferential with the cylinder, the E-plane direction corresponds to the plane $\theta = 90°$ while the orthogonal H-plane occurs in the plane $\phi = 0°$ through the centre of the patch.

The patch antenna that is described in Figure 5.4 for a planar geometry is now mounted on a cylinder of radius $R_0 = 50\lambda$. This patch has dimensions $w = \lambda/4$, $\ell = 0.8\lambda$, $\varepsilon_r = 2.54$ and $h = 0.1\lambda$. The principal plane patterns were calculated from Eq. 5.14 and are shown plotted in Figure 5.7. There are significant differences in the patterns compared with the same patch close to a ground plane. In particular, the patterns are seen to contain significant ripples at all levels due to the interference between the field radiated directly by the patch and creeping waves that are excited on the cylinder. The H-plane pattern is similar to the planar case as it is along the axis of the cylinder and apparently this appears similar to a ground plane. Nevertheless, there is still interference from nearby creeping waves. More will be said about the creeping wave in connection with mutual coupling on conformal surfaces in Chapter 8.

5.4 Problems

P5.1 Show that the radiation pattern of a two-element array of identical slots symmetrically located along the x-axis a distance s in each direction from the origin as shown in Figure 5.3 is $2\cos(s\sin\theta\cos\phi) \times$ (slot pattern).

P5.2 Suppose a rectangular patch antenna has dimensions $h = 0.12\lambda$, $w = 0.45\lambda$ and $\ell = 0.87\lambda$ is mounted on a dielectric substrate with relative permittivity of 3. It is fed from a microstrip line with a centre conductor of width 0.15λ. The conductor thickness of the input line and the patch is $10^{-3}\lambda$.
   a. Find the direction (θ, ϕ) of the radiation maximum.
   b. Identify the E-plane and compute the half-power beamwidth.
   c. Identify the H-plane and compute the half-power beamwidth.

P5.3 For the patch antenna described in P5.2, calculate the characteristic impedance of the patch and, therefore, estimate the input reflection coefficient viewed by the microstrip. Estimate the maximum gain of the patch.
   Answer: $Z_{op} = 58\Omega$, $\Gamma_i = -24.55$ dB and $G_{max} = 6.71$ dBi.

P5.4 An expression for the conductance G_r in the transmission line model of a patch antenna shown in Figure 5.5 is given in Eq. 5.12.
   a. Verify this expression and
   b. Calculate the conductance of the patch antenna described in P5.2.
   Answer: (b) $G_r = 0.013$ S.

P5.5 From the equivalent circuit for the patch antenna shown in Figure 5.5, obtain an approximate expression for the input admittance when it is fed from one end.

$$\text{Answer: } Y_{in} = G_r + jB + Y_{op}\frac{G_r + j(B + Y_{op}\tan\beta\ell)}{Y_{op} + j(G_r + jB)\tan\beta\ell}, \text{ where } B = \omega C_e.$$

P5.6 Repeat exercise P5.5 this time with a feed placed at a distance ℓ_1 from one end of the parallel plate section.

P5.7 Plot the radiation pattern at a frequency of 9 GHz for a patch of dimensions 1 cm × 1 cm on a substrate of thickness 1 mm and dielectric constant 2.4 that is mounted on a cylinder of radius $R_0 = 100$ cm.

References

Deschamps, G. and Sichak, W. (1953): 'Microstrip microwave antennas', Proc. 3rd Symp. USAF Antenna R&D Program, October 1953.

Gupta, K.C., Garg, R. and Bahl, I.J. (1979): 'Microstrip lines and slotlines', Artech House, Inc., Dedham, MA.

Jackson, D.R. (2007): 'Microstrip antennas' (Chapter 7). In: Volakis, J.L. (eds) 'Antenna engineering handbook', 4th ed., McGraw-Hill, New York.

James, J.R., Hall, P.S. and Wood, C. (1982): 'Microstrip antenna theory and design', Peter Peregrinus Ltd., London, UK.

Munson, R.E. (1974): 'Conformal microstrip antennas and microstrip phased arrays', IEEE Trans. Antennas Propag., Vol. **AP-22**, No. 1, pp. 74–78.

Sainati, R.A. (1996): 'CAD of microstrip antennas for wireless communications', Artech House, Norwood, MA.

Wait, J.R. (1959): 'Electromagnetic radiation from cylindrical structures', Pergamon Press, London, UK.

6

Reflector Antennas

6.1 Introduction

In transmission, a reflector antenna concentrates energy received from another antenna, called a feed, into a narrow beam of radiation. In reception, the reflector re-directs the impinging field and concentrates it in a smaller volume, called the focal region, where it can be collected. Figure 6.1 shows, in cross-section, several basic reflector configurations. Most reflector antennas are designed to maximize the signal in one direction, that is, for the examples given in Figure 6.1 parallel to the ray paths at the aperture. In some cases, for example, in satellite antennas, it may be desirable to design the antenna for approximately constant gain over a chosen extended angular region. This is often achieved by 'shaping' the main reflector profile or using an array of horns as the feed.

The front-fed paraboloid in Figure 6.1a is the most common type of reflector configuration. Use of a subreflector in a Cassegrain configuration in Figure 6.1b can give improved performance and is widely used in large earth stations. The two reflector profiles can be defined to enhance the gain while at the same time reducing antenna noise temperature. The unwanted random signals ('noise') that an antenna receives from the sky and the earth is expressed as an equivalent antenna noise temperature T_a (in Kelvin). A major contribution to T_a arises from feed 'spillover' at the edges of the reflectors. Spillover past the main reflector edge is especially important, since this allows energy to be received from the earth which is a good radiator of noise. In a front-fed paraboloid, the feed sees the 'hot' earth directly through its sidelobes, while in a Cassegrain, the feed sees the 'cold' sky. The ratio of gain to overall temperature, denoted by G/T_{sys} (expressed in dB K^{-1}), is an important parameter for receiving antennas. Here, $T_{sys} = T_a + T_{loss} + T_{rx}$ where T_a arises from feed spillover, T_{loss} is the noise temperature due to losses in the feed and reflector and T_{rx} is the contribution from

Fundamentals of Aperture Antennas and Arrays: From Theory to Design, Fabrication and Testing,
First Edition. Trevor S. Bird.
© 2016 John Wiley & Sons, Ltd. Published 2016 by John Wiley & Sons, Ltd.
Companion website: www.wiley.com/go/bird448

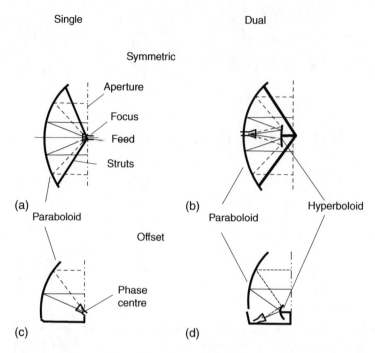

Figure 6.1 Reflector antenna configurations. For receiving, the ray paths are reversed. (a) Symmetric paraboloid. (b) Cassegrain. (c) Offset paraboloid and (d) Offset Cassegrain

the receiver that is connected to the antenna. A Cassegrain geometry can be designed to maximize G/T_{sys}.

A disadvantage of any axisymmetric reflector system is the blockage created by the feed or subreflector and associated strut supports. These obstructions have a deleterious effect on the antenna gain and, more importantly on the sidelobe levels, particularly for small antennas ($D/\lambda < 150$). Blockage can be avoided by using offset-fed reflector configurations as illustrated in Figure 6.1c and d. Although the lack of symmetry creates a number of design problems, these antennas are capable of performance that is usually superior to that of their axisymmetrical counterparts. One problem is that the offset-paraboloid Figure 6.1c has high cross-polarization. This is substantially reduced by means of a subreflector which is adjusted to a correct offset angle α (Fig. 6.1d). In this chapter details are given of the front-fed paraboloid along with some properties of the offset-parabolic reflector and Cassegrain antennas.

6.2 Radiation from a Paraboloidal Reflector

A fundamental property of the paraboloid in transmission is that it converts a wave with a spherical wave-front from a source, which is situated at the geometric focus, O, into a wave emanating from the aperture with a plane wave-front. This is possible because the path length of any ray from the focus, which arrives perpendicular to the aperture, Figure 6.2, is constant and equal to $2f$, where f is the focal length. As a result, $\rho + s_r = 2f$.

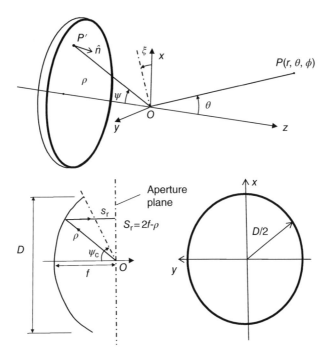

Figure 6.2   Geometry of paraboloidal reflector

In spherical polar co-ordinates (ρ, ψ, ϕ) that are defined at the focus of the paraboloid, the radial distance is

$$\rho = \frac{2f}{1 + \cos \psi} = f \sec^2 \frac{\psi}{2}. \tag{6.1}$$

From the focus, the reflector rim subtends a half-cone angle given by

$$\psi_c = 2 \arctan \left[\frac{1}{4(f/D)} \right]. \tag{6.2}$$

On the surface of the paraboloid, S, the outward unit normal in the various co-ordinate systems defined in Figure 6.2 are given by

$$\hat{n} = -\frac{1}{2\sqrt{\rho f}} (\hat{x}x + \hat{y}y) + \sqrt{\frac{f}{\rho}} \hat{z}. \tag{6.3a}$$

$$\hat{n} = -\sin \frac{\psi}{2} (\hat{x} \cos \xi + \hat{y} \sin \xi) + \cos \frac{\psi}{2} \hat{z} \tag{6.3b}$$

$$\hat{n} = -\hat{\rho} \cos \frac{\psi}{2} + \hat{\psi} \sin \frac{\psi}{2}. \tag{6.3c}$$

Consider a feed antenna located at the focus with the reflector is in its far-field region. Therefore, the field from the feed is represented by

$$\mathbf{E_f}(\rho,\psi,\xi) = E_o \mathbf{F}(\psi,\xi)\frac{e^{-jk\rho}}{\rho}$$

$$= E_o \left[\hat{\psi}F_\psi(\psi,\xi) + \hat{\xi}F_\xi(\psi,\xi)\right]\frac{e^{-jk\rho}}{\rho} \tag{6.4a}$$

$$\mathbf{H_f} = \frac{1}{\eta_o}\hat{\rho}\times\mathbf{E_f}, \tag{6.4b}$$

where E_o is a constant scale factor. An approximate surface current on the reflector is given by $J_s \approx 2\,\hat{n}\times\mathbf{H_f}|_S$. The vector function $\mathbf{F}(\psi,\xi)$ gives the spatial distribution of the field (i.e. the radiation characteristics) due to the feed. A special case of this feed function is when the power pattern is axisymmetric, in which case $F_\psi(\psi,\xi) = P(\psi)\cos(\xi-\xi_o)$ and $F_\xi(\psi,\xi) = P(\psi)\sin(\xi-\xi_o)$ where $P(\psi)$ is the radiation pattern and ξ_o is the reference polarization direction relative to the initial line. In many practical cases, the reflector diameter is very large in terms of wavelength; typically $D/\lambda > 100$. In that instance, for large reflectors, geometric optics (GO) can be used, which simplifies the reflector analysis. Geometric optics is used in the next section to find approximate aperture fields.

6.2.1   Geometric Optics Method for a Reflector

In a homogeneous medium, waves described by geometric optics propagate in straight lines, as verified in Section 2.1.7. The ray paths for input and reflected rays are as shown in Figure 6.3. At a perfect conductor of the type illustrated, the boundary conditions require zero net tangential electric field at the surface as well as continuity in the tangent plane of the normal components of the electric field. Again referring to Figure 6.3, suppose $\mathbf{E_i}$ is the incident electric field and $\mathbf{E_r}$ is the reflected field. The boundary conditions on S require that

$$(\mathbf{E_i} + \mathbf{E_r})\times\hat{n} = 0 \tag{6.5a}$$

and

$$(\mathbf{E_i} - \mathbf{E_r})\cdot\hat{n} = 0, \tag{6.5b}$$

where $\hat{n}$ is the normal at the reflector. Taking the cross-product of Eq. 6.5a with $\hat{n}$ results in

$$(\mathbf{E_i} + \mathbf{E_r}) - \hat{n}\left[(\mathbf{E_i} + \mathbf{E_r})\cdot\hat{n}\right] = 0.$$

This is further simplified by means of Eq. 6.5b to

$$\mathbf{E_r} = 2\hat{n}\,(\hat{n}\cdot\mathbf{E_i}) - \mathbf{E_i}. \tag{6.6}$$

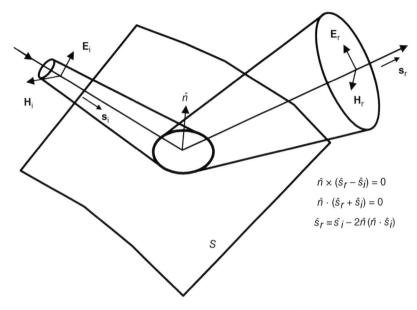

Figure 6.3   Reflection at a conducting surface S

Equation 6.6 gives the resulting reflected electric field from a conducting surface in terms of the incident field. This is incomplete because an additional phase factor is required to account for the path length from the reflector to the aperture as shown in Figure 6.2, which is given by $s_r = 2f - \rho$. As a result, according to GO, a feed located at the focus o radiating an electric field $\mathbf{E}_f$ produces an aperture field, given by

$$\mathbf{E}_a = [2\hat{n}(\hat{n} \cdot \mathbf{E}_f) - \mathbf{E}_f]e^{-jk(2f - \rho)} \tag{6.7a}$$

$$\mathbf{H}_a = \frac{1}{\eta_o}\hat{z} \times \mathbf{E}_a, \tag{6.7b}$$

where $\mathbf{E}_f$ is the electric field due to the feed that is incident at the reflector surface. In spherical polar co-ordinates relative to the feed, the normal to the reflector is given by Eq. 6.3c.

Using Eqs. 6.4a and 6.3c in Eq. 6.7a results in

$$\mathbf{E}_a = -E_o\left[\hat{\rho}F_\psi \sin\psi + \hat{\psi}F_\psi \cos\psi + \hat{\xi}F_\xi\right](1 + \cos\psi)\frac{\exp(-jk2f)}{2f}. \tag{6.8}$$

Alternatively, in rectangular components,

$$\mathbf{E}_a = -E_o\left[\hat{x}\left(F_\psi \cos\xi + F_\psi \sin\psi\right) + \hat{y}\left(F_\psi \sin\xi + F_\xi \cos\xi\right)\right]$$
$$\times (1 + \cos\psi)\frac{\exp(-jk2f)}{2f}. \tag{6.9}$$

From these aperture fields, the field radiated by the paraboloid is obtained from Eqs. 3.20 are

$$E_\theta(r,\theta,\phi) = \frac{jk}{4\pi} \frac{e^{-jkr}}{r} (1 + \cos\theta)(N_x \cos\phi + N_y \sin\phi) \tag{6.10a}$$

$$E_\phi(r,\theta,\phi) = \frac{jk}{4\pi} \frac{e^{-jkr}}{r} (1 + \cos\theta)(-N_x \sin\phi + N_y \cos\phi), \tag{6.10b}$$

where

$$\mathbf{N}(\theta,\phi) = \int_0^{2\pi} d\xi \int_0^{D/2} \mathbf{E_a}(t,\xi) \exp(jwt\cos(\phi-\xi)) t\, dt \tag{6.11}$$

with $w = k\sin\theta$ and $t = \rho\sin\psi$. When the aperture field is axisymmetric, the integral over ξ can be completed by means of Eq. B.3 allowing Eq. 6.11 to be simplified to

$$\mathbf{N}(\theta,\phi) = 2\pi \int_0^{D/2} \mathbf{E_a}(t) J_0(wt) t\, dt, \tag{6.12}$$

where J_0 is the zero-order Bessel function of the first kind. Eq. 6.12 is axisymmetric as it is now independent of ϕ. To investigate the above results a little further, some specific feed antenna examples are considered firstly the half-wave dipole and then circular waveguides and horns.

6.2.1.1   Dipole Feed

One of the simplest feeds to fabricate and, therefore, one of the most frequently used feeds is the half-wave dipole. An attractive feature of this feed is that the input transmission line can also be used to support the feed at the focus. The approximate electric field radiated by a thin half-wave dipole that is oriented parallel to the x-direction is given by

$$\mathbf{E_f} = E_o \frac{e^{-jkr}}{r} A(\theta,\phi) [\hat{\theta}\cos\theta\cos\phi - \hat{\phi}\sin\phi], \tag{6.13}$$

where $A(\theta,\phi) = \sin((\pi/2)\cos\theta)$.
  In the E-plane ($\phi = 0$ or π), the field is

$$\mathbf{E_f} = \hat{\theta} E_o \frac{e^{-jkr}}{r} \sin\left(\frac{\pi}{2}\cos\theta\right)\cos\theta \tag{6.14a}$$

and in the H-plane ($\phi = \pm\pi/2$), it is

$$\mathbf{E_f} = \mp\hat{\phi} E_o \frac{e^{-jkr}}{r} \sin\left(\frac{\pi}{2}\cos\theta\right). \tag{6.14b}$$

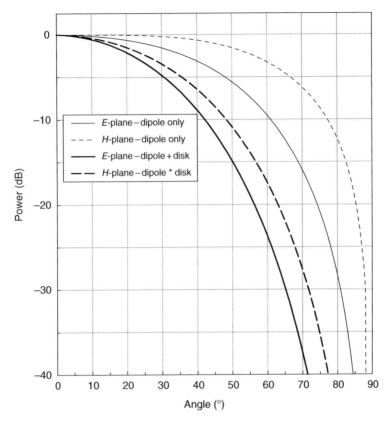

Figure 6.4 Half-wave dipole radiation patterns and dipole in front of a conducting plate (dark line)

The radiation patterns corresponding to Eqs. 6.14 are plotted in Figure 6.4.

The aperture field created by a half-wave dipole located at the focus of the paraboloid is found from Eqs. 6.9 and 6.13 to be

$$\mathbf{E_a} = -E_oA(\psi,\xi)\left[\hat{x}\left(\cos\psi\cos^2\xi + \sin^2\psi\right) + \hat{y}\cos\xi\sin\xi\right]$$
$$\times (1+\cos\psi)\frac{\exp(-jk2f)}{2f}.$$

Clearly it is seen that the radiation pattern is strongly influenced by the pattern function $A(\psi,\xi)$. In the E-plane (x–z plane), the field is

$$\mathbf{E_a}(x',0) = -\hat{x}E_o(1+\cos\psi)\cos\psi\sin\left(\frac{\pi}{2}\cos\psi\right)\frac{e^{-jk2f}}{2f} \tag{6.15a}$$

and in the H-plane (y–z plane), it is

$$\mathbf{E_a}(0,y') = -\hat{x}E_o(1+\cos\psi)\sin\left(\frac{\pi}{2}\cos\psi\right)\frac{e^{-jk2f}}{2f}. \tag{6.15b}$$

It is seen from Eqs. 6.15 that the aperture field in the principal planes is simply the feed radiation in that plane divided by the distance to the reflector from the feed, namely, $\rho = 2f/(1 + \cos \psi)$. This latter factor reduces the aperture field towards the edge of the aperture and, therefore, it is more 'tapered' towards the reflector rim than the feed radiation pattern alone would indicate as the latter is usually measured on a sphere of constant radius. However, there is significant radiation in all directions and a loss of efficiency.

In practice the dipole has a backing plate as illustrated in Figure 4.17a to reduce rear-directed radiation and the back lobe. If a backing plate is employed and the dipole is located a distance s from the plate, from image theory (see Figure 2.2a) there is now a pair of dipoles spaced $2s$ apart. Therefore, the feed radiation is approximately equivalent to that of a two element array, in which case Eq. 6.13 is then multiplied by the factor $2j \sin(ks \cos \psi)$. Typically a spacing of $s = 0.18 - 0.25\lambda$ is selected to achieve the desired performance, which also makes the feed radiation more directive as shown in Figure 6.4. The presence of a backing plate improves the input match and provides an additional variable (s) with which to account for the thickness of the dipole element and the characteristic impedance of the input line. This line can be a coaxial cable (with a balun transformer) (Silver, 1946) where typically the characteristic impedance is close to 50 Ω or it could be an open-wire transmission line where typically the dipole elements need to be longer to achieve an acceptable match.

The polarization of the field in the aperture has distinctive characteristics and this is illustrated in Figure 6.5. The field is polarized mainly in the x-direction, but there is also a cross-polarized

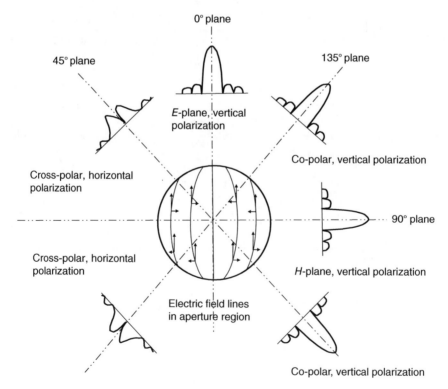

Figure 6.5 Electric field in the aperture of a paraboloid due to a linearly polarized feed (after Jones, 1954)

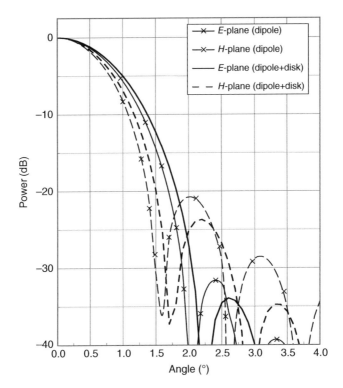

Figure 6.6 Principal plane radiation patterns of a paraboloid with $D = 1.5$ m, $f/D = 0.33$ and a half-wave dipole feed and one with a backing reflector (dipole + disk). Frequency 10 GHz

component in the orthogonal y-direction. This cross-polar component takes power from the co-polar direction and could interfer with adjacent systems operating in the y-directed polarization. It is shown in Figure 6.5 that the maxima of the cross-polarized field occurs at 45° to the principal directions (either x or y directions) (Jones, 1954).

The field radiated by the paraboloid with a dipole feed is obtained by substituting Eqs. 6.15 in Eqs. 6.10 via Eq. 6.12. In this case, the transform of the aperture field is best found numerically. Since the field in the E-plane is more tapered than for the H-plane, the far-zone radiation pattern is broader in the E-plane than in the H-plane. Furthermore, the radiation pattern side-lobes are lower in the E-plane than in the H-plane due the $\cos \theta$ factor. Principal plane radiation patterns are shown in Figure 6.6 for a 1.5 m ($D = 50 \lambda$) reflector with $f/D = 0.33$ at 10 GHz clearly demonstrates these properties. Also, shown in Figure 6.6 are the corresponding radiation patterns of the same reflector with a half-wave dipole backed by an ideal ground plane to approximate a disk reflector. The spacing between the dipole and the backing reflector was chosen to be $s = 0.25\lambda$. The principal plane patterns of the reflector with the dipole and disk are more comparable and there is a significant improvement in the gain (>2 dB).

6.2.1.2 Circular Waveguides and Horn Feeds

Smooth wall and corrugated circular waveguides or horns are widely used as feeds for reflectors, and are capable of higher performance than the dipole. Some properties of circular feeds were described in Section 4.4. For horns excited in the TE_{11} or HE_{11} modes, the radiated field has a single period in azimuth and, in terms of Eq. 6.4a,

$$F_\psi(\psi, \xi) = A(\psi) \cos \xi; \quad F_\xi(\psi, \xi) = -B(\psi) \sin \xi, \tag{6.16}$$

where $A(\psi)$ and $B(\psi)$ are pattern functions. As a result, the field created in the aperture of the paraboloid due to a circular feed is

$$\mathbf{E}_a = -E_o \left[\hat{x} \left(A(\psi) \cos^2 \xi + B(\psi) \sin^2 \psi \right) + \hat{y} \cos \xi \sin \xi (A(\psi) - B(\psi)) \right]$$
$$\times (1 + \cos \psi) \frac{\exp(-jk2f)}{2f}. \tag{6.17}$$

In common with the half-wave dipole, this aperture field has, in general, two non-zero field components present in regions away from the two principal planes. This is because Eq. 6.16 is not pure polarized as discussed in Section 4.4.5. However, if $A(\psi) = B(\psi)$, the aperture field produced by a circular horn is purely polarized and the radiation pattern is axisymmetric about the z-axis. For an axisymmetric feed, Eq. 6.17 simplifies to

$$\mathbf{E}_a = -\hat{x} E_o A(\psi)(1 + \cos \psi) \frac{\exp(-jk2f)}{2f}, \tag{6.18}$$

which produces a linearly polarized aperture field. This aperture field can be used in Eq. 6.12 so that the transformation integral becomes

$$N_x(u, v) = -E_o \pi \frac{e^{-jk2f}}{f} \int_0^{D/2} J_0(wt) A(\psi)(1 + \cos \psi) t dt. \tag{6.19}$$

In general, Eq. 6.19 must be evaluated numerically although there are several types of feeds that have distributions with special functions that enable a closed-form solution. It is seen that the integrand function

$$I(\psi) = A(\psi)(1 + \cos \psi) \tag{6.20}$$

determines the aperture illumination. Uniform illumination corresponds to $I(\psi) = \text{constant}$ and could be achieved if the feed function, $A(\psi)$, had an inverse taper to compensate for the free-space path loss. In particular, when $I(\psi) = 1$ Eq. 6.19 simplifies to

$$N_x(u, v) = -E_o \pi \left(\frac{D}{2} \right)^2 \frac{e^{-jk2f}}{f} \left[\frac{2J_1(wD/2)}{wD/2} \right], \tag{6.21}$$

where Eq. B.5 has been used to evaluate the integral. The function $2 J_1(x)/x$ is plotted in Figure 3.4. Near the antenna boresight (the z-axis in Figure 6.2), the beam of the reflector is narrow and $\theta \approx 0°$. Therefore $(1 + \cos \theta)/2 \approx 1$ so that the radiated fields near boresight are approximately

$$E_\theta(r,\theta,\phi) \approx \frac{jk}{2\pi} \frac{e^{-jkr}}{r} N_x(\theta,\phi) \cos \phi \qquad (6.22a)$$

and

$$E_\phi(r,\theta,\phi) \approx -\frac{jk}{2\pi} \frac{e^{-jkr}}{r} N_x(\theta,\phi) \sin \phi. \qquad (6.22b)$$

In the case of a uniformly illuminated aperture, Eqs. 6.21 and 6.22 show that the function $2J_1(x)/x$ determines the width of the main lobe of the electric and magnetic fields and the side-lobe levels. As a result, the half-power beamwidth is $\sim 58.4\lambda/D$ degrees and the first sidelobe is ~ 17.6 dB below the peak (see Figure 3.4). With tapered aperture illumination functions $A(\psi)$ the beamwidth is wider, and the first sidelobe level is lower than for uniform illumination. To illustrate this, radiation patterns are shown in Figure 6.7 for a feed with an axisymmetric Gaussian pattern function given by

$$A(\psi) = \exp\left(-\alpha\psi^2\right) \qquad (6.23)$$

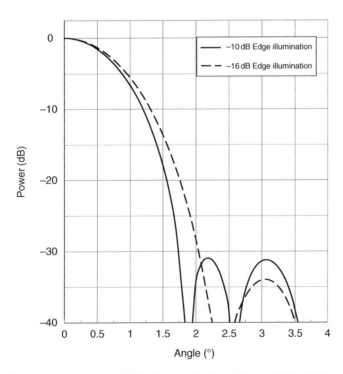

Figure 6.7 Radiation pattern of paraboloid reflector with $D = 1.5$ m and $f/D = 0.33$ at a frequency of 10 GHz. The feed pattern is a Gaussian function chosen to give a selected edge illumination at $\psi_c = 74.2°$

where α is a constant that is chosen to give a specified feed illumination at the reflector edge $\psi = \psi_c$. As has been seen in Section 4.4.4, the main lobe of the Gaussian function, Eq. 6.23, is a good approximation to the feed function of corrugated waveguide at the balanced hybrid condition, which is given in Eq. 4.61. If the field illumination at the reflector rim is $E_{dB} = -20\log_{10}(|A(\psi_c)|)$ dB, then from Eq. 6.23,

$$\alpha = \frac{E_{dB}}{\psi_c^2(20\log_{10}e)}. \tag{6.24}$$

The results given in Figure 6.7 are for the same conditions as for the dipole feed example above. Patterns are shown for reflector edge illumination levels of -10 and -16 dB. Main lobe broadening is evident particularly in the latter case as compared to the half-wave dipole (see Figure 6.6).

6.2.2   Edge Taper and Edge Illumination

As the feed radiates a spherical wave, the distance to the reflector rim, or edge, is greater than to the centre of the paraboloid as shown in Figure 6.8. It has been shown that, in any plane, the radiation pattern is closely related to the aperture illumination in that plane and particularly how it rolls-off, or tapers, at the edge. Therefore, it is common practice to refer to the level of the field at the reflector edge to provide a 'rule-of-thumb' description of beamwidth and sidelobe levels. Two different terms are used, often interchangeably, and these are now defined. The first term is 'edge illumination'; that is the level of illumination directed at the edge. Edge

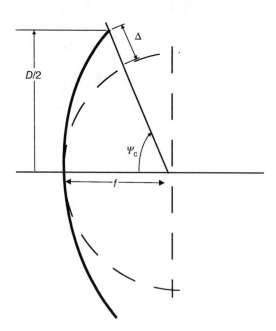

Figure 6.8   Edge taper

illumination is the ratio of the field strength radiated by the feed in the direction of the edge and its level at the reflector vertex, where both are measured on a circle of radius equal to the focal length. The second term is edge taper. This is the ratio of the feed field strength at the actual reflector edge and its level at the vertex. The difference between edge illumination and edge taper is the free-space loss due to the distance Δ (shown in Figure 6.8) from the sphere to the reflector. If E is the edge illumination, by definition, edge taper is

$$T = E \frac{f}{f + \Delta}$$

$$= \frac{E}{L_e},$$

(6.25)

where L_e is the edge taper loss factor. The free-space loss reduces the amplitude of the spherical wave as it propagates towards the edge. The loss is

$$L_e = \frac{f + \Delta}{f}.$$

(6.26)

For a paraboloid, the extra distance is

$$\Delta = \frac{f}{(4f/D)^2}$$

(6.27)

and, therefore,

$$L_e = 1 + \frac{1}{(4f/D)^2}.$$

(6.28)

It is usual practice to express Eq. 6.25 in dB, namely,

$$(\text{Edge taper}, \text{dB}) = (\text{edge illumination}, \text{dB}) - (\text{path loss}, \text{dB}).$$

(6.29)

Some values of L_e for different (f/D) ratios of a paraboloid are listed in Table 6.1.

Table 6.1   Edge taper loss factor versus parabolic reflector f/D

f/D	L_e (dB)
0.2	8.17
0.25	6.02
0.3	4.58
0.33	3.94
0.4	2.86
0.5	1.94

As an example, consider the half-wave dipole feed example described in Section 6.2.1.1 for a parabolic reflector with $f/D = 0.33$. At the reflector edge at $\psi_c = 74.29°$, the edge illumination is -13.2 dB in the E-plane and 0 dB in the H-plane. The path loss for this reflector is $L_e = 4$ dB. The respective edge tapers are, therefore, -17.2 and -4 dB.

A term related to edge taper that is used in signal processing applications in which the aperture illumination is tailored to achieve a desired outcome is apodization. It was originally used in optics to refer to the modification of the central illumination of a lens to suppress secondary maxima, that is, sidelobes, to improve the dynamic range of a telescope. Often the central illumination is non-linear and approaches zero at the edges of the aperture. While the term apodization occurs occasionally in the antenna area, it is more commonly used in other related imaging areas such as optics, audio and photography.

6.2.3   Induced Current Method

In the previous sections, the radiation from a paraboloidal reflector has been described in terms of the aperture field method. Another approach that is widely used for reflector antenna analysis is the induced current method, which is also a more accurate method. In this approach, the radiated field is determined from an electric surface current set up on the reflector by the feed. The problem is to find this current and generally numerical methods must be used.

An approximate approach, which yields good results, is to assume that, at any point on the reflector, the current that is induced is the same as on an infinite plane conductor. Thus, if the feed radiates a magnetic field $\mathbf{H_f}$ at the reflector, this surface current is given by

$$\mathbf{J_s} = 2\hat{n} \times \mathbf{H_f}|_{\text{reflector } \Sigma}, \tag{6.30}$$

where $\hat{n}$ is the unit outward normal to the reflector surface Σ. This representation of the surface current is called the physical optics approximation. The factor of two occurs in Eq. 6.30 because, it will be recalled, the total magnetic field at the reflector is twice the incident field.

Equation 6.30 may be used in Eq. 3.24 to calculate the radiated fields. There is no magnetic surface current at the reflector and, therefore, the radiated electric field is

$$\mathbf{E}(r,\theta,\phi) = -\frac{jk\eta_o}{2\pi}\frac{e^{-jkr}}{r}[\mathbf{F}(\theta,\phi) - \hat{r}(\mathbf{F}(\theta,\phi)\cdot\hat{r})], \tag{6.31}$$

where

$$\mathbf{F}(\theta,\phi) = \int_\Sigma \mathbf{J_s}\exp(jk\hat{r}\cdot\mathbf{r}')dS'. \tag{6.32}$$

The primed co-ordinates relate to co-ordinates on the reflector surface. Notice that the second term inside the square brackets of Eq. 6.31 cancels out a radial vector component that is introduced by the first term. For example, in rectangular components, the vector in the square braces becomes

$$\hat{x}(F_x - F_r \sin\theta\cos\phi) + \hat{y}(F_y - F_r \sin\theta\sin\phi) + \hat{z}(F_z - F_r \cos\theta),$$

where $F_r = \mathbf{F} \cdot \hat{r}$. The contributions from the radial terms in the above directly cancel identical terms in F_x, F_y and F_z when these are expressed in spherical polar co-ordinates as given in the vector identities in Appendix A.2.

The fields predicted by the induced current method (given by Eq. 6.31) and the aperture field method (Eq. 6.10) are very similar for large reflectors ($D > 100\lambda$), but there are important differences. These differences affect the co-polar radiation pattern the least but become significant several beamwidths from boresight. The main differences are as follows:

a. The z-component of the current in Eq. 6.31 gives a term in the far-field that is not predicted by the aperture field method.
b. The phase function in the two methods is different due to the use of different path lengths.

Both (a) and (b) become important as θ increases from boresight. The z-component in (a) is particularly important for accurate prediction of the antenna cross-polarization. However, if the reflector is reasonably large and the feed introduces a significant amount of its own cross-polarization, the effect of (a) is small. As a result of the limitations mentioned, the induced current method is usually preferred to the aperture field method in reflector calculations. Historically, the latter has the advantage that the integral over the aperture (in Eq. 6.11) may be evaluated directly by means of the fast Fourier transform (FFT) algorithm. However, research in the 1980s showed that the current transform Eq. 6.32 can also be evaluated by means of the FFT, after some modification (see, e.g. Franceschetti & Mohsen, 1986). One such approach is to express the current transform, as a series of Fourier transforms. To do this, in the phase of Eq. 6.32 let $\hat{r} \cdot \mathbf{r'} = ux' + vy' + \cos\theta z'$. In addition, on the reflector surface Σ, the z' co-ordinates are expressed as a function of the co-ordinates x' and y'. This allows Eq. 6.32 to be written as an integral over an aperture A consisting of the area projected by Σ onto the $x'-y'$ plane; thus,

$$\mathbf{F}(\theta,\phi) = \int_A \mathbf{J}_s W(x',y') \exp(jk(ux' + vy' + \cos\theta z'))dx'dy' \tag{6.33}$$

with

$$W(x',y') = \sqrt{1 + \left(\frac{\partial z'}{\partial x'}\right)^2 + \left(\frac{\partial z'}{\partial y'}\right)^2}. \tag{6.34}$$

The exponential in the integrand, $\exp(jk\cos\theta z')$, can be expanded in its Taylor series. With this substitution, Eq. 6.33 can be expressed as follows:

$$\mathbf{F}(\theta,\phi) = \sum_{p=0}^{\infty} \frac{1}{p!}(jk\cos\theta)^p \mathbf{F}_p(u,v), \tag{6.35a}$$

where

$$\mathbf{F}_p(u,v) = \int_A \mathbf{J}_s W(x',y')z'^p \exp(jk(ux' + vy'))dx'dy'. \tag{6.35b}$$

Each Fourier transform $\mathbf{F}_p$ may now be evaluated for each value of p by means of the efficient FFT algorithm (see, e.g. Brigham, 1974). In practice, it has been found that very good accuracy is achieved with most common types of reflectors by taking only the first two terms of the series; that is $p = 0$ & 1 in Eq. 6.35a.

6.2.3.1   Radiation from Symmetrical Reflectors with General Profile

A common situation is to have a main reflector with a general profile that is symmetric about the z-axis as illustrated in Figure 6.9. Consider such a symmetrical reflector that is excited by a feed located on its axis of symmetry a distance f_o from the vertex. Suppose that the profile can be specified by the function $\rho(\psi)$, where ρ is the radial distance from the origin to the reflector surface and ψ is the angle from the axis to a point on the reflector. The feed is located a distance $\Delta z = d - f_o$ from the origin where $d = \rho(0)$ is the distance from the origin to the reflector vertex.

The normal to the reflector is given by

$$\hat{n} = \frac{1}{\sqrt{\rho^2 + \rho_\psi^2}} [-n_a(\hat{x}\cos\xi + \hat{y}\sin\xi) + \hat{z}n_z], \qquad (6.36)$$

where ρ_ψ of $\rho(\psi)$ and the coefficients are

$$n_a = \rho_\psi \cos\psi - \rho\sin\psi \text{ and } n_z = \rho_\psi \sin\psi + \rho\cos\psi.$$

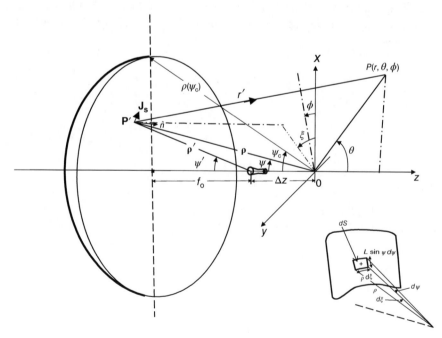

Figure 6.9   Geometry of radiation from a prime focus reflector with a general profile. Surface element shown in bottom right

The feed radiates an electric field in the primed co-ordinate system (ρ', ψ', ξ') given by

$$\mathbf{E_f} = \frac{e^{-jk\rho'}}{\rho'} A(\psi')(\hat{\psi} \cos \xi' - \hat{\xi} \sin \xi')$$

and

$$\mathbf{H_f} = \frac{1}{\eta_o}(\hat{\rho}' \times \mathbf{E_f}),$$

where the pattern function $A(\psi')$ can be arbitrary, $\rho' = \sqrt{\rho^2 + \Delta z^2 - 2\rho \Delta z \cos \psi}$, $\psi' = \sin^{-1}\left(\frac{\rho}{\rho'} \sin \psi\right)$, $\xi' = \xi$ and $\rho' = \rho - \hat{z}\Delta z$.

Following physical optics assumptions, the current induced on the reflector by this feed is

$$\mathbf{J_s} \approx \frac{2}{\eta_o} \hat{n} \times (\hat{\rho} \times \mathbf{E_f})$$

$$= -\frac{2A(\psi)}{\eta_o} \frac{\exp(-jk\rho)}{\rho L}[\hat{x}'(-n_a \sin \psi \sin^2 \xi + n_z(\cos^2 \xi + \cos \psi \sin^2 \xi)$$

$$+ \hat{y}' \cos \xi \sin \xi (n_a \sin \psi + n_z(1 - \cos \psi)) + \hat{z}' n_a \cos \xi].$$

The incremental surface element for a general surface is $dS = \rho L \sin \psi \, d\psi \, d\xi$, where $L = \sqrt{\rho^2 + \rho_\psi^2}$ is the segment length as shown in the inset to Figure 6.9. This cancels with the term in the denominator of $\mathbf{J_s}$. Substituting this and the current into Eqs. 6.31 and 6.32 results in far-field components

$$E_\theta(r,\theta,\phi) = \frac{jk}{2\pi} \frac{e^{-jkr}}{r}[\cos\theta(F_x(\theta,\phi)\cos\phi + F_y(\theta,\phi)\sin\phi) - F_z(\theta,\phi)\sin\theta] \qquad (6.37a)$$

$$E_\phi(r,\theta,\phi) = -\frac{jk}{2\pi} \frac{e^{-jkr}}{r}[-F_x(\theta,\phi)\sin\phi + F_y(\theta,\phi)\cos\phi]. \qquad (6.37b)$$

The transforms can be evaluated in the ξ-direction in closed form using Eq. B.3. Thus,

$$F_x(\theta,\phi) = 2\pi \int_0^{\psi_c} d\psi \rho \sin\psi \left[\frac{\rho}{\rho'} A(\psi')\right] J_0(w) \exp[-jk(\rho' + \rho \cos\theta \cos\psi)] \qquad (6.38a)$$

$$F_y(\theta,\phi) = 2\pi \sin 2\phi \int_0^{\psi_c} d\psi \sin\psi \left[\frac{\rho}{\rho'} A(\psi')\right] J_2(w)) \left(\rho_\psi \cos\left(\frac{\psi}{2}\right) - \rho \sin\left(\frac{\psi}{2}\right)\right)$$

$$\times \sin\left(\frac{\psi}{2}\right) \exp[-jk(\rho' + \rho \cos\theta \cos\psi)] \qquad (6.38b)$$

$$F_z(\theta,\phi) = -4\pi j \cos\phi \int_0^{\psi_c} d\psi \left[\frac{\rho}{\rho'} A(\psi')\right] J_1(w)\rho \sin^2\left(\frac{\psi}{2}\right) \times \exp[-jk(\rho' + \rho \cos\theta \cos\psi)],$$

$$(6.38c)$$

where $w = k\rho \sin\theta$ and J_n is the ordinary Bessel function of order n. The first quantity in the square brackets on the right-side of Eqs. 6.38 reduces to $A(\psi)$ when $\Delta z = 0$ and also $f_0 = f$, the reflector focal length. The integrals in Eqs. 6.38 are readily integrated providing $\rho(\psi)$ and its derivative are known either in closed form or through interpolation of a set of data points.

As an example of the former, consider a paraboloid defined by $\rho = 2f/(1 + \cos\psi)$ and $\rho_\psi = \rho \tan(\psi/2)$. The surface element for a paraboloid is, therefore, expressed as $dS = \rho^2 \sec(\psi/2) \sin\psi\, d\psi\, d\xi$. With these substitutions and the same feed defined above, Eqs. 6.38 reduce to

$$F_x(\theta,\phi) = 2\pi f \int_0^{\psi_c} d\psi A(\psi) J_0(w) \tan\left(\frac{\psi}{2}\right) \exp[-jk\rho(1 + \cos\theta\cos\psi)] \qquad (6.39a)$$

$$F_y(\theta,\phi) = 0 \qquad (6.39b)$$

$$F_z(\theta,\phi) = -4\pi jf \cos\phi \int_0^{\psi_c} d\psi A(\psi) J_1(w) \tan^2\left(\frac{\psi}{2}\right) \times \exp[-jk\rho(1 + \cos\theta\cos\psi)]. \qquad (6.39c)$$

The far-fields then follow from Eqs. 6.37. For instance, Figure 6.10 shows the radiation pattern of a paraboloid with dimensions $D = 100\lambda$ and $f/D = 0.4$ and a Gaussian feed function with an edge illumination of -10 dB. Other examples of the use of Eqs. 6.38 using discrete data points in the calculation of radiation pattern are given in the next section and later in this chapter in Section 6.9 in relation to shaped reflectors.

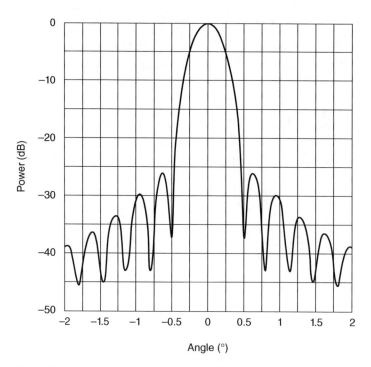

Figure 6.10   Principal radiation pattern of a paraboloid with dimensions $D = 100\lambda$ and $f/D = 0.4$ with a feed having an axisymmetric Gaussian pattern function that results in a -10 dB edge illumination

6.2.3.2 Spherical Reflector

Another reflector type that is used in a variety of applications because of its ease of use in scanned beams is the spherical reflector that is illustrated in Figure 6.11 (Li, 1959). The surface is obtained by taking a section of a sphere of radius R_o. Unlike the paraboloid, the spherical reflector does not have a perfect focus. Typically, a focus is taken on the line of symmetry at a distance $f_o = R_o/2$ from the reflector vertex. However, this does not take account of other rays near the axis of symmetry, and as a result, the best feed location tends to be closer to the reflector, the actual distance depending on the requirement of the antenna. For example, to keep the phase error in the aperture to within $\pm\lambda/16$, the diameter, D, of the aperture should not exceed $D = 256\lambda(f_o/D)^3$. Common methods for feeding a spherical reflector are an array of dipoles or waveguides, a correcting concave reflector, which results in a Gregorian-corrected dual-reflector configuration, or alternatively a line source provided by a travelling wave feed along the axis, which radiates a field towards the reflector.

To examine the spherical reflector geometry further, consider a possible location of a single feed as shown in Figure 6.11. The distance from the focal point to the aperture plane at $z = 0$ is given by

$$L = FP' + P'A = \sqrt{s^2 + \left[\sqrt{\rho^2 - s^2} - (\rho - f_o)\right]^2} + \sqrt{\rho^2 - s^2}, \qquad (6.40)$$

where $\rho = R_o$ the radius of the spheroid, s is the radial distance from the z-axis and f_o is the distance from the vertex to the focal point at F. The path difference between a paraxial ray and a non-axial ray is given by

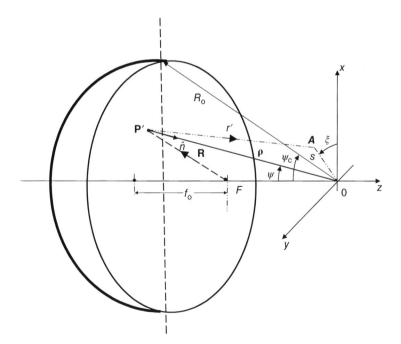

Figure 6.11 Spherical reflector geometry

$$\Delta = R_o + f_o - L.$$

This path difference in wavelengths is

$$\frac{\Delta}{\lambda} = \frac{R_o}{\lambda}\left(1 + \frac{f_o}{R_o} - \sqrt{\left(\frac{s}{R_o}\right)^2 + \left[\sqrt{1 - \left(\frac{s}{R_o}\right)^2} - \left(1 - \frac{f_o}{R_o}\right)\right]^2} - \sqrt{1 - \left(\frac{s}{R_o}\right)^2}\right). \quad (6.41)$$

The total phase error over a prescribed aperture is least when the phase error at the aperture edge is zero. If the aperture radius is $D/2$, which describes a cone of angle $\psi_c = \sin^{-1}(D/2R_o)$ at O, the optimum focal length when $\Delta/\lambda = 0$ is given by

$$f_o = \frac{1}{4}\left(R_o + \sqrt{R_o^2 - \left(\frac{D}{2}\right)^2}\right). \quad (6.42)$$

While this is the optimum location for a uniform illumination, it is not necessarily the most suitable location for a tapered feed pattern when the feed needs to be moved further away from the reflector vertex. Correspondingly, there is a given aperture dimension for a minimum total phase error. Thus, the phase error tolerance limits the aperture size. It has been determined that the maximum allowable total phase error $(\Delta/\lambda)_{max}$ for a given aperture diameter in wavelengths D/λ is

$$\left(\frac{\Delta}{\lambda}\right)_{max} = \frac{(D/\lambda)^4_{max}}{75\pi(R_o/\lambda)^3}. \quad (6.43)$$

The radiation pattern of a spheroid can be obtained by means of the expressions given by Eqs. 6.37 and 6.38. As an example, a spheroid is chosen with a radius of $R_o = 50\lambda$. According to Eq. 6.43, the maximum diameter of the spherical segment for a phase error of $\lambda/16$ is $D/\lambda = 36.384$. Assuming an approximate uniform illumination, Eq. 6.42 gives a focal distance of $f_o = 24.121\lambda$. At this distance, the feed-cone angle is $\theta_o = 56.78°$. A feed pattern function given by $A(\psi) = \cos^6\psi$ was chosen for this reflector to provide an edge taper of -8.21 dB. The resulting radiation pattern that was computed is shown in Figure 6.12, and the maximum gain is 37.64 dBi, which results in an aperture efficiency of 41.2%. These results were obtained from a discrete data representation of the reflector and are comparable with those obtained from the closed-form expressions for the spheroid. The first sidelobe level is seen to occur at -25 dB, and the maximum cross-polar level is -44.3 dB. This sidelobe level is significantly lower than what might be expected from the edge taper with a paraboloid of the same diameter.

6.2.4   Receive-Mode Method

In the previous section, the radiation pattern of the reflector was obtained by assuming that the feed antenna was transmitting. The reflector can also be easily analysed by adopting a receive-mode method. In the receive-mode approach, the reflector antenna is assumed under

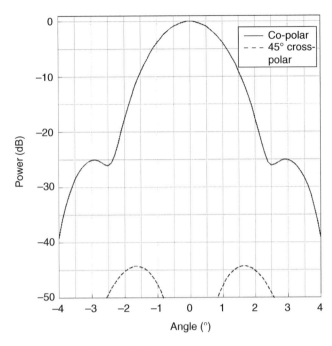

Figure 6.12   Radiation patterns of a spherical reflector with $D = 36.834\lambda$, $R_o = 50\lambda$ and $f_o = 24.121\lambda$, edge illumination -6.41 dB

illumination from a distant point source. The illumination is by a plane wave from a fixed direction. The wave can be linearly or circularly polarized. For a distributed source, the illumination could be created from a superposition of plane waves from all relevant angles of incidence. Recalling reciprocity, the receive-mode method should give the same results as in transmission. Its advantage is that for a fixed incident beam the feed need not enter the calculation until after the current on the nearest reflector has been computed. In addition, with little difficulty, the same computer program can determine both the focal region fields, knowledge of which is important for designing the feed excitation and the far-field radiation patterns. In the latter calculation, the methods of Section 3.7 can be used. When there are two or more reflectors, it is usual to integrate over the reflector nearest the feed as the latter's surface is usually the smallest one of the two. Both calculations are feasible once the current on the relevant reflector is known. The current is often stored as this is usually the most complicated calculation. The radiation efficiency can then be compared relatively quickly for a variety of feeds or feed element locations.

The field finally reaching the feed is usually subject to several approximations. It is common practice to ignore either a direct component into the feed from the incident field or an indirect diffracted component from the subreflector. Similarly, multiple reflections from the subreflector, or feed, which reach the main reflector, are excluded. This latter assumption eliminates the reaction between the reflectors and the reflectors and feed. On the whole, practice with symmetrical reflectors has shown that these assumptions do not significantly affect the main beam and the first few sidelobes. In the case of the dual offset reflector, these approximations are likely to have less effect.

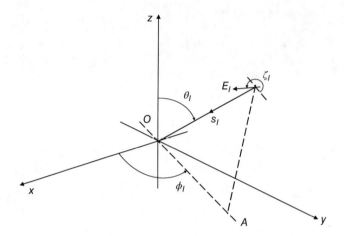

Figure 6.13   Linearly polarized plane wave incident in direction s_I

The direction of arrival of an incident linearly polarized plane wave can be described by three angles θ_I, ϕ_I and ζ_I as shown in Figure 6.13. Suppose s_I is the beam direction and $\mathbf{E}_I$ is the electric field associated with the wave. As may be verified from Figure 6.13

$$s_I = -(\hat{x}\cos\phi_I + \hat{y}\sin\phi_I)\sin\theta_I - \hat{z}\cos\theta_I. \tag{6.44}$$

As well, $\mathbf{E}_I$ has a polarization angle ζ_I, which is defined relative to an initial line that is in the $z-OA$ plane which is defined by ϕ_I. A field defined in the co-polarized direction is

$$\mathbf{E}_I = \hat{x}(\cos\zeta_I\cos\phi_I\cos\theta_I - \sin\zeta_I\sin\phi_I)$$
$$+ \hat{y}(\cos\zeta_I\sin\phi_I\cos\theta_I + \sin\zeta_I\cos\phi_I) - \hat{z}\cos\zeta_I\sin\theta_I. \tag{6.45a}$$

An orthogonal cross-polarized field can similarly be defined by simply replacing $\zeta_I \rightarrow \zeta_I + \pi/2$. That is,

$$\mathbf{E}_{Ix} = -\hat{x}(\sin\zeta_I\cos\phi_I\cos\theta_I + \cos\zeta_I\sin\phi_I)$$
$$- \hat{y}(\zeta_I\sin\phi_I\cos\theta_I - \cos\zeta_I\cos\phi_I) + \hat{z}\sin\zeta_I\sin\theta_I. \tag{6.45b}$$

The field arriving at the point $P(r_p, \theta_p, \phi_p)$ that is scattered from a paraboloid is shown in Figure 6.14. The reflector is oriented with its axis of symmetry along the z-axis. The field can be computed using the induced current method or by some other technique such as the geometrical theory of diffraction (GTD). In the latter, the resulting field consists of field contributions arising from specular reflection that is a geometric optics contribution plus partial fields that are due to diffraction from several points Q_D^k ($k = 1, 2$) on the reflector rim. All points are determined from Fermat's principle of finding the least path length from incidence to the reflection or diffraction points. The field obtained close to the focus is the focal region field. This field is an indicator of the best feed aperture distribution with which to excite the reflector. It may also be used to calculate the radiation pattern by means of the power coupling theorem

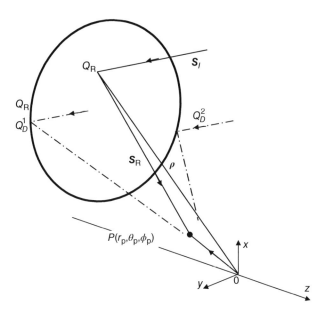

Figure 6.14   Reflection point QR and sites of edge diffraction Q_D^k ($k = 1, 2$) on an offset paraboloid

(Wood, 1980) or field correlation as described in Section 3.7. The version of the power coupling theorem given here includes the assumption that physical optics is applicable at the reflector or a subreflector. In addition, it is assumed that the scattered field does not influence the feed
radiation. With these assumptions, the fraction of the power in the field scattered from the main
reflector that is coupled into the feed is given by Eq. 3.64 as

$$\eta(\theta,\phi) = \frac{\left| \displaystyle\iint_{\text{Srefl}} (\mathbf{E}_f \cdot \mathbf{J}_{\text{refl}}(\theta,\phi)) dS \right|^2}{16 P_f P_{\text{inc}}}, \tag{6.46}$$

where $\mathbf{E}_f$ is the electric field radiated by the feed independently of the reflector, P_f is the power
radiated by the feed and P_{inc} is the power in the wave incident on the reflector. The current on
the reflector is

$$\mathbf{J}_{\text{reff}} = 2Y_o \hat{n} \times (\mathbf{s}_I \times \mathbf{E}_I), \tag{6.47}$$

where Y_o is the wave admittance of the incident plane wave, $\hat{n}$ is the normal to the reflector in
rectangular co-ordinates and for the paraboloid it is given by Eq. 6.3. The incident electric field
$\mathbf{E}_I$ is given by Eqs. 6.45a and 6.45b for determining the co- and cross-polar efficiencies, respectively. Eq. 6.46 represents the fraction of the power radiated by the feed in the direction of the
incoming wave. Its maximum value is the peak antenna efficiency or beam efficiency and when
the antenna radiates multiple pencil beams, its maximum value in each beam direction is the
beam efficiency. Far-field radiation patterns can be determined by evaluating Eq. 6.46 over
solid angles around each incident wave direction. The integrals in Eq. 6.46 are usually

evaluated numerically using methods such as trapezoidal quadrature or the Gauss-Legendre quadrature rule (Fröberg, 1974).

As an example of the possible simplification of Eq. 6.46, consider the special case of a symmetrical paraboloid. This reflector is fed by a Huygens source with a radiated field of

$$\mathbf{E}_f = A(\theta)\left(\hat{\theta}\cos\phi - \hat{\phi}\sin\phi\right)\exp\frac{-jkr}{r}$$

where $A(\theta)$ is the Gaussian function given by Eq. 6.23. The radiated power is

$$P_f = \frac{\pi}{\eta_o}\int_0^{\pi/2} d\theta |A(\theta)|^2,$$

while the power in the incident plane wave is $P_{inc} = (\pi D)^2/\eta_o$, where D is the diameter of the paraboloid. Assume an incident plane wave that is parallel to the z-axis (i.e. $\theta_I = 0°$ and $\phi_I = 0°$) that is polarized parallel to the x-axis (i.e. $\zeta_I = 0°$). The maximum efficiency given by Eq. 6.46 is

$$\eta_{max} = \frac{\int_0^{\pi/2} d\psi |A(\psi)\cos(\psi/2)|^2}{2(\pi D)^2 \int_0^{\pi/2} d\psi |A(\psi)|^2}.$$

For a paraboloid with $D = 100\,\lambda$ and $f/D = 0.35$ that is fed with a Huygens source with an edge illumination of -16 dB, Eq. 6.23, the maximum efficiency predicted by power coupling is 77%. The induced current method gives 74% for the same geometry.

6.3 Focal Region Fields of a Paraboloidal Reflector

An understanding of the fields excited in the vicinity of the focus of a reflector by an incident signal is very useful for designing suitable feed antennas. By making the aperture fields of the feed a good match to the focal region fields, high gain and low cross-polarization may be achieved. To calculate the focal region fields for far-field operation either in reception or transmission, the reflector is illuminated by a uniform plane wave from infinity. In this section, the fields in the focal region (at $z = 0$ in Figure 6.15) are found by means of the induced current method. Other approximate methods such as GTD can be used. A high frequency approximation will also be described later in this section.

A paraboloidal reflector is illuminated by a plane wave linearly polarized in the x-direction, as illustrated in Figure 6.15. To make the result more general, the wave is incident at an angle to the axis of symmetry and its electric field is

$$\mathbf{E}_i = E_o(\hat{x}\cos\theta_i - \hat{z}\sin\theta_i)e^{jk(x\sin\theta_i + z\cos\theta_i)}, \tag{6.48}$$

where θ_i is the angle of incidence relative to the negative z-direction. This field induces the current on the reflector surface as follows:

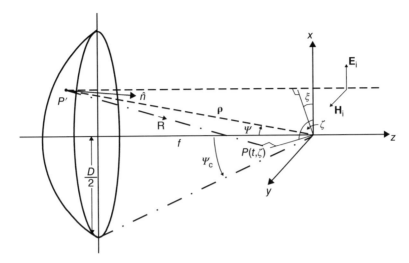

Figure 6.15 Geometry for focal region field analysis for on-axis incidence θ_i.

$$\mathbf{J_s} = 2\hat{n} \times \mathbf{H_i}\big|_{\text{reflector } \Sigma}$$

$$= 2\hat{n} \times \frac{1}{\eta_o}\left[(-\hat{x}\sin\theta_i - \hat{z}\cos\theta_i) \times \mathbf{E_i}\right],$$

(6.49)

where $\hat{n}$ is given by Eq. 6.3. In rectangular co-ordinates the current is expressed as follows:

$$\mathbf{J_s} = \frac{2E_o}{\eta_o} e^{jk(x\sin\theta_i + z\cos\theta_i)}\left(\hat{x}\cos\frac{\psi}{2} + \hat{z}\sin\frac{\psi}{2}\cos\xi\right).$$

(6.50)

As focal plane is not generally in the far-field of the reflector and, therefore, Eqs. 3.8 are required to evaluate the fields. Only the electric field in the focal region is considered. Eq. 3.8a gives

$$\mathbf{E}_F(r,\theta,\phi) = \frac{jk\eta}{4\pi}\int_{\Sigma}\frac{e^{-jkR}}{R}\left[\mathbf{J_s} - \hat{R}(\mathbf{J_s}\cdot\hat{R})\right]dS'.$$

(6.51)

As shown in Figure 6.15 for on-axis incidence θ_i, the vector $\mathbf{R} = \mathbf{t} - \boldsymbol{\rho}$ is from P' on the reflector to P in the focal region and $\mathbf{t}$ is a vector in the focal plane. Near the focus $|\mathbf{t}| \ll |\boldsymbol{\rho}|$, and this allows approximations to be made in a similar fashion to estimating the far-fields. Accordingly, let

$R = |\mathbf{R}| \approx |\boldsymbol{\rho}| = \hat{\rho}\cdot\mathbf{t}$ in the phase function of the integrand and
$\hat{R} \approx \hat{\rho}$ inside the square brackets in the amplitude function of the integrand.

The amplitude function in the integral of Eq. 6.51 is now

$$\frac{1}{R}\left[\mathbf{J_s}-\hat{R}(\mathbf{J_s}\cdot\hat{R})\right]dS' \approx [\mathbf{J_s}-\hat{\rho}(\mathbf{J_s}\cdot\hat{\rho})]\rho d\psi d\xi$$

$$= -\frac{2fE_0}{\eta}e^{jk(x\sin\theta_i + z\cos\theta_i)}\left\{\hat{x}\left(1-\tan^2\frac{\psi}{2}\cos 2\xi\right)\sin\psi\right.$$
(6.52)

$$\left.-\hat{y}\sin\psi\tan^2\frac{\psi}{2}\sin 2\xi + \hat{z}2\sin\psi\tan\frac{\psi}{2}\cos\xi\right\}\rho d\rho d\psi.$$

Also, the combined phase function is given by

$$\Phi' = R - (\cos\theta_i z + \sin\theta_i x)$$

$$\approx (x_F\cos\xi + y_F\sin\xi)\sin\psi + 2f\sin\theta_i\tan\frac{\psi}{2}\cos\xi + f\cos\theta_i\sec^2\frac{\psi}{2}-2f$$
(6.53)

$$= t(\psi)\sin\psi\cos(\zeta-\xi) + f\cos\theta_i\sec^2\frac{\psi}{2}-2f,$$

where $t(\psi) = \sqrt{(x_F + f\sin\theta_i\sec^2(\psi/2))^2 + y_F^2}$. With these approximations, Eq. 6.51 gives (Minnett & Thomas, 1968)

$$E_{Fx} = -\frac{jkfE_0}{2\pi}\int_0^{2\pi}d\xi\int_0^{\psi_c}\left(1-\tan^2\frac{\psi}{2}\cos 2\xi\right)\sin\psi e^{-jk\Phi'}d\psi$$
(6.54a)

$$E_{Fy} = \frac{jkfE_0}{2\pi}\int_0^{2\pi}d\xi\int_0^{\psi_c}\sin\psi\tan^2\frac{\psi}{2}\sin 2\xi e^{-jk\Phi'}d\psi$$
(6.54b)

$$E_{Fz} = -\frac{jkfE_0}{\pi}\int_0^{2\pi}d\xi\int_0^{\psi_c}\sin\psi\tan\frac{\psi}{2}\cos\xi e^{-jk\Phi'}d\psi.$$
(6.54c)

The integration with respect to ξ can be completed by means of Eq. B.3 allowing Eqs. 6.54 to be reduced to

$$E_{Fx}(x_F, y_F, \theta_i) = A_0(x_F, y_F) + A_2(x_F, y_F)$$
(6.55a)

$$E_{Fy}(x_F, y_F, \theta_i) = B_2(x_F, y_F)$$
(6.55b)

$$E_{Fz}(x_F, y_F, \theta_i) = -2jA_1(x_F, y_F),$$
(6.55c)

where

$$A_n(x_F, y_F) = \kappa\int_0^{\psi_c}J_n(kt\sin\psi)\tan^n\frac{\psi}{2}\sin\psi\cos n\zeta\exp\left(-jka_z\tan^2\frac{\psi}{2}\right)d\psi$$
(6.56a)

and

$$B_n(x_F, y_F) = \kappa\int_0^{\psi_c}J_n(kt\sin\psi)\tan^n\frac{\psi}{2}\sin\psi\sin n\zeta\exp\left(-jka_z\tan^2\frac{\psi}{2}\right)d\psi$$
(6.56b)

in which $a_z = 2f \sin^2(\theta_i/2)$; $\kappa = -jkfE_o \exp(-jk2f \cos^2\theta_i/2)$ and $\tan \zeta = y_F/(x_F + f \sin \theta_i \sec^2(\psi/2))$.

For on-axis incidence ($\theta_i = 0$), Eqs. 6.55 become

$$E_{Fx}(x_F, y_F, 0) = \Lambda_0(t) + \Lambda_2(t) \cos 2\zeta \tag{6.57a}$$

$$E_{Fy}(x_F, y_F, 0) = \Lambda_2(t) \sin 2\zeta \tag{6.57b}$$

$$E_{Fz}(x_F, y_F, 0) = -2j\Lambda_1(t) \cos \zeta, \tag{6.57c}$$

where

$$\Lambda_n(t) = \kappa \int_0^{\psi_c} J_n(kt \sin \psi) \tan^n \frac{\psi}{2} \sin \psi \, d\psi \tag{6.57d}$$

and $\kappa = -jkfE_o e^{-jk2f}$. Further insight to these equations is possible by now considering a paraboloid with a long focal length. Suppose the angle to the rim, ψ_c, is small allowing the functions in Eqs. 6.57 to be approximated as follows

$$\Lambda_0(t) \approx \kappa' \left[2\frac{J_1(U)}{U} \right] \tag{6.58a}$$

$$\Lambda_1(t) \approx \kappa' \left[\psi_c \frac{J_2(U)}{U} \right] \tag{6.58b}$$

$$\Lambda_2(t) \approx 0, \tag{6.58c}$$

where $U = kt \sin \psi_c$ and $\kappa' = 2\kappa \sin^2(\psi_c/2)$. Note that $\sin(\psi_c/2) \approx \tan(\psi_c/2) \approx D/2f$ so that $\kappa' \approx -jkE_o e^{-jk2f} D^2/8f$ and therefore

$$E_{Fx}(t, \zeta) \approx \kappa' \left[2\frac{J_1(U)}{U} \right] \tag{6.59a}$$

$$E_{Fy}(t, \zeta) \approx 0 \tag{6.59b}$$

and

$$E_{Fz}(t, \zeta) \approx -j2\kappa' \left[\psi_c \frac{J_2(U)}{U} \right]. \tag{6.59c}$$

Therefore, the dominant field amplitude in the focal plane is

$$|E_{Fx}| \approx E_o \left(\frac{kD^2}{8f} \right) \left[2\frac{J_1(kt\psi_c)}{kt\psi_c} \right]. \tag{6.60}$$

Equation 6.60 is the scalar solution obtained by the mathematician and Astronomer Royal George Biddell Airy in the early part of the nineteenth century when he investigated the distribution of light in the focal region of a lens. He observed that the focal region field consists of bright ('Airy') and dark rings. Eq. 6.60 shows that these rings occur when $J_1(U) = 0$, resulting in a first dark ring when $U = 3.832$.

Even for paraboloids with a short focal length, the field in the focal region has similar characteristics through the expressions given in Eqs. 6.57. Figure 6.16 shows the field distribution in the focal plane of the Parkes radio telescope ($\psi_c = 63°$ and $f/D = 0.41$). In the central region, the field is nearly linearly polarized. It changes sign at the first zero, which occurs at a radius of $t_1 = 0.610\lambda / \sin \psi_c = 0.55\lambda$, which is indicated as a dotted line in Figure 6.16. For optimum performance, a feed should have an aperture field distribution that closely matches the focal fields. A corrugated waveguide operating in the HE_{11} mode (see Eq. 4.61a) provides a good match to the focal field when the waveguide, a, is approximately equal to t_1 in both the co- and cross-polar directions. The TE_{11} mode of circular waveguide is also quite a good feed for a paraboloid, but it does not have zero cross-polarization as required in Figure 6.16b. For maximum gain, its radius should be slightly greater than t_1 (see Figure 6.21). This is because the TE_{11} mode aperture field is more uniformly polarized near the centre of the waveguide than at its walls and is, therefore, a better match to the focal fields. A circular waveguide with a diameter of about 1.1λ would be a reasonable option. If a beam is required off-axis, a small array can also be used (Poulton & Bird, 1988).

6.3.1 Asymptotic Representation of the Scattered Field*

The field scattered from the offset reflector to the focal region can be approximated for large reflectors or for high frequencies by means of asymptotic methods. The basic technique was described in Section 3.8.1. The focal fields or radiation pattern can then be obtained either by correlating this field with the field of the horn on the aperture or using the field to approximate the current on a subreflector if one is present. Consider the situation illustrated in Figure 6.17. The magnetic field scattered to a point P from paraboloid due to plane wave incidence is given by

$$\mathbf{H}(\mathbf{R}) = \frac{jk}{4\pi} \iint_S (\mathbf{J}_S \times \mathbf{s}_R) \frac{\exp(-jks_R)}{s_R^2} dS, \tag{6.61}$$

where $\mathbf{J}_S = (2/\eta_o)[\hat{n} \times (\hat{s}_I \times \mathbf{E}_I)] \exp(-jk\hat{s}_I \cdot \boldsymbol{\rho})$.

$\mathbf{E}_I$ is the incident electric field;
$\hat{s}_I$ is a unit vector in the direction of the incident plane wave;
$\hat{s}_I = -(\hat{x} \cos \phi_I + \hat{y} \cos \phi_i) \sin \theta_I + \hat{z} \cos \theta_I$;
$\boldsymbol{\rho} = \rho(\hat{x} \cos \xi \sin \psi + \hat{y} \sin \xi \sin \psi + \hat{z} \cos \psi)$ is a vector from the focus to the reflector surface and for a paraboloid $\rho = 2f/(1 + \cos \psi)$;
$\hat{n}$ is the normal to the surface of the reflector

$$\hat{n} = \frac{1}{\sqrt{\rho^2 + \rho_\psi^2}} [(\hat{x} \cos \xi + \hat{y} \sin \xi)(\rho_\psi \cos \psi - \rho \sin \psi) - \hat{z}(\rho_\psi \sin \psi + \rho \cos \psi)];$$

$\mathbf{s}_R = |\boldsymbol{\rho} - \mathbf{R}|$ is a vector from the source point to P and $\hat{s}_R = \mathbf{s}_R / s_R$ is a unit vector and s_R is the magnitude; and $\mathbf{R}$ is a vector from O to P the observation point

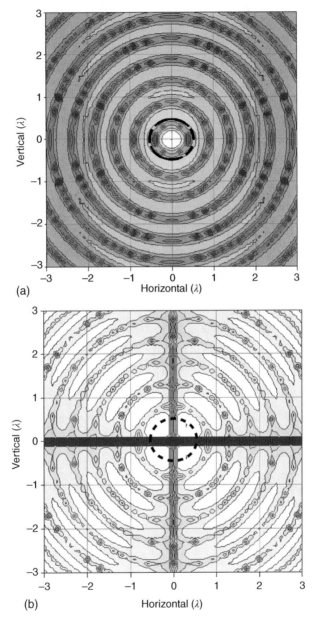

(a)

(b)

Figure 6.16   Contour plots of the electric field amplitude in the focal plane of a paraboloid with diameter $D = 300\lambda$ and $f/D = 0.41$ which corresponds to a half-cone angle $\psi_c = 63°$ (a) Co-polar $|E_x|$; and (b) cross-polar $|E_y|$. The dashed circle at the centre has a radius of 0.55λ

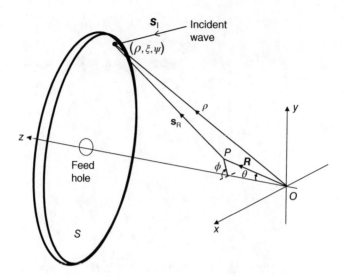

Figure 6.17   Incident plane wave on a shaped symmetrical reflector

$$\mathbf{R} = R(\hat{x}\cos\phi\sin\theta + \hat{y}\sin\xi\sin\theta + \hat{z}\cos\theta).$$

For convenience, this field is expressed as follows:

$$\mathbf{H}(\mathbf{R}) = \iint_S \mathbf{F}(\mathbf{R}|\boldsymbol{\rho})\exp(jkg(\mathbf{R}|\boldsymbol{\rho}))dS, \tag{6.62}$$

where $\mathbf{F}(\mathbf{R}|\boldsymbol{\rho}) = \dfrac{jk}{2\pi\eta_o}\dfrac{[\hat{n}\times(\hat{s}_I\times\mathbf{E}_I)]\times\mathbf{s}_R}{s_R^2}$ and $g(\mathbf{R}|\boldsymbol{\rho}) = -(\hat{s}_I\cdot\boldsymbol{\rho} + s_R)$.

This field can be evaluated asymptotically by methods described in Section 3.8.1. In brief, there will, typically, be three contributions to the asymptotic solution, namely, a stationary point from the surface of the reflector corresponding to specular reflection, two or more edge contributions due to stationary points on the peripheries due to edge diffraction and a contribution from any discontinuities on the boundary of S.

In the case of an axisymmetric reflector, the components of the integral in Eq. 6.62 are integrals of the form

$$I = \int_0^{2\pi} d\xi \int_{\psi_e(\xi)}^{\psi_u} d\psi\, f(\xi,\psi)\exp(jkg(\xi,\psi)), \tag{6.63}$$

where $f(\xi,\psi)$ is a component of Eq. 6.62. $\psi = \psi_e(\xi)$ is the inner boundary of the surface S and corresponds to shadowing by a subreflector in a dual reflector or a feed in a prime focus configuration. $\psi = \psi_u$ defines the outer rim of the reflector.

As has been mentioned, there are potentially three different types of critical points relating to the domain S. The critical point of the first kind, or spectral point, at (ξ_1, ψ_1) on the reflector surface is defined by

$$\left.\frac{\partial g}{\partial \xi}\right|_{\substack{\xi=\xi_1 \\ \psi=\psi_i}} = g_\xi(\xi_1,\psi_1)=0=g_\psi(\xi_1,\psi_1). \tag{6.64a}$$

As well, there may be up to four critical points of the second kind, or edge points, from the two boundaries of S, which are defined by

$$\left.\frac{\partial g}{\partial \xi}\right|_{\substack{\xi=\xi_1 \\ \psi=\psi_c}} = g_\xi(\xi_2,\psi(\xi_2))=0 \tag{6.64b}$$

on the inner and outer boundaries of S. Finally, there could potentially be two critical points of a third kind due to a discontinuously turning tangent on the inner boundary. However, for simplicity, it is assumed here that $\psi_c(\xi)$ is a smooth function with no discontinuities. The critical points of the first kind yield a geometric optics type of contribution to Eq. 6.63 and those of the second kind give diffraction contributions from the edges. For large reflectors with a relatively small inner hole defined by $\psi_c(\xi)$, the contributions to Eq. 6.63 from the inner boundary edge points are small compared with the other terms.

When the functions f and g are expanded in their Taylor series about the critical points, as described in Section 3.8.1, and at each critical point the first term of the asymptotic expansion of Eq. 6.63 is taken into account, an asymptotic representation of this equation is given by

$$I \sim \sigma I_1(\xi_1,\psi_1) - \sum_i \varepsilon_{ui} I_2(\xi_{u2},\psi_2) + \sum_j \varepsilon_{uj} I_2(\xi_{\ell2},\psi(\xi_{\ell2})). \tag{6.65}$$

The subscripts u and ℓ refer to the upper and lower limits of ψ and $i,j=1,2$ is a summation over the critical points of a second kind on the two boundaries. To define the other variables in Eq. 6.65, let

$$V(\xi,\psi) = g_\psi - \left(\frac{g_\xi g_{\xi\psi}}{g_{\xi\xi}}\right) \tag{6.66a}$$

$$U(\xi,\psi) = \frac{\Delta}{g_{\xi\xi}} = g_{\psi\psi} - \left(\frac{g_{\xi\psi}^2}{g_{\xi\xi}}\right), \tag{6.66b}$$

where g_ψ, g_ξ and so on refer to derivatives of g with respect to ψ, ξ and

$$\Delta = \begin{vmatrix} g_{\psi\psi} & g_{\xi\psi} \\ g_{\psi\xi} & g_{\xi\xi} \end{vmatrix}$$

is the determinant of the Hessian matrix of $g(\xi,\psi)$. For a general critical point denoted by q_i, representing boundary q of S, point i (which is set equal to 1 for the stationary point) then in Eq. 6.65

$$\varepsilon_{qi} = \operatorname{sgn}\frac{V(\xi_{qi},\psi_{qi})}{U(\xi_{qi},\psi_{qi})} \tag{6.67a}$$

and also

$$\sigma = \begin{cases} \varepsilon_{ui}; & \text{when } \text{sgn}(V(\xi_{u1},\psi_{u1})) = \text{sgn}(V(\xi_{u2},\psi_{u2})) \ \& \ \text{sgn}(V(\xi_{\ell 1},\psi_{\ell 1})) \neq \text{sgn}(V(\xi_{\ell 2},\psi_{\ell 2})) \\ 0; & \text{otherwise} \end{cases}$$

(6.67b)

is a step function that distinguishes whether there is a geometric optics contribution to the integral based on a test of the derivatives of g at the edge points. This test can be made before a search is undertaken to find a stationary point on S and this can help shorten computation time. In addition,

$$I_1(\xi_1,\psi_1) = \sqrt{\pi}W(\xi_1,\psi_1)\exp\left(\frac{j\mu\pi}{4}\right)$$

$$I_2(\xi_{qi},\psi_{qi}) = W(\xi_{qi},\psi_{qi})F_\mu\left(\sqrt{\frac{k}{2|U|}}|V|\right),$$

where

$$\mu = \text{sgn}(U)$$

and

$$W(\xi_{qi},\psi_{qi}) = \frac{2}{k}\sqrt{\frac{\pi}{\Delta}}f(\xi_{qi},\psi_{qi})\exp\left(j\,\text{sgn}(g_{\psi\psi})\frac{\pi}{4}\right)$$

$$\times \exp\left[jk\left(g(\xi_{qi},\psi_{qi}) - \frac{1}{2}\left(\frac{g_\xi^2}{g_{\xi\xi}} + \frac{V^2}{U}\right)\right)\right].$$

$F_\mu(z)$ is a Fresnel integral, which is defined in Appendix E. It can have a positive or negative exponential in the argument of the integral depending on the sign of μ. A particular feature of the asymptotic solution given above is that the critical points need only be approximate. The reason is that Eq. 6.65 includes compensating terms that involve the derivative g_ξ and g_ψ, which are set to zero in some asymptotic forms. In the present formulation, the stationary point (ξ_1, ψ_1) is a simultaneous solution of g_ξ and g_ψ such that $|g_\xi|, |g_\psi| < \varepsilon_1$ while the edge points are solutions to Eq. 6.64b with $|g_\xi| < \varepsilon_2$ where ε_1 and ε_2 are assumed to be small. Approximate solutions of the critical points in the range $0.001 < \varepsilon_1, \varepsilon_2 < 1$ have been tried and only small variations to the overall value of the integral have been observed. The critical points can be found by a variety of root finding methods. For example, the two-dimensional Newton–Raphson (Dixon, 1972) is particularly apt for finding the stationary point on the reflector in present application. The i-th Newton–Raphson step towards a minimum of $|g|$ given by Eq. 6.64a is given by

$$\Phi_{i+1} = \Phi_i - \left(\bar{\bar{H}}^{-1}\bar{J}\right)_i$$

where $\Phi_i = \begin{bmatrix} \xi_i \\ \psi_i \end{bmatrix}$, $\bar{J}_i = \begin{bmatrix} g_\xi \\ g_\psi \end{bmatrix}_i$, and $\bar{\bar{H}}^{-1} = \frac{1}{\Delta}\begin{bmatrix} g_{\xi\xi} & g_{\psi\xi} \\ -g_{\xi\psi} & g_{\psi\psi} \end{bmatrix}$ is the inverse Hessian matrix.

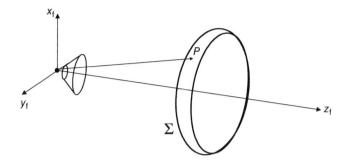

Figure 6.18   Feed illuminating subreflector

As mentioned earlier, the present solution for the scattered field can be used to find the focal region fields or the radiation of a reflector. In a dual reflector application, the radiation pattern or focal region fields can be obtained from the physical optics currents induced on the subreflector as shown in Figure 6.18. The current at point P on the surface Σ is given by $\mathbf{J}_s(r,\theta,\phi) = 2\hat{n}_p \times \mathbf{H}(r,\theta,\phi)$ where $\mathbf{H}$ is given by the asymptotic form of Eq. 6.62 and $\hat{n}_p$ is the normal at P. The radiation from a dual reflector could be obtained by field correlation from Eq. 6.46 by integrating the asymptotic field with the electric field from the feed $\mathbf{E}_f$ over the subreflector surface as indicated. The feed field needs to be transformed from the local feed co-ordinate system to the global co-ordinate (x, y, z) system. In the simpler prime focus application, field correlation can also be used, but in this case, the integration of the asymptotic solution should be done over the aperture of the feed with both the electric and magnetic fields present as given by the alternative Eq. 3.62.

6.4   Blockage

The feed and its support structure in a front-feed paraboloid scatter energy away from the aperture producing a shadow as illustrated in Figure 6.19. To a first approximation, shadowing, or blockage as it is known, may be included in field calculations by eliminating the blocked parts from the integration over the aperture or reflector surface. For example, in the aperture field method, blockage from a feed of diameter D_o is included by removing a circle of radius $D_o/2$ from the centre of the aperture. Eq. 6.11 would then be replaced by

$$\mathbf{N}(\theta,\phi) = \int_0^{2\pi} d\xi \int_{D_o/2}^{D/2} \mathbf{E}_a(t,\xi)\exp(jwt\cos(\phi-\xi))t\,dt. \tag{6.68}$$

The effect of blockage on the radiation pattern is demonstrated by the one-dimensional example illustrated in Figure 6.20. The Fourier transform of the aperture distribution in this figure is

$$\begin{aligned}
E &= \int_{-\infty}^{\infty} f(x')e^{j2\pi ux'} \\
&= \int_{-a/2}^{a/2} e^{j2\pi ux'}\,dx' = \int_{-b/2}^{b/2} e^{j2\pi ux'}\,dx' \\
&= aS(\pi ua) - bS(\pi ub).
\end{aligned} \tag{6.69}$$

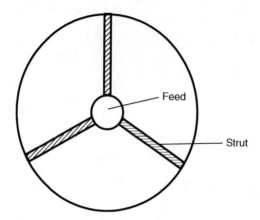

Figure 6.19 Aperture blockage by feed and supporting struts

Aperture distribution

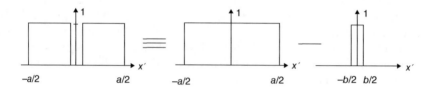

Far-field intensity

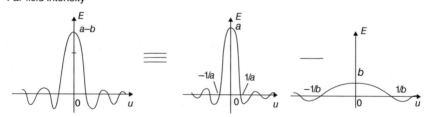

Figure 6.20 Illustration of blockage effects in terms of Fourier distributions

The result is illustrated in Figure 6.20. It is seen that the subtracted transform due to the central blockage is wider than the transform due to the main aperture distribution. Therefore, the central lobe and the even-numbered sidelobes of the main aperture are out of phase with the blockage. Therefore, blockage reduces the on-axis gain and increases the nearby odd-numbered sidelobes extending out from the central beam. As well it decreases the even-numbered sidelobes. Blockage can also have an important effect on the sidelobes far from the main lobe, sometimes increasing them to unacceptably high levels. This will depend on the size and shape of the blockage as well as the illumination.

6.5 Reflector Antenna Efficiency

Under normal operating conditions, a reflector antenna has maximum gain for a uniform, equiphase aperture distribution. Then from Eq. 3.40, a reflector with a diameter D has maximum gain of

$$G_o = \left(\frac{\pi D}{\lambda}\right)^2. \tag{6.70}$$

If the reflector illumination is tapered, as it usually is in practice to keep sidelobes at acceptable levels, the maximum is less than that given Eq. 6.70. To accommodate this, in Section 3.5.6 maximum gain[1] was defined as

$$G_{max} = \eta_a G_o, \tag{6.71}$$

where η_a is the aperture efficiency. Eq. 6.71 neglects power losses in the reflector system due to feed spillover, mismatch, conductor losses and so on. These power losses may be accounted for in the calculation of gain by modifying Eq. 6.71 as is now demonstrated for reflector spillover.
    The power density radiated by the feed is

$$P_f = \frac{1}{2\eta} |E_f|^2 \tag{6.72}$$

giving a total radiated power

$$P_T = \int_0^{2\pi} d\xi \int_0^\pi P_f \rho \sin\psi \, d\psi. \tag{6.73}$$

Not all this power is intercepted by the reflector. Some of it falls outside the reflector causing a power loss. This power loss is called spillover. The power collected by a reflector subtending an angle ψ_c is

$$P_c = \frac{1}{2\eta} \int_0^{2\pi} d\xi \int_0^{\psi_c} |E_f|^2 \rho \sin\psi \, d\psi. \tag{6.74}$$

Therefore, the power loss due to spillover is

$$P_s = P_T - P_c = \frac{1}{2\eta} \int_0^{2\pi} d\xi \int_{\psi_c}^\pi |E_f|^2 \rho \sin\psi \, d\psi. \tag{6.75}$$

---

[1] Higher gains than Eq. 6.70 are possible with non-uniform aperture phase distributions. This supergain phenomenon can be difficult to achieve and obtain in practice because of losses and often the improvement is very narrowband.

The spillover efficiency can be defined as

$$\eta_s = \frac{P_c}{P_T} = 1 - \frac{P_s}{P_T}.$$

(6.76)

By means of Eqs. 6.73 and 6.75, the spillover efficiency is

$$\eta_s = \frac{\int_0^{2\pi} d\xi \int_0^{\psi_c} |\mathbf{E_f}|^2 \rho \sin \psi \, d\psi}{\int_0^{2\pi} d\xi \int_0^{\pi} |\mathbf{E_f}|^2 \rho \sin \psi \, d\psi}.$$

(6.77)

Ideally, η_s should be close to 1 although typically it is $0.6 < \eta_s < 0.95$ depending on the feed taper.

Let the power density in the far-field region of the reflector be

$$P_T = \frac{1}{2\eta_o} |\mathbf{E}|^2.$$

(6.78)

By means of Eq. 3.48, the gain function becomes

$$G(\theta, \phi) = \frac{4\pi r^2 P_r}{P_T}$$

$$= \eta_s \frac{4\pi r^2 P_r}{P_c}.$$

(6.79)

Therefore, the maximum gain is

$$G_{max} = \eta_a \eta_s \left(\frac{\pi D}{\lambda}\right)^2 = \eta_a \eta_s G_0.$$

(6.80)

Equation 6.80 is the extension of Eq. 6.71 to account for spillover.

For a paraboloidal reflector and a feed with an axisymmetric pattern $A(\psi)$, the power density on boresight is found from Eqs. 6.19 and 6.78, to be given by

$$P_r = \frac{2}{\eta_o} \left(\frac{kf}{r}\right)^2 \left| \int_0^{\psi_c} A(\psi) \tan \psi \, d\psi \right|^2.$$

(6.81)

To obtain this result from Eq. 6.19 the substitution $t = \rho \cos \psi$ was made. The power collected by the reflector from the feed is

$$P_c = \frac{\pi}{\eta_o} \int_0^{\psi_c} |A(\psi)|^2 \sin \psi \, d\psi.$$

(6.82)

Hence,

$$G_{\max} = \eta_s \left(\frac{\pi D}{\lambda} \right)^2 2\cot^2 \frac{\psi_c}{2} \frac{\left| \int_0^{\psi_c} A(\psi) \tan (\psi/2) d\psi \right|^{12}}{\int_0^{\psi_c} |A(\psi)|^2 \sin\psi d\psi}. \tag{6.83}$$

The aperture efficiency is, therefore,

$$\eta_a = 2\cot^2 \frac{\psi_c}{2} \frac{\left| \int_0^{\psi_c} A(\psi) \tan (\psi/2) d\psi \right|^{12}}{\int_0^{\psi_c} |A(\psi)|^2 \sin\psi d\psi}. \tag{6.84}$$

In a similar fashion, efficiency factors may also be defined to account for feed mismatch, η_f, and reflector and feed conductor losses, η_c. These efficiencies are then incorporated in an overall efficiency factor

$$\eta_T = \eta_a \eta_s \eta_f \eta_c. \tag{6.85}$$

The maximum antenna gain is, therefore, expressed as

$$G_{\max} = \eta_T G_o. \tag{6.86}$$

The theoretical aperture efficiency of a paraboloid with a circular waveguide feed is plotted in Figure 6.21 as a function of feed radius. Curves are given for a typical range of feed radii for a reflector of diameter 100λ for various half-cone angles, ψ_c. The circular waveguide operates in the TE_{11} mode and results are given for two types of aperture terminations: namely, the waveguide terminates at an infinite metallic flange, and secondly the circular waveguide has no flange and the waveguide walls are infinitely thin. Radiation characteristics of the first type were discussed in Section 4.4. For the second type of circular feed, an exact solution is available (Weinstein, 1969) which, in contrast with the E–H model (Eq. 4.37), accurately represents the currents on thin waveguide walls. Efficiency values presented in Figure 6.21 include mismatch loss at the feed aperture and correspond to Eq. 6.85 with $\eta_c = 1$. Blockage, however, is not included, but its effect on the efficiency in most instances will be small. Figure 6.21 indicates there is an optimum feed diameter for maximum efficiency for every half-cone angle. Efficiency increases as ψ_c decreases because the pattern narrows in one plane which lowers spillover loss but the beam is asymmetric and the cross-polarization will be high. In undertaking design with these feeds, there is a compromise between the competing requirements of high gain, sidelobe level, pattern symmetry and cross-polarization.

A third type of circular waveguide feed, which is in common use in satellite earth stations, is illustrated in Figure 1.1b. It has an aperture flange that contains a number of ring-slots or corrugations. The feed pattern function has good axial symmetry, and low cross-polarization can be achieved by optimizing the slot width and depth as well as the slot spacing relative to the aperture (James, 1979). A design strategy that maximizes reflector antenna efficiency results in a feed

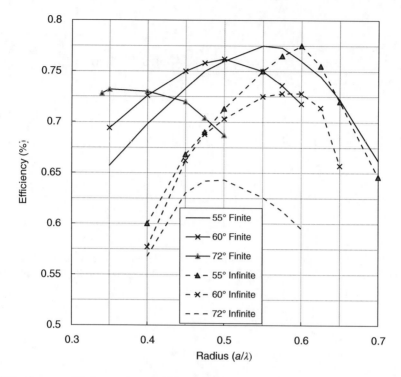

Figure 6.21   Aperture efficiency of paraboloidal reflector of diameter 100 λ versus radius a of a circular waveguide feed. Parameter is ψ_c (degree). Solid: thin-wall waveguide; dashed: --- Infinite flange

consisting of central waveguide and ring-slots that approximately coincide with the Airy rings in the focal plane. As in the case of the circular waveguide feeds in Figure 6.21, there is an optimum reflector half-cone angle, ψ_c, that maximizes reflector efficiency, for a given feed. This is shown in Figure 6.22 for a waveguide of radius 0.37 λ. A maximum efficiency occurs in this case because of the trade-off between energy loss due to spillover and the uniformity of illumination of the reflector. As ψ_c increases, spillover efficiency increases but the reflector is less uniformly illuminated. In the plots shown in Figure 6.22, the flange has a varying number of rings-slots. The lowest feed cross-polarization occurs for three ring-slots with a slot spacing of 0.05 λ, width 0.13 λ and depth 0.26 λ. Curve (a) in Figure 6.22 gives the paraboloid efficiency for this feed. The maximum efficiency reduces slightly with a single ring-slot, as shown by curve (b). In common with the three ring case the slot dimensions were chosen to minimize feed cross-polarization.

The remaining curves in Figure 6.22, labelled (c)–(f), illustrate the importance of conditions at the flange. For a waveguide of radius 0.37 λ, they show that the efficiency falls as the flange width, σ (see Figure 6.22), reduces to zero. This waveguide is obviously not optimum with a small flange. Higher efficiencies are possible with thin wall waveguides, as shown in Figure 6.21, when the waveguide radius is increased to about 0.5–0.6 λ. Coincidently, this size also gives lowest feed cross-polarization.

Maximum efficiency for the three slot-ring case (curve (a) in Figure 6.22) occurs when $\psi_c = 57°$. Principal plane radiation patterns in this case are plotted in Figure 6.23. Blockage is not included and the peak is plotted relative to a uniformly illuminated aperture, giving

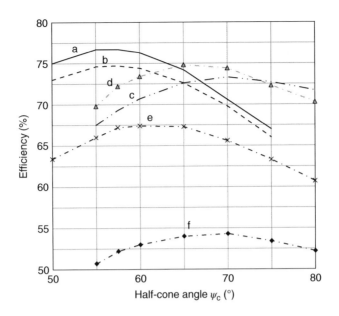

Figure 6.22 Aperture efficiency of a paraboloidal reflector of diameter $100\,\lambda$ with a flanged circular waveguide feed of radius $0.37\,\lambda$. (a) Flange with three ring-slots, spacing $0.05\,\lambda$, width $0.13\,\lambda$ and depth $0.26\,\lambda$. (b) Flange with one ring-slot spacing $0.18\,\lambda$, width $0.13\,\lambda$, and depth $0.26\,\lambda$. (c) Infinite flange ($\sigma = \infty$). (d) Flange-width $\sigma = 1\lambda$. (e) Flange-width $\sigma = \lambda/4$. (f) Thin-wall waveguide ($\sigma \ll \lambda$)

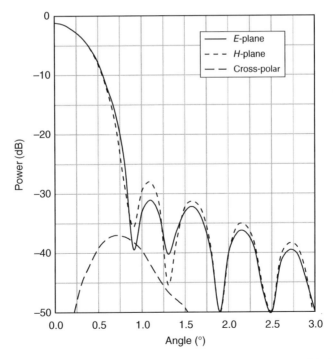

Figure 6.23 Radiation patterns of a paraboloidal reflector and a circular waveguide feed with a flange containing 3 ring-slots (see Figure 6.15). Reflector parameters: $\psi_c = 57°$, $D = 100\,\lambda$. Solid line: *E*-plane; short dash: *H*-plane; long dash: 45° plane cross-polar

an efficiency of −1.16 dB (76.6%). The peak cross-polarization in the 45° plane is −35.8 dB. This is 5.4 dB below the peak cross-polar level of the feed in isolation. This is fairly typical for the peak cross-polarization in the radiation pattern of a paraboloid. Depending on the half-cone angle and the feed cross-polar peak, the paraboloid has lower peak cross-polarization by about 3–5 dB in the 45° plane.

Although efficiencies of 75% and higher are predicted in Figures 6.21 and 6.22, practical efficiencies for the cases considered seldom exceed 60–65%; that is, there is usually a loss of gain of about 1–0.6 dB in implementation. This can occur due to manufacturing errors, surface finish and various ohmic losses. The effect of surface errors on the antenna gain is considered in the next section.

6.6   Reflector Surface Errors

In the manufacture of a reflector, various systematic and random errors occur causing the final surface to depart from the ideal shape. Systematic errors may be minimized by proper attention to detail during the design and construction phases. The latter type of error is determined mainly by the accuracy of manufacture and gives the upper limit of performance when all systematic error is eliminated. Random error modifies the aperture field and, if the surface error is small, this results in a random aperture phase error.

Random error modifies the aperture field as will be demonstrated. If the surface error is relatively small, a random aperture phase error results that can be approximated by

$$\mathbf{E}'_a = \mathbf{E}_a e^{-j\alpha}, \tag{6.87}$$

where α is a small random variable and $\mathbf{E}_a$ is the aperture field with no reflector surface errors. Since $\alpha \ll 1$, let $\exp(-j\alpha) \approx 1 - \alpha^2/2 + j\alpha$. Without any loss of generality, consider a paraboloid with a feed having an axisymmetric pattern. With a small phase error, the aperture efficiency Eq. 6.84 is modified to

$$\eta'_a = \frac{1}{2\pi^2} \cot^2 \frac{\psi}{2} \frac{\left| \int_0^{2\pi} d\xi \int_0^{\psi_c} A(\psi) \tan(\psi/2)(1 - (\alpha^2/2) + j\alpha) d\psi \right|^{12}}{\int_0^{\psi_c} |A(\psi)|^2 \sin\psi \, d\psi} \tag{6.88}$$

$$\approx \eta_a \left(1 - \overline{(\alpha^2)} + (\overline{\alpha})^2 \right),$$

η_a is the aperture efficiency with no surface error, while $\overline{\alpha^2}$ and $\overline{\alpha}$ are the mean-square phase error and mean phase error weighted by the compound aperture illumination function $A(\psi)\tan(\psi/2)$. The weighted mean square phase deviation is

$$\overline{(\alpha^2)} = \overline{(\alpha - \overline{\alpha})^2} = \frac{\left| \int_0^{2\pi} d\xi \int_0^{\psi_c} A(\psi) \tan\frac{\psi}{2}(\alpha - \overline{\alpha})^2 d\psi \right|^{12}}{\int_0^{2\pi} d\xi \int_0^{\psi_c} |A(\psi)|^2 \tan\frac{\psi}{2} d\psi} \tag{6.89}$$

$$= \overline{(\alpha^2)} - (\overline{\alpha})^2.$$

Therefore,

$$\eta_a' \approx \eta_a \left(1 - \overline{(\alpha^2)}\right) \approx \eta_a \exp\left(-\bar{\Delta}^2\right). \qquad (6.90)$$

Other more sophisticated models of surface errors have been developed. One, in particular, that is in wide use is due to Ruze (1966). This model is valid for large errors that are Gaussian distributed. If the errors are completely correlated in small regions of the aperture with diameter much less than D, the aperture efficiency is

$$\eta_a' \approx \eta_a \exp\left(-\overline{(\alpha^2)}\right). \qquad (6.91)$$

$\bar{\delta}^2$ is the mean square error of the Gaussian distribution. For small errors Eq. 6.91 is similar to Eq. 6.90 with the exception that $\bar{\Delta}^2$ is a weighted mean. The two means converge for small errors and uniform illumination.

A useful parameter in practice is the rms surface error defined by

$$\varepsilon = \frac{\lambda}{4\pi} \sqrt{\bar{\delta}^2}. \qquad (6.92)$$

The exponential factor in Eqs. 6.91 and 6.90 indicates that an rms surface error of $\lambda/37$ results in a 0.5 dB loss, while an error of $\lambda/24$ gives a loss of 1.19 dB.

An important implication of Eq. 6.91 for several applications and especially for radio astronomy is that there is a maximum operating frequency for a given reflector surface error beyond which any further increase in frequency causes the gain to decrease. Initially, gain increases as the square of the frequency until reflector errors take over causing the gain to decrease. In the presence of surface errors the maximum gain occurs at the frequency

$$f_{max} = \frac{c}{4\pi\varepsilon}, \qquad (6.93)$$

where a loss of 4.3 dB is incurred. For example, a reflector with rms surface error of 0.5 mm has a maximum operating frequency of 48 GHz.

6.7   Offset-fed Parabolic Reflector

As has been seen in the previous section, blockage by the feed and feed support struts reduces the gain and increases the sidelobe level. This can be overcome by adopting the offset-fed configurations as illustrated in Figure 6.24 for the paraboloid (Rudge & Adatia, 1978). A further advantage of offset parabolic antenna is that the interaction between the feed and reflector is quite small. Offset-fed reflectors are widely used as satellite and radar antennas because the often complicated feed systems can be placed close to bulky feed networks.

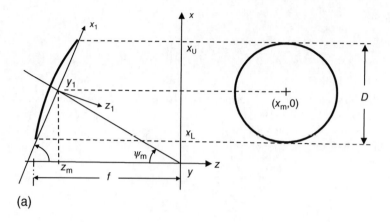

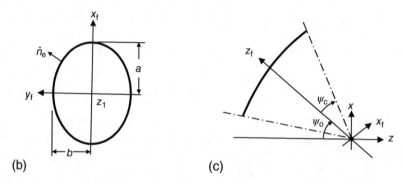

Figure 6.24   The offset paraboloid. (a) Projected aperture. (b) Elliptical rim. (c) Rotated co-ordinate system $\{X_f\}$

The offset parabolic reflector is formed by rotating the feed through an angle ψ_o as shown in Figure 6.24, and illuminating only part of the paraboloid. The feed and reflector are contained within a cone of half-angle ψ_c with the feed at its apex and the reflector rim lying on its surface. The projection of the rim onto the x–y plane (Figure 6.24a) is a circle of diameter

$$D = \frac{4f \sin \psi_c}{\cos \psi_o + \cos \psi_c} \tag{6.94}$$

and centre $(x_m, 0)$, where

$$x_m = \frac{2f \sin \psi_o}{\cos \psi_o + \cos \psi_c}. \tag{6.95}$$

f is the focal length of the original paraboloid and ψ_o is the angle of rotation ('offset angle') of the axes about the focus. The effective focal length of an offset parabolic reflector is defined to be

$$f_{eff} = \frac{2f}{1 + \cos \psi_o},$$ (6.96)

which is the distance from the focus to the vertex in the offset geometry. The inverse of Eqs. 6.94 and 6.95 is sometimes useful. That is,

$$\psi_{o,c} = \tan^{-1}\left(\frac{x_m + D/2}{2f}\right) \pm \tan^{-1}\left(\frac{x_m - D/2}{2f}\right).$$

Feed blockage is avoided if the largest feed extremity in the positive x-direction is less than the clearance distance, x_L, between the paraboloid's rim and the z-axis, where

$$x_L = 2f \tan\left(\frac{\psi_o - \psi_c}{2}\right).$$ (6.97)

The reflector rim lies on the projected cone and is an ellipse, Figure 6.24b having major and minor axes of length

$$a = \frac{D}{2 \sin \gamma}; \quad b = \frac{D}{2},$$

where $\gamma = \tan^{-1}(2f/x_m)$. The centre of the ellipse is $(x_m, 0, z_m)$ wherein

$$x_m = f\left[\frac{\sin^2 \psi_o + \sin^2 \psi_c}{(\cos \psi_o + \cos \psi_c)^2} - 1\right].$$ (6.98)

In terms of the spherical polar co-ordinates (ρ, ψ, ξ) in the rotated co-ordinate system $\{X_f\}$ relative to the focus, a point P on the paraboloid has rectangular co-ordinates

$$x_{fp} = \rho \sin \psi \cos \xi; \quad y_{fp} = \rho \sin \psi \sin \xi; \quad z_{fp} = \rho \cos \psi,$$ (6.99)

where

$$\rho = \frac{2f}{1 - \cos \xi \sin \psi \sin \psi_o + \cos \psi \cos \psi_o}.$$ (6.100)

The global co-ordinates of this point are

$$x_p = x_{fp} \cos \psi_o + z_{fp} \sin \psi_o;$$

$$y_p = -y_{fp};$$ (6.101)

$$z_p = x_{fp} \sin \psi_o - z_{fp} \cos \psi_o.$$

The geometric optics approximation to the aperture field can be obtained by the method described in Section 6.2.1 for the symmetrical paraboloid. The main geometrical difference is feed rotation although this does not alter the paraboloid's basic properties; namely, the distance from the focus to the aperture plane at $z = 0$ is $2f$, and the z-component of the aperture field is zero as, in general, because the output wave is planar. These properties allow the aperture field expressions to be simplified. For the incident field from the feed expressed by Eq. 6.4, the components of the electric field in the projected aperture are from Eqs. 6.7

$$E_{ax} = g_0 \left[c_1 F_\psi(\psi, \xi) + d_1 \Gamma_\xi(\psi, \xi) \right] \qquad (6.102a)$$

$$E_{ay} = g_0 \left[-d_1 F_\psi(\psi, \xi) + c_1 F_\xi(\psi, \xi) \right], \qquad (6.102b)$$

where

$$c_1 = \sin \psi \sin \psi_0 + \cos \xi (1 + \cos \psi \cos \psi_0),$$

$$d_1 = \sin \xi (\cos \psi + \cos \psi_0), \text{ and}$$

$$g_0 = \exp(-jk2f)/2f.$$

The far-fields can be obtained from Eqs. 6.102 by applying Eq. 3.26 to the aperture plane shown in Figure 6.24a. To do this, define polar co-ordinates (t, ζ) centred on $(x_m, 0)$ such that $x = x_m + t \cos \zeta$, $y = t \sin \zeta$ and $z = (x^2 + y^2 - 4f^2)/4f$ then carry out the integration over the projected circle, Figure 6.24. To obtain the angle co-ordinates relative to the rotated feed co-ordinates, use the standard co-ordinate relations to express $\psi = \sin^{-1}(z_f/\rho)$ and $\xi = \tan^{-1}(y_f/x_f)$, where $\rho = \sqrt{x_f^2 + y_f^2 + z_f^2}$. For a feed with an axisymmetric pattern (i.e. $F_\psi(\psi, \xi) = A(\psi) \cos \xi$ and $F_\xi(\psi, \xi) = -A(\psi) \sin \xi$), the aperture fields will be symmetric about the vertical (x) axis. Study of these aperture fields shows that while they may be symmetric about the x-axis, by contrast with the symmetrical paraboloid, there is cross-polarization. This is maximum in the plane of asymmetry (i.e. $\xi = 90°$, $270°$), and it occurs because the feed rotation causes the illumination of the reflector to be no longer linearly polarized. For a general feed, the plane of maximum cross-polarization normally occurs between the 90° and 45° planes, depending on the level of feed cross-polarization. Special feeds have been designed to cancel the cross-polarization in the offset paraboloid (Rudge & Adatia, 1978). The radiation field can be obtained in the usual way by substituting Eqs. 6.102 into Eqs. 6.10. As in the case of the symmetrical paraboloid, the resulting integral transform, **N**, can be evaluated by numerical integration, or by means of the FFT (Brigham, 1974).

The method of physical optics can also usefully applied to the offset reflector as is briefly outlined below. A feed antenna is assumed to radiate an electric field given by

$$\mathbf{E}_f(\psi, \xi) = \left(\hat{\psi} E_{f\psi}(\psi, \xi) - \hat{\xi} E_{f\xi}(\psi, \xi) \right) \exp[-jk \, \rho(\psi, \xi)]/\rho(\psi, \xi),$$

where the co-ordinates (ρ, ψ, ξ) are in the co-ordinate system $\{X_f\}$. The reflector rim subtends an angle ψ_c to the axis z_f as shown in Figure 6.24c. The feed illumination induces a current on the surface given by

$$\mathbf{J}_s(\psi,\xi) = 2\hat{n}(\psi,\xi) \times \frac{1}{\eta_0}\left[\hat{\rho} \times \mathbf{E}_f(\psi,\xi)\right]$$

$$= 2\hat{n}(\psi,\xi) \times \frac{\exp(-jk\rho)}{\eta_0\rho}\left[(\hat{x}\cos\psi_0 + \hat{z}\sin\psi_0)\left(E_{f\xi}(\psi,\xi)\cos\psi\cos\xi - E_{f\psi}(\psi,\xi)\sin\xi\right)\right.$$

$$\left. -\hat{y}\left(E_{f\xi}(\psi,\xi)\cos\psi\sin\xi + E_{f\psi}(\psi,\xi)\cos\xi\right) - (\hat{x}\sin\psi_0 - \hat{z}\cos\psi_0)E_{f\xi}(\psi,\xi)\sin\psi\right]$$

where $\hat{n}(\psi,\xi) = -\left(1/(2\sqrt{\rho(\psi,\xi)}f)\right)(\hat{x}x_p + \hat{y}y_p) + \sqrt{(f/(\rho(\psi,\xi)))}\hat{z}$ in which x_p, y_p are given by Eqs. 6.101. That is,

$$\mathbf{J}_s(\psi,\xi) = 2\hat{n}(\psi,\xi) \times \frac{\exp(-jk\rho)}{\eta_0\rho}\left[\hat{x}\left(\cos\psi_0\left(E_{f\xi}\cos\psi\cos\xi - E_{f\psi}\sin\psi_0\right)\right.\right.$$

$$- \sin\psi_0 E_{f\xi}\sin\psi) - \hat{y}\left(E_{f\xi}\cos\psi\sin\xi + E_{f\psi}\cos\xi\right) \tag{6.103}$$

$$+ \hat{z}\left(\sin\psi_0\left(E_{f\xi}\cos\psi\cos\xi - E_{f\psi}\sin\psi_0\right) + \hat{z}\cos\psi_0 E_{f\xi}\sin\psi\right].$$

Equation 6.103 is then substituted into Eq. 6.32 and the integrals evaluated over the feed angular co-ordinates as follows:

$$\mathbf{F}(\theta,\phi) = \int_0^{2\pi}\int_0^{\psi_c} \mathbf{J}_s(\psi,\xi)\exp[jk\Phi(\psi,\xi)]\rho^2(\psi,\xi)\Gamma(\psi,\xi)\sin\psi\,d\psi\,d\xi,$$

where

$$\Gamma(\psi,\xi) = \sqrt{1 + \left[\rho\left(\frac{\cos\xi\cos\psi\sin\psi_0 + \sin\psi\cos\psi_0}{2f}\right)\right]^2},$$

and

$$\Phi(\psi,\xi) = \left(x_{fp}\cos\psi_0 + z_{fp}\sin\psi_0\right)\sin\theta\cos\phi - y_{fp}\sin\theta\sin\phi + \left(x_{fp}\sin\psi_0 - z_{fp}\cos\psi_0\right)\cos\theta.$$

The field components are then obtained from Eq. 6.31.

Due to the asymmetry, the radiation pattern of the offset paraboloid is often assessed from two-dimensional plots, and Figure 6.25 illustrates typical examples. The figure shows contour plots of patterns for an offset reflector obtained from the method of physical optics. The reflector is defined by $\psi_0 = 40°$, $\psi_c = 30°$ and $D = 100\,\lambda$, and results are given with two different feeds. The contours are in dB, in increments above the −60 dBi level. The reference field polarization is parallel to the x (vertical) direction', that is, $\phi_o = 0$ in Eq. 3.45. In Figure 6.25a, the radiation patterns given are for an axisymmetric feed, which has a Gaussian pattern function with an edge illumination of −10 dB (see Eq. 6.23). In Figure 6.25b the patterns are for an offset reflector with an asymmetric feed that has different E- and H-plane patterns both of which are Gaussian functions, giving edge illuminations of −10 dB and −16 dB, in the two planes, respectively. For the axisymmetric feed, in Figure 6.25a, the gain is 48.92 dBi (efficiency 76.0%), and for Figure 6.25b, the peak gain reduces to 48.74 dBi (efficiency 75.8%) when the principal

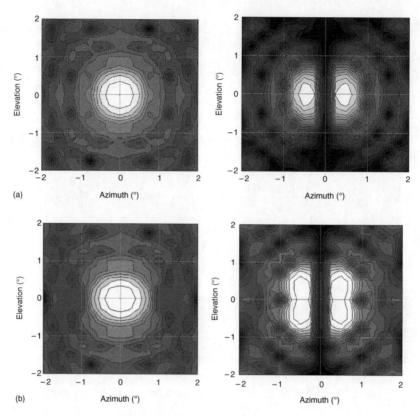

Figure 6.25 Radiation patterns of an offset reflector antenna with $\psi_o = 40°$, $\psi_o = 30°$ and $D = 100\lambda$. Contours are in dBi, at increments above −60 dB. (a) Axisymmetric feed pattern, −10 dB edge illumination. Cross-polar peak −25.1 dB; (b) Asymmetric edge illumination: −10 dB in E-plane and −16 dB in H-plane. Cross-polar peak at −25.8 dB

plane patterns of the feed are different. In the first case the main beam is almost circular with a HPBW of 0.65°. The narrower H-plane feed pattern in the second case broadens the reflector's H-plane pattern giving a HPBW in this plane of 0.73°, while maintaining virtually the same E-plane HPBW. A useful estimate of the HPBW of a pattern cut of an offset parabolic reflector is

$$\text{HPBW(deg.)} = (0.9E_{dB} + 58)\left(\frac{\lambda}{D}\right), \tag{6.104}$$

where E_{dB} is the edge illumination in dB in the same plane.

When the principal plane feed patterns are different, the peak cross-polarization occurs near the 45° plane and usually is at a higher level than in the axisymmetric case. However, the maximum cross-polar level in the plane of asymmetry (horizontal) is approximately the same for both types of feed.

The focal region fields of an offset paraboloid can be obtained by the method described in Section 6.3. For on-axis plane wave incidence with the electric field polarized parallel to the plane of symmetry (z–x plane), the focal field components in the $\{X_f\}$ plane are given by (Bem, 1969)

$$E_{fx} = A_0(U) + 2j \tan \frac{\psi_o}{2} B_1(U) \cos \zeta' \tag{6.105a}$$

$$E_{fy} = -2j \tan \frac{\psi_o}{2} B_1(U) \sin \zeta' \tag{6.105b}$$

$$E_{fz} = -2j B_1(U) \cos \zeta', \tag{6.105c}$$

where

$$A_0(U) = 2\kappa' \frac{J_1(U)}{U} \tag{6.106a}$$

$$B_1(U) = \kappa' \psi_c \frac{J_2(U)}{U} \tag{6.106b}$$

in which $U = kt' \sin \psi_c$, and $\kappa' \approx (jkD^2 E)/8f_{\text{eff}} e^{-jk(f_{\text{eff}} + t' \sin \zeta')}$. t' and ζ' are polar co-ordinates in the $\{X_f\}$ plane (located at $z_f = 0$ in Figure 6.24) of the offset parabola. The solution for the field polarized parallel to the y-axis is obtained from Eqs. 6.105 by interchanging the co-ordinates x and y, changing the sign of A_0 in Eq. 6.105a and replacing ζ' with $\zeta' + \pi/2$. It is observed that Eqs. 6.105 reduce to Eqs. 6.57 when the offset angle is zero ($\psi_o = 0$). Also, as the offset angle is increased, the quadrature term of the principal field component increases, as does the cross-polar field component.

As an example, consider the focal region fields of an offset parabola with $D = 50\lambda$, $\psi_o = 40°$ and $\psi_c = 30°$. The amplitude contour plots are given Figure 6.26 for a wave incident parallel to the z-axis and polarized in the x–z plane. The co-polar component has highly circular Airy rings, which occur in long focal length reflectors, while the cross-polar component has lobes that peak in the plane of offset.

In the design of offset parabolic reflectors, the concept of the 'effective paraboloid' is helpful for establishing initial design information prior to more detailed analysis. The offset reflector is assumed identical to a symmetrical paraboloid of the same diameter that is given by Eq. 6.94 with an effective focal length f_{eff} given by Eq. 6.96. For the effective paraboloid, the focal length to diameter ratio is given by f_{eff}/D. As an example, consider the design of an offset paraboloid to produce a beam in the far-field at the elevation and azimuth angles θ_b and ϕ_b. Often required is an estimate of where to place the feed to transmit to or receive from this direction. Assume θ_b is close to reflector boresight and consider an incoming ray from (θ_b, ϕ_b). In the effective paraboloid the reflected ray makes an angle β with respect to the axis, where

$$\beta \approx \theta_b \left(1 + \frac{1}{32(f_{\text{eff}}/D)^2} \right). \tag{6.107}$$

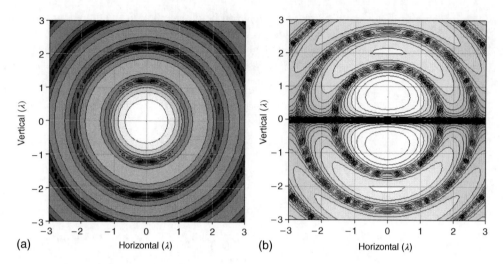

Figure 6.26 Amplitude of the focal region fields of an offset parabolic reflector with $D = 50\lambda$, $\psi_o = 40°$ and $\psi_c = 30°$. The incident field is polarized parallel to the x-axis. (a) Principal component $|E_x|$. (b) Cross-polar component $|E_y|$

Equation 6.107 is a formula obtained from considerations of a paraboloidal reflector. An estimate for the location of a single feed in the focal plane is

$$x_{oeff} = f_{eff} \sin \beta \cos \phi_b; \quad y_{oeff} = f_{eff} \sin \beta \sin \phi_b. \tag{6.108}$$

6.8 Cassegrain Antenna

6.8.1 Classical Cassegrain

The Cassegrain antenna (see Figure 6.1b) had its origins in an optical telescope that was invented during the seventeenth century. In its classical form it consists of a paraboloidal main reflector and a smaller hyperboloidal subreflector, the geometry of which is shown in Figure 6.27. Some geometrical relationships for the hyperboloid are summarized in Table 6.2. The Cassegrain geometry is shown in Figure 6.28.

A hyperboloid with an eccentricity e has both real and virtual foci, labelled F and F' in Figure 6.27, and it is symmetric about the axis FF'. The profile from either foci is defined in the first row of Table 6.2. In the Cassegrain, the virtual focus F' of the hyperboloid is placed coincident with the paraboloid's focus, while the feed is placed at the real focus F. If a source of spherical waves is located at the focus, the hyperboloid reflects the wave to the paraboloid in such a way that the wave appears to emanate from a source located at F'.

The equivalent parabola approximation is also useful for the analysis of the Cassegrain. This is possible because of the properties of the hyperboloid and is illustrated in Figure 6.28. As in the case of the offset paraboloid, the feed remains at the focus F, but both reflectors are replaced by another paraboloid with a longer focal length, f'. The equivalent parabola has the same diameter as the main reflector and the feed half-cone angle, θ_c, at the subreflector rim is the

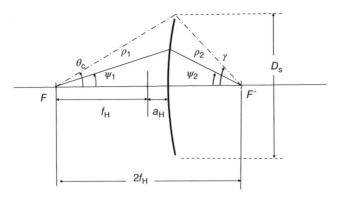

Figure 6.27 Geometry of a hyperboloid reflector

Table 6.2 Geometrical relationships of hyperboloid with reference to Figure 6.27

Profile	$\rho_1 = \dfrac{-e\beta}{1 - e\cos\psi_1}$	$\rho_2 = \dfrac{e\beta}{1 + e\cos\psi_2}$
Intermediate	$\tan\dfrac{\psi_1}{2} = M\tan\dfrac{\psi_2}{2}$	$\alpha = \dfrac{2\beta}{D_s}$
Cone angles	$\cot\theta_c + \cot\gamma = \dfrac{f_H}{D_s}$	$\theta_c = \tan^{-1}\left(-\dfrac{1}{\alpha}\right) + \cos^{-1}\left(\dfrac{\alpha}{e\sqrt{\alpha+1}}\right)$
Eccentricity	$M = \dfrac{e+1}{e-1}, e = \dfrac{f_H}{a_H}$	$\beta = f_H\left(1 - \dfrac{1}{e^2}\right)f_H + a_H = \dfrac{e\beta}{e-1}$

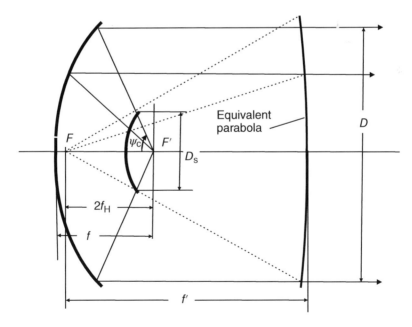

Figure 6.28 The Cassegrain and the equivalent parabola

half-cone angle at the rim of the equivalent paraboloid. This is used to show that the effective focal length of the equivalent paraboloid for the Cassegrain is

$$f' = Mf. \tag{6.109}$$

where M is the magnification factor which is defined in Table 6.2 and for a Cassegrain it is typically greater than 1 (usually 2.5–5). Cassegrain antennas have, therefore, similar properties to long focal length paraboloids. The equivalent parabola can be used to estimate the Cassegrain antenna efficiency and the sidelobe levels, the focal region fields (from Eq. 6.57) and the effect of scanning the beam off-axis through Eq. 6.107 (in which f' replaces f_{eff}). Although a useful design aid, the equivalent paraboloid is not a substitute for more detailed analysis, which is required to achieve best performance.

The main advantages of a Cassegrain over a single reflector antenna are that the feed can be situated close to the main reflector and to the receiver. In earth station antennas, the feed spillover is directed towards the cold sky. A disadvantage of the classical symmetrical Cassegrain (Figure 6.1b) is the decrease in antenna efficiency due to blockage and diffraction by the subreflector and the subreflector supports. However, if the reflector profiles are shaped, the impact of subreflector blockage can be reduced. Reflector shaping applied to both surfaces allows the aperture illumination to be selected to enhance performance. This makes the symmetrical shaped Cassegrain superior in every respect to front-fed reflectors and is a major reason for their widespread use in large earth stations. Blockage can be reduced by tailoring the feed illumination or avoided entirely with the offset Cassegrain configuration that is illustrated in Figures 6.1d and 6.29. Feed spillover at the subreflector is an important contributor to the far-out sidelobes of all types of Cassegrain antennas and, to minimize this, the feed sidelobes should be small. Typically, subreflector edge illumination needs to be −16 to −20 dB in order to keep the spillover contributions to the sidelobes of the Cassegrain at an acceptably low level.

The properties of Cassegrain antennas can be analysed by means of the techniques described in Section 6.2.1. Geometric optics (GO) ray tracing can be applied to both reflectors to find the aperture field. This approach is not accurate for small subreflectors (diameter less than about 30 λ) because of diffraction at the rim. GO is sufficiently accurate for most purposes when the diameter of the reflectors is several hundred wavelengths. Considerable improvement in accuracy results for smaller reflectors when diffraction is included through methods such as the GTD (James, 1986). Accurate results are also possible by applying the physical optics approximation at both reflectors or combining the techniques of GTD and physical optics.

6.8.2 Offset Cassegrain Antenna

Blockage in the classical symmetrical Cassegrain impacts the performance, particularly in reduced gain and increased sidelobes, although cross-polarization can remain low due to axial symmetry. By adopting the offset Cassegrain configuration shown in Figure 6.29, both gain and sidelobe performance can be improved. Cross-polarization can increase due to the offset geometry but, through the selection of the feed and subreflector rotation angles, it can be significantly reduced below the level of a single offset.

A GO analysis of the offset Cassegrain configuration shown in Figure 6.29 provides an approximate description of the aperture fields. For a feed with an axisymmetric pattern with

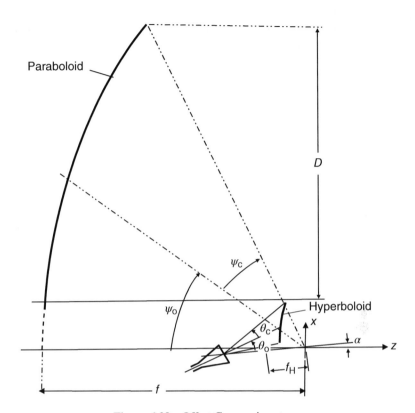

Figure 6.29 Offset Cassegrain antenna

the electric field polarized parallel to the plane of symmetry (x–z plane), the aperture fields are given by

$$E_{ax}(\psi',\zeta') = g_o/D(\psi',\zeta')\left[A\sin\psi'\cos\zeta' + B\left(\sin^2\zeta' + \cos\psi'\cos^2\zeta'\right) + C(1+\cos\psi')\right]$$
$$(6.110a)$$

$$E_{ay}(\psi',\zeta') = -g_o/D(\psi',\zeta')\sin\zeta'[A\sin\psi' + B(\cos\psi'-1)\cos\zeta'],\qquad (6.110b)$$

where

$$g_o = F(\psi') - \exp\frac{(-2jkf + a_H)}{2fl},$$
$$D(\psi',\zeta') = B\cos\psi' + A\sin\psi'\cos\zeta' + B\cos\psi' + C(1+\cos\psi'),$$
$$A = L\sin\alpha\cos\theta_o - \sin\theta_o(K+\cos\alpha)$$
$$B = L\sin\alpha\sin\theta_o + \cos\theta_o(K+\cos\alpha) - C,$$

$C = 1 + K\cos\alpha$, $K = (1-M^2)/(1+M^2)$, $L = 2M/(1+M^2)$, where M and other hyperboloid subreflector parameters are defined in Table 6.2. Notice that $L^2 + K^2 = 1$. As well (ψ', ξ') are the elevation and azimuth angles of a spherical co-ordinate system that is located at the feed phase centre and $F(\psi')$ is the feed pattern. Eq. 6.110 are in identical form as for a single offset reflector (Eq. 6.102) except that in the latter the following replacements are required: $A = \sin\theta_o$, $B = 1 - \cos\theta_o$, $C = -1$ and $g_o = F(\psi')\exp(-2jkf)/2f$. The aperture fields of a symmetrical Cassegrain are obtained by setting $\phi_o = 0$, $\theta_o = 0$ and $\alpha = 0$.

Unlike the single-offset paraboloid, GO predicts the offset Cassegrain can have zero cross-polarization. From an inspection of Eq. 6.110 this occurs when the feed and hyperboloid offset angles satisfy the condition

$$\tan\frac{\theta_o}{2} = M\tan\frac{\alpha}{2}. \tag{6.111}$$

When Eq. 6.111 is satisfied, the aperture field is axisymmetric also as illustrated in Figure 6.30 for an offset Cassegrain with $D = 150\lambda$, $f = 166.79\lambda$, $\psi_o = 59.553°$, $\psi_c = 20.511°$, $\theta_o = 28°$, $\theta_c = 11°$, $e = 2.4575$, $f_H = 41.054\lambda$ and $\alpha = 12.045°$. When the feed pattern is asymmetric,

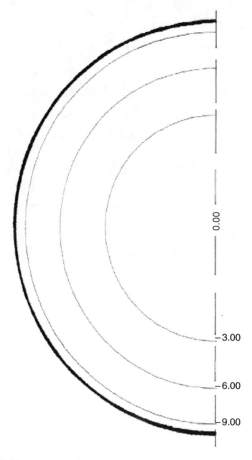

Figure 6.30 Principal field component in aperture plane of offset Cassegrain antenna with $\theta_0 = 28°$, $\theta_c = 11°$, $f = 166.79\lambda$, $e = 2.475$, $f_H = 41.054\lambda$, $\alpha = 12.054°$ obtained by geometric optics. Feed is a Huygen's source having a Gaussian feed pattern with a 10 dB beamwidth of 22°. The electric field is polarized parallel to plane of symmetry

(a)   (b)

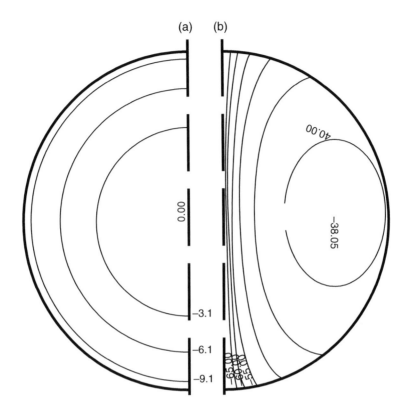

Figure 6.31   Principal (a) and cross-polar (b) field components in aperture plane of an offset Cassegrain antenna with $\theta_0 = 28°$, $\theta_c = 11°$, $f = 166.79\lambda$, $e = 2.475$, $f_H = 41.054\lambda$, $\alpha = 23.054°$ obtained by geometric optics. The feed has 10 dB beamwidths in E- and H-planes of 22° and 24°, respectively

Eq. 6.111 corresponds to the condition for zero cross-polarization in the plane of asymmetry (the y-axis). Furthermore, for this case the principal component of the aperture field is elliptical and maximum cross-polarization occurs between the 45° and 90° planes. Figure 6.31 shows the aperture fields of an antenna that satisfies Eq. 6.111, but where the feed pattern is asymmetric. The geometry in this case is identical to the previous one except that $\alpha = 23.054°$. The feed pattern is a Gaussian function which gives a subreflector edge illumination of −2.5 and −2.1 dB, respectively, in the E- and H-planes. As could be expected, the co-polar contours are almost uniform and peak cross-polarization occurs in 90° and 270° planes.

The radiation patterns can be obtained by substituting the aperture fields into Eqs. 3.20. A sequence of co-ordinate changes are required from the aperture plane through to the local co-ordinate system of the feed. An example of the principal plane patterns obtained in the plane of asymmetry is shown in Figure 6.32 for an offset Cassegrain with a geometry given by $D = 84.9\lambda$, $\psi_0 = 38.5°$, $\psi_c = 25.7°$, $e = 2.8$, $\theta_0 = 22.8°$, $\theta_c = 15.2°$, $f_H = 12.028\lambda$ and $\alpha = 6.0°$. Note that the co-polar pattern in the 90° plane is the H-plane pattern. This antenna is pictured in Figure 1.1j. The beamwidth and first sidelobe levels agree approximately with measured results (Bird & Boomars, 1980) although the measured cross-polarization is higher than shown in Fig. 6.32. The differences are most likely due to incorrect alignment of the subreflector and feed.

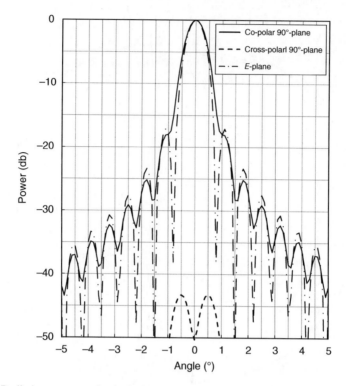

Figure 6.32 Radiation patterns of a dual-offset reflector antenna computed from geometric optics. $D = 84.9\lambda$, $\psi_o = 38.5°$, $\psi_c = 25.7°$, $e = 2.8$, $\theta_o = 22.8°$, $\theta_c = 15.2°$, $f_H = 12.028\lambda$ and $\alpha = 6.0°$. Feed has Gaussian pattern with -3 dB subreflector edge illumination

The GO formulation described above does not take into account diffraction from the subreflector, which, in particular, increases the cross-polarization. For example, in the plane of asymmetry (i.e. $\phi = \pm 90°$), a rigorous analysis of the antenna described in Figure 6.30 by means of physical optics at the subreflector and GTD at the main reflector (Bird & Boomars, 1980) has shown that the cross-polarization varies as demonstrated in Figure 6.33 (Bird, 1981). The subreflector diameter in each case is given approximately by $D_S \approx 2(f_H + a_H)\tan\theta_c$. For the curves given in Figure 6.33, the results correspond to subreflectors with major axial lengths of 40λ, 20λ and 10λ. These result in subreflector diameters of about $D_S \approx 9.7\lambda$, $D_S \approx 5.8\lambda$ and $D_S \approx 3.8\lambda$, respectively.

6.9   Shaped Reflectors

While simple surfaces such as the paraboloid, cylinder and spheroid are in common use for reflectors, shaped reflectors can often provide significantly improved performance. In this section, the shaping of the reflector surface is described in order to achieve a prescribed radiation pattern. Three techniques are outlined. The first and oldest technique is based on geometrical optics and was initially used in the 1940s (Silver, 1946). It creates a set of coupled differential equations, which need to be solved to determine the reflector profile. The second approach is

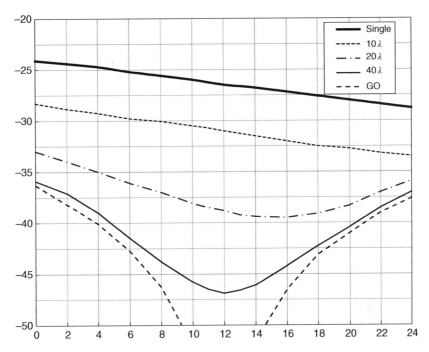

Figure 6.33   Maximum cross-polar level relative to peak co-polar in the plane of asymmetry of a dual offset reflector antenna versus feed offset angle α in degrees. The parameter is the hyperboloid major axial length in wavelengths. Also shown are the corresponding values for a single offset reflector and the geometric optics (GO) result (Bird, 1981)

more recent and is based on computer numerical optimization techniques. This latter approach is potentially more useful because it can include the effects of diffraction as well as limitations due to the reflector structure or the feed antenna. More accurate methods can also be employed such as physical optics or even some numerical methods, such as the method of moments, as long as their implementation is fast enough for use with a standard optimizer. A third shaping technique that is briefly outlined employs an algorithm specially developed for reflector shaping, which is fast and is based on the method of successive projections.

6.9.1   Reflector Synthesis by Geometric Optics

A geometric optics technique is presented for the shaping of a single reflector to achieve maximum gain over an angular range. This approach is often referred to as reflector synthesis in the literature. It can be extended to two (Galindo, 1964) or more reflectors by continued application of methods of geometric optics.

Consider the geometry shown in Figure 6.34. The z_1-axis is taken as the axis of rotation of the reflector whose profile is to be determined and F is its focus. The reflector has a maximum dimension in the vertical direction given by x_{1max}. Let $\rho_1(\theta_1)$ be the radial distance from

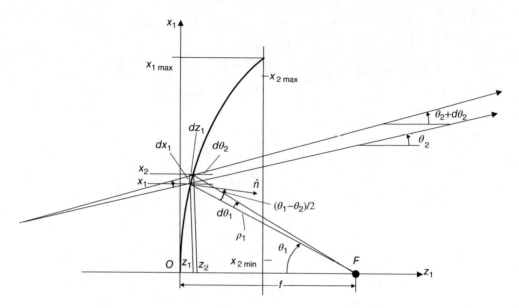

Figure 6.34   Ray paths at a shaped axisymmetric reflector fed from the focus F

F to a point on the reflector at an elevation angle of θ_1. A ray from a source at the focus undergoes reflection and exits at an angle θ_2 to the z_1-axis. The second law of reflection requires the angle between the normal to the reflector surface and the incident and reflected rays to be $(\theta_2 - \theta_1)/2$ as illustrated in Figure 6.23. An incremental application of Snell's law results in the differential equation

$$\frac{dx_1}{dz_1} = \frac{1}{\rho_1}\frac{\partial \rho_1}{\partial \theta_1} = \tan\left(\frac{\theta_1 - \theta_2}{2}\right). \tag{6.112}$$

Integrating both sides of Eq. 6.112 with respect to θ_1 gives

$$\int_0^{\theta_1} \frac{1}{\rho_1}\frac{\partial \rho_1}{\partial \theta_1} d\theta_1 = \int_0^{\theta_1} \tan\left(\frac{\theta_1 - \theta_2}{2}\right) d\theta_1$$

resulting in

$$\ln\left[\frac{\rho_1(\theta_1)}{f}\right] = \int_0^{\theta_1} \tan\left(\frac{\theta_1 - \theta_2}{2}\right) d\theta_1. \tag{6.113}$$

The solution to Eq. 6.113 results in the profile of the reflector. To achieve this, a relationship between θ_1 and θ_2 is required.

As a simple example, suppose $\theta_2 = \theta_b = $ constant, which means the exit ray is at a constant angle to the z_1-axis. Substituting this relation into Eq. 6.113 results in $\ln[(\rho_1(\theta_1))/f] = 2\ln|\sec[(\theta_1 - \theta_b)/2]| - 2\ln|\sec(\theta_b/2)|$, which gives the profile as $\rho_1(\theta_1) = f(1 + \cos\theta_b)/[1 + \cos(\theta_1 - \theta_b)]$. This shape gives a beam at an angle θ_b for a feed located at the focus F. When the exit ray is parallel to the z-axis, that is, $\theta_b = 0$, the profile simplifies to Eq. 6.1, namely, $\rho_1(\theta_1) = 2f/(1 + \cos\theta_1)$.

In general, the angles θ_1 and θ_2 should be chosen to enable the energy incident on the reflector to disperse into the secondary radiation pattern. This can be achieved by ensuring conservation of energy from the feed into the aperture field. This energy is contained in the angular wedges $d\theta_1$ and $d\theta_2$ shown in Figure 6.34. Therefore, if $P(\theta_1)\,d\theta_1$ is the energy incident from a primary feed located at F and $I(\theta_2)\,d\theta_2$ is the power emanating from the reflector where $I(\theta)$ is the power density per unit solid angle in this output field, conservation of energy requires that

$$P(\theta_1)\sin\theta_1\,d\theta_1 = I(\theta_2)\sin\theta_2 d\theta_2.$$

Integrating this requirement over the angles subtended in the input and output leads to

$$\int_0^{\theta_1} P(\theta_1)\sin\theta_1\,d\theta_1 = \int_{\theta_{2\min}}^{\theta_2} I(\theta_2)\sin\theta_2 d\theta_2 = K\int_{x_{2\min}}^{x_1} I(x_2')x_2'dx_2', \tag{6.114}$$

where $\theta_2 = \tan^{-1}(x_2 - x_1)/(z_2 - z_1)$ and K is a constant. The angles $\theta_{2\min}$ and $\theta_{2\max}$ are the minimum and maximum angles over which $I(\theta_2)$ has been specified. $\theta_{2\max}$ can be chosen from zero to several beamwidths. The constant K is found by evaluating Eq. 6.114 at the upper limits of the angular range, thus $\theta_1 = \theta_{1\max}$ and $\theta_2 = \theta_{2\max}$. This results in

$$K = \frac{\displaystyle\int_0^{\theta_{1\max}} P(\theta_1)\sin\theta_1\,d\theta_1}{\displaystyle\int_{\theta_{2\min}}^{\theta_{2\max}} I(\theta_2)\sin\theta_2\,d\theta_2}. \tag{6.115}$$

Equation 6.115 is then substituted into Eq. 6.114 to give

$$\frac{\displaystyle\int_0^{\theta_1} P(\theta_1)\sin\theta_1\,d\theta_1}{\displaystyle\int_0^{\theta_{1\max}} P(\theta_1)\sin\theta_1\,d\theta_1} = \frac{\displaystyle\int_{\theta_{2\min}}^{\theta_2} I(\theta_2)\sin\theta_2 d\theta_2}{\displaystyle\int_{\theta_{2\min}}^{\theta_{2\max}} I(\theta_2)\sin\theta_2 d\theta_2}. \tag{6.116}$$

Equation 6.116 is the relationship that is required between θ_2 and θ_1 which can be used in conjunction with Eq. 6.113 to determine $\rho_1(\theta_1)$. In principle, either $P(\theta_1)$ or $I(\theta_2)$ could be specified by theoretical or measured data from which a relationship can be obtained.

A feed pattern that is quite useful for many practical feeds is $P(\theta_1) = \cos^n\theta_1$ where n is the power of the cosine-shaped radiation pattern. For this pattern function Eq. 6.116 gives

$$\frac{\displaystyle\int_0^{\theta_1} \cos^{n+1}\theta_1 \sin\theta_1\,d\theta_1}{\displaystyle\int_0^{\theta_{1\max}} \cos^{n+1}\theta_1 \sin\theta_1\,d\theta_1} = \frac{\displaystyle\int_{\theta_{2\min}}^{\theta_2} I(\theta_2)\sin\theta_2 d\theta_2}{\displaystyle\int_{\theta_{2\min}}^{\theta_{2\max}} I(\theta_2)\sin\theta_2 d\theta_2}.$$

Evaluating the integrals on the left-side gives

$$\frac{1-\cos^{n+1}\theta_1}{1-\cos^{n+1}\theta_{1\,max}}=\frac{\displaystyle\int_{\theta_{2\,min}}^{\theta_2}I(\theta_2)\sin\theta_2d\theta_2}{\displaystyle\int_{\theta_{2\,min}}^{\theta_{2\,max}}I(\theta_2)\sin\theta_2d\theta_2}. \tag{6.117}$$

In general, Eq. 6.117 results in a transcendental equation of the form $F(\theta_2(\theta_1))-0$ where F is an arbitrary function, which can be solved iteratively for θ_2. As an example assume $I(\theta_2)=1$ for the angular range $\theta_{2\,min}\le\theta_2\le\theta_{2\,max}$. With these specifications in Eq. 6.117 followed by carrying out the integrations and reorganizing the result is

$$\theta_2(\theta_1)=\cos^{-1}\left[\cos\theta_{2\,min}+\left(\cos\theta_{2\,max}-\cos\theta_{2\,min}\right)\frac{1-\cos^{n+1}\theta_1}{1-\cos^{n+1}\theta_{1\,max}}\right]. \tag{6.118}$$

This relation can be used in conjunction with Eq. 6.113 to determine the reflector profile.

Consider the design of the profile of reflector with diameter $80\,\lambda$ and focal length $f=32\lambda$ which is required to give a field-of-view of about $\pm4°$, which is a typical shaped beam requirement. A cosine-shaped feed pattern function with $n=1.5$ was chosen to give an edge illumination of about $-10\,dB$. Suppose a uniform illumination function is needed over the symmetrical field-of-view defined by $\lambda/20D\le\theta\le5.6\lambda/D$. The shaped reflector profile obtained from a solution of Eq. 6.113 along with Eq. 6.118 is shown in Figure 6.35a and the resulting radiation pattern obtained from physical optics is given in Figure 6.35b. The peak value of the gain over

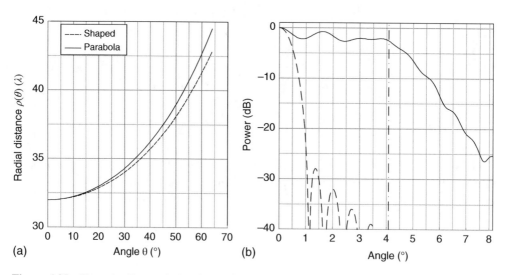

Figure 6.35 Shaped reflector design for uniform illumination over $\lambda/20D\le\theta\le5.6\lambda/D$ where $D=80\lambda$, $f/D=0.4$, $n=1.5$. (a) Reflector profile radial distance versus angle shaped reflector compared with parabola. (b) Radiation patterns of both reflectors. Dashed curves: parabola; solid curves: shaped

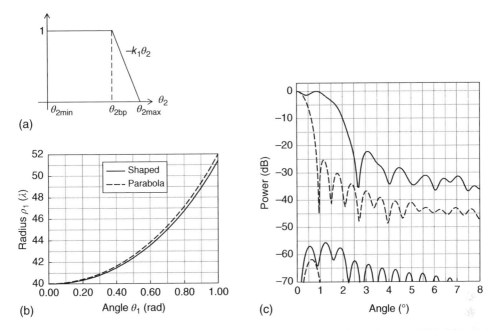

Figure 6.36 Shaped reflector designed for uniform illumination over $2\lambda/D$ where $D = 100\lambda$, $f/D = 0.4$, $n = 1.5$ and sidelobes < -20 dB. (a) Envelope of desired beam. (b) Reflector profile radial distance versus angle-shaped reflector compared with parabola. (c) Radiation patterns of both reflectors. Dashed curves: parabola; solid curves: shaped

the field of view is 28.75 dBi. The profile of a parabolic reflector with the same diameter, focal length $f/D = 0.4$ and same feed pattern is also shown in Figure 6.35a. Its radiation pattern has a maximum gain of 47.07 dBi and is shown in Figure 6.35b. The shape of the two profiles in Figure 6.35a is similar although they diverge as the edge is approached. Observe that when $\theta_{2\max}$ is reduced, the synthesized profile approaches the parabolic profile.

As a further example, consider the design of a reflector to achieve the radiation pattern envelope shown in Figure 6.36a with a cosine to the power n feed pattern. It can be shown that Eq. 6.117 gives

$$\left(\frac{1 - \cos^{n+1}\theta_1}{1 - \cos^{n+1}\theta_{1\max}}\right)\left[F_1(\theta_{2bp}) + F_2(\theta_{2\max})\right] = \begin{cases} F_1(\theta_{2bp}) + F_2(\theta_2); & \theta_2 > \theta_{2bp} \\ F_1(\theta_2); & \theta_2 \leq \theta_{2bp} \end{cases}, \qquad (6.119)$$

where $F_1(\theta) = -\cos\theta + \cos\theta_{2\min}$ and $F_2(\theta) = -k_1(\sin\theta - \sin\theta_{2bp} - \theta\cos\theta + \theta_{2bp}\cos\theta_{2bp})$, θ_{2b} is the breakpoint angle and $-k_1$ is the slope of the outer envelope as shown in Figure 6.36a. The relationship for θ_2 in terms of θ_1 is obtained as the root of Eq. 6.119. Suppose the reflector is required to have a diameter $D = 100\lambda$ and focal length $f = 40\lambda$. Also, the extent of the output field is $\theta_{2bp} = 2\lambda/D$ with sidelobes less than -20 dB. To achieve the latter let $k_1 = -100$. The reflector profile that is obtained is shown in Figure 6.36b and the resulting radiation pattern is shown in Figure 6.36c.

Equation 6.117 could be used in reverse to estimate a desired pattern function needed to achieve a desired radiation pattern. For example, in the design of a reflector for a radar application it is desirable to reduce reflections from the ground. Such an illumination function is $I(\theta_2) = \operatorname{cosec}^2 \theta_2$. In normal use, when this function is substituted into Eq. 6.116, it can be shown that

$$\operatorname{cosec}\theta_2 = \operatorname{cosec}\theta_{2\,\mathrm{min}} - \left(\operatorname{cosec}\theta_{2\,\mathrm{min}} - \operatorname{cosec}\theta_{2\,\mathrm{max}}\right) \frac{\displaystyle\int_0^{\theta_1} P(\theta_1)\sin\theta_1\,d\theta_1}{\displaystyle\int_0^{\theta_{1\,\mathrm{max}}} P(\theta_1)\sin\theta_1\,d\theta_1}. \tag{6.120}$$

The feed pattern can be in the form of measured data or be represented as a modal summation through which θ_2 can be expressed in terms of θ_1. Eq. 6.113 along with Eq. 6.120 can be used to synthesize the reflector profile.

Shaped beams in two-dimensions can also be created from shaped reflectors that are designed by means of Eqs. 6.113 and 6.116 as illustrated in Figure 6.37. This is providing the reflector contour lies in any of the transverse plane passing through the point O as shown in Figures 6.34 and 6.37. For example, an elliptical shaped beam could be created with a feed with an axisymmetric pattern by shaping the reflector contour in selected transverse planes. In

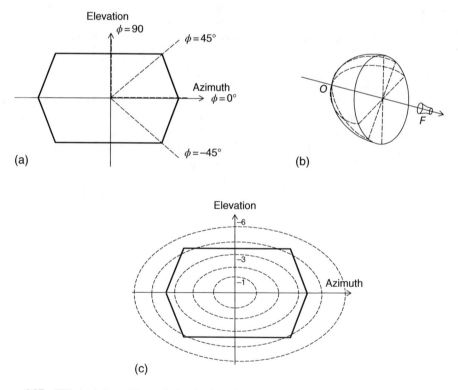

Figure 6.37 Elliptical-shaped beam design in elevation and azimuth with elliptically contoured reflector and circular feed horn. (a) Coverage region. (b) Transverse segments of reflector. (c) Elliptical beam produced

Figure 6.37a these planes occur at 45° intervals in the azimuthal direction about O. By this approach, Eq. 6.118 could be applied to each section as illustrated in Figure 6.37. A sequence of reflector profiles is produced and these form a continuous reflector in the azimuth direction. When illuminated by the original axisymmetric feed, an elliptical beam results.

The approach described above for a single reflector is readily extended to a dual reflector (Galindo, 1964). A second differential equation results from GO requirements on the second reflector and this differential equations couples with the one from the first. It is found that in the integral expression for conservation of energy for the second reflector, which is equivalent to Eq. 6.114, when its upper limit is chosen to be positive, a Cassegrain-type solution results. When this upper limit is negative, a Gregorian-type reflector geometry is created with a concave subreflector.

6.9.2   Reflector Synthesis by Numerical Optimization

The geometric optics approach described above has limitations on accuracy as well as utility as the number of physical constraints and additional requirements increase. As with a profiled horn design that was described in Section 4.5.3, the synthesis problem can be broadened by means of numerical methods. There are several ways of doing this both directly and indirectly. There are direct improvements for reflector shaping such as the inclusion of diffraction, blockage and accurate feed models. One such technique is the method of successive projections (Poulton & Hay, 1991). The indirect approach is to use a numerical representation of the surface and to use this with accurate radiation and feed models to meet the various system requirements. Reflector synthesis with standard optimization methods is described initially, and this is followed by a short overview of successive projections.

A basic requirement in reflector synthesis is that the reflector surface should be represented numerically. One approach that has proved very effective and accurate for reflector synthesis and computation is to use basis-spline or B-spline functions that are briefly outlined in Section 4.5.3 for an axisymmetric surface. The degree of the spline function can be selected as required although, in reflector synthesis, third order has been found sufficient in both accuracy and efficient in computation time. The reflector surface $z = f(x,y)$ is represented by Eq. 4.79 (de Boor, 1978) in this case written

$$f(x,y) = \sum_{i=0}^{m}\sum_{j=0}^{n} \alpha_{ij} N_i(x) N_j(y), \qquad (6.121)$$

where $N_i(x)$ and $N_j(x)$ are standard cubic B-spline functions which have $m+1$ and $n+1$ control points, respectively. The expansion coefficients α_{ij} are the unknowns here and as in Section 4.5.3 they are determined through optimization. A B-spline polynomial in the variable x is a piecewise function of degree $p_x = 3$. It is defined over a range $t_1 \leq x \leq t_m$, with $m = p_x + 2$. The points where $x = t$ are called knots or break-points, which are arranged in ascending order. The number of knots is the minimum for the degree of the B-spline, which has a non-zero value only in the range between the first and last knot. Each piece of the function is a polynomial of degree p_x between and including adjacent knots. The surface given by Eq. 6.121 has a set of $(m+1)(n+1)$ control points, which is in common with other interpolation methods, except that the major difference is the surface does not generally pass through the central control points. Expressions for the polynomial pieces are easily generated by means of recursion formulae (de Boor, 1978). If there is more than one reflector, the remaining surfaces are expressed in a similar form to Eq. 6.121 where the coefficients of each series for the new reflector surfaces are included in the optimization.

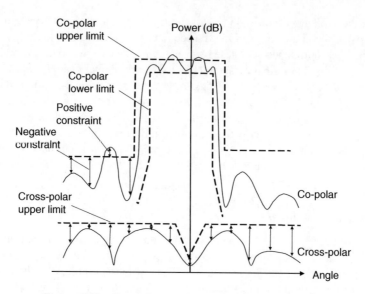

Figure 6.38   Constraints on the far-field radiation patterns

There are several ways of progressing from here. One way is to use Eq. 6.121 in conjunction with an analysis method and to adjust the unknown coefficients to achieve the system requirements based on the pattern and feed mismatch. Within the optimizer, the reflector-shaping procedure could use a transmit-mode radiation-pattern analysis based on physical-optics, which uses a numerical integration of the current. This could be performed within a gradient search algorithm for optimum reflector shapes via the expansion coefficients such as α_{ij}. This was used earlier to design a dual-reflector feed for a radio telescope (Granet et al., 1997). A similar approach is used in the design of array feeds where the array coefficients are found instead of the reflector surface coefficients. Both cases could employ either constrained or unconstrained variables that are optimized with a standard numerical optimizer. This technique for arrays will be described in Chapter 7. The intention of this approach is to use the power of standard numerical optimization.

In order to specify pattern constraints, consider a cut through the far-field as shown in Figure 6.38. At point P in the radiation pattern, define the following limits and weight parameters on these limits:

c_{up} = the maximum co-polar level
c_{lp} = the minimum co-polar level
x_{up} = the maximum cross-polar level
w_{up} = the weighting factor on co-polar maximum
w_{lp} = the weighting factor on co-polar minimum

In the same way, specifications on the input match and gain can also be included through constraints and weight parameters. For example, constraints could be placed on the input reflection coefficient of the feed over a range of frequencies. At a frequency k, specify

γ_{uk} = the maximum feed reflection level;
w_{rk} = the weighting factor on reflection coefficient.

All constraints can be included in a single index as follows (Bandler & Charalambous, 1972). If the co- and cross-polar efficiencies of the reflector antenna over the field of view are optimized, respectively, η_{cp} and η_{xp} are both obtained via Eq. 6.46. A number of positive discrete differences can be formed as follows:

$$\Delta_k = \begin{cases} -w_{lk}(\eta_{ck}-c_{lk}); & k=p; & p=1,2,...,N \\ w_{uk}(\eta_{ck}-c_{uk}); & k=N+p; & p=1,2,...,N \\ w_{xk}(\eta_{ck}-c_{lk}); & k=2N+p; & p=1,2,...,N \\ w_{rk}(\Gamma_{uk}-\gamma_{uk}); & k=3N+p; & p=1,2,...,N_f \end{cases}, \qquad (6.122)$$

where N is the number of sample points in the radiation pattern and N_f is the number of frequency points. The objective of the optimization is to find the vector of coefficients for which $\Delta_k \leq 0$ for all $k=1,2,...,3N+N_f$. A single performance index that incorporates all constraints is the least p-th index (Bandler & Charalambous, 1972)

$$I = H\left[\sum_{k\in\kappa}\left(\frac{\Delta_k}{H}\right)^p\right]^{1/p}, \qquad (6.123)$$

where $H=\max(\Delta_k)$, $p= \mathrm{sgn}(\Delta_k)q$ and κ is the set of specifications

$$\kappa = \begin{cases} \text{select all } \Delta_k & \text{if } H \leq 0 \\ \text{select only positive } \Delta_k & \text{if } H > 0 \end{cases}.$$

Any integer index q can be chosen in p although $q=2, 4, 10$ and 100 have been found most useful in antenna designs. Eq. 6.123 can be minimized with most standard optimization methods (Dixon, 1972). One such method that has proved reliable in several antenna applications is based on gradient search and numerically calculated differentials (Fletcher, 1972). Techniques such as the genetic algorithm (Goldberg, 1989) and particle swarm optimization (Kennedy & Eberhart, 1995) can be very effective at the commencement of the search for the minimum.

The method of successive projections is a general iterative technique that can be used to determine a common intersection point among a number of conflicting requirements or sets. Related to aperture antennas it has been used to obtain array excitation coefficients (Poulton, 1986), to adjust the shape of a reflector antenna from measured data and in specifying the shapes of single and dual reflector antennas (Hay, 1999). The aim is to achieve a reflector shape that produces a beam that has directivity constrained between lower and upper bounds, respectively, G_L and G_U. The approach is illustrated in Figure 6.39.

The first step is to specify a collection of sets of $\{X\}$, which consist of complex valued functions on a rectangular region containing the projected aperture, a weighting function w constrained to have a unit magnitude (i.e. $|w|=1$), and a weight with continuous phase, for example, $w = \exp(jp)$; $p \in \{P\}$. The next step is to provide an approximate solution for the reflector. It is usual to choose a suitable function that represents a smooth reflector, which is of the required size that radiates over a specified coverage region, that has approximate directivity bounds, and phase centres that lie between the feed and the reflector. The method that follows is illustrated in Figure 6.39. This iterative process designs a suitable smooth reflector by first commencing with a stepped reflector that satisfies the directivity bounds. This is done

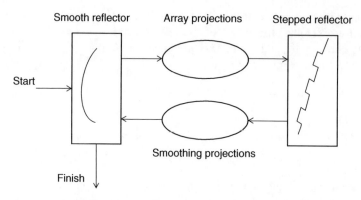

Figure 6.39 Method of successive projections

by dividing the last reflector surface obtained into an array of elements and then projecting these onto sets within the space of complex weights for the radiation patterns of the elements. These are often obtained by Fourier transforms. In this way, a vector $\mathbf{\Phi}_i$ is determined, which is a vector of the elementary field components involving the integral of the PO currents, to obtain the far-field radiation at points $i = 1, 2, \cdots, M$ in a number of specified far-field directions (θ_i, ϕ_i). To evaluate $\mathbf{\Phi}_i$ the integral may be approximated by summing integrated samples on a rectangular grid in the x–y plane. In this process, weights are applied that represent a stepped weight distribution for the current on the surface of the last obtained reflector surface. The phase of the weight distribution represents the surface of the stepped reflector relative to the last reflector, and the intersection of the collection of sets represents weights that satisfy both the specified directivity bounds and also the constraint on the magnitude of each weight equals unity. Next, these functions are projected onto the first two sets of requirements. The formula to do this is simple to compute and, therefore, the process can be very fast. A sequence of iterations $\{\mathbf{x}^n\}$ is generated in the following form:

$$\mathbf{x}^{n+1} = \left(\frac{y_1}{|y_1|}, \frac{y_2}{|y_2|}, \cdots, \frac{y_N}{|y_N|} \right) \tag{6.124}$$

$$\mathbf{y} = \mathbf{x}^n + r \left(\frac{\sqrt{G_i}}{|F_i^n|} - 1 \right) \left(\frac{F_i^n}{\mathbf{\Phi}_i \cdot \mathbf{\Phi}_i^*} \right) \mathbf{\Phi}_i^*, \quad i = 1, \ldots, N, \tag{6.125}$$

where $N = 4P_x P_y$, $r \sim 4$ is a relaxation factor, $F_i^n = \mathbf{x}^n \cdot \mathbf{\Phi}_i$ where $n = 1, 2, \ldots$ is the iteration number and,

$$G_i = \begin{cases} G_L & \text{if } |F_i|^2 \le G_L \\ G_U & \text{if } |F_i|^2 \ge G_U \end{cases}.$$

The parameters P_x and P_y are limits on the Fourier series that are used to ensure any solution is sufficiently smooth. In each direction, the series have $2P_x + 1$ and $2P_y + 1$ terms, which are chosen so that the minimum of the harmonic periods is usually taken to about two times the side length of an element in the stepped phase distributions. A smooth reflector is found by

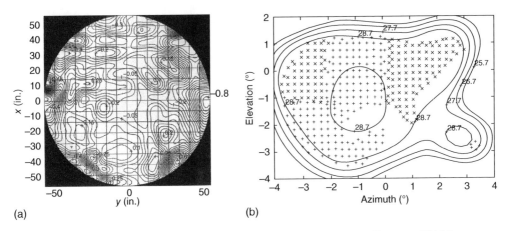

Figure 6.40 Shaped reflector designed by successive projections aperture diameter of 112 in. (a) Difference δz of the initial and final reflector surfaces in inches. (b) One shaped beam for a composite coverage requirement (indicated by + and ×) for continental USA showing directivity at 4.2 GHz (Hay, Private communication). *Source*: Reproduced with permission from CSIRO

projecting the stepped weight distribution onto a set of weight functions w each with sufficiently smooth phase and unity magnitude. This projection comprises a finite-term Fourier series. If the smooth reflector is a suitable solution to the problem, then the iterative process terminates; otherwise, it repeats until convergence is achieved. To optimize a directivity pattern with respect to upper or lower bounds, the bounds are tightened until they can be tightened no further.

An example of a shaped reflector design, Figure 6.40b, shows a beam shaped that was designed to cover continental USA at C-band transmit and receive frequencies (Hay, Private communication). An offset reflector was chosen for the antenna with a projected circular aperture diameter of 112 in. (i.e. 2845 mm). The reflector surface was designed using the method of successive projections to achieve the directivities of 27.7 dBi in zone I (shown in Figure 6.40b) and 28.7 dBi in zone II (indicated by × in Figure 6.40b). The starting reflector had a parabolic surface and following the synthesis process the difference in the height in the axial direction, δz, between the initial and final reflector surfaces is shown in Figure 6.40a. The radiation pattern given by the shaped reflector at 4.2 GHz is shown in Figure 6.40b.

6.10   Problems

P6.1 For a parabolic reflector show that the half-cone angle ψ_c is given by

$$\psi_c = 2\tan^{-1}\left(\frac{1}{4f/D}\right),$$

where f is the focal length and D is the diameter of the reflector.

P6.2 Using geometric optics, obtain the aperture field of a paraboloidal reflector excited by an elementary source with

a. Radiated far-fields

$$E_\theta = E_1 \cos\theta \cos\phi \, \frac{e^{-jkr}}{r},$$

$$E_\phi = 0 = E_r.$$

b. A second elementary source is available with radiated fields

$$E_\phi = -E_2 \cos\theta \sin\phi \, \frac{e^{-jkr}}{r},$$

$$E_\theta = 0 = E_r.$$

Find the aperture fields that this source produces.

c. Hence, find the total aperture field from a source with both contributions and the complex amplitudes required to yield zero cross-polarization in the aperture.

P6.3 A half-wave dipole illuminates a 3 m parabolic reflector antenna at a frequency of 10 GHz. If the reflector has an $f/D = 0.433$, what is the level in the E- and H-planes of the:

a. edge illumination; and

b. edge taper?

P6.4 Verify that the field radiated by a half-wave dipole given initially by Eq. 6.13 when placed a distance d in front of a large conducting plate now has an amplitude approximately given by $A(\theta,\phi) = 2j \sin((\pi/2)\cos\theta)\sin(kd\cos\theta)$.

P6.5 At 3 GHz the total input power to a feed antenna situated at the focus of 3 m paraboloid is 1 W. Measurements have shown that the efficiency of the feed is 82%. The reflector has a calculated spillover efficiency of 98% and an aperture efficiency of 63%. Calculate the antenna gain and the power density at a receiver situated 5 km away.

P6.6 This problem verifies some equations for the normal to a paraboloidal surface.

a. Show that the equation for the normal to general surface, Eq. 6.36, reduces to Eq. 6.3b for a paraboloid.

b. Given the equation for a paraboloid in Eq. 6.1, show that the normal to the surface in spherical polar co-ordinates is given by Eq. 6.3c.

P6.7 Use Eqs. 6.39 to obtain an expression for the fields radiated by a paraboloidal reflector that is fed by a corrugated waveguide and operates in its HE_{11} mode at the balanced hybrid condition.

P6.8 Blockage of a reflector by a feed or a subreflector is examined in this problem. Radiation from a circular aperture of diameter Σ is blocked by a centrally placed object of diameter D. Assume a uniform linearly polarized field distribution in the aperture.

a. Find the radiated field. You may need the identity:

$$\int_0^{2\pi} \exp(ju\cos\phi)\,du = 2\pi J_0(u)$$

b. Demonstrate the effect of central blockage is to (i) reduce the gain, (ii) to increase the odd numbered sidelobes and (iii) to decrease the even numbered sidelobes.

c. Suggest a way of reducing the impact of blockage.

P6.9   A feed is moved axially a distance s from the focus towards the vertex of a paraboloid reflector. Assuming s is very much smaller than the focal length (i.e. $s \ll f$) and the feed uniformly illuminates the reflector, determine the dominant aberrations produced by the axial feed movement.

P6.10   A feed for a paraboloidal reflector has an axisymmetric pattern and pattern function

$$A(\psi) = \begin{cases} \cos^n\psi; & 0 \le \psi \le \pi/2 \\ 0; & \text{elsewhere} \end{cases}.$$

where $n = 1$. Obtain expressions for
a. the aperture efficiency, and
b. the spillover efficiency. Hence
c. determine the reflector half-cone angle that gives maximum gain with this feed, assuming there are no other losses.

P6.11   Show that the spillover efficiency, η_s, of a feed with an axisymmetric pattern and pattern function given in P6.10, which illuminates a reflector with cone angle ψ_c, is given by $\eta_s = 1 - \cos^{2n+1}\psi_c$.

P6.12   A half-wave dipole antenna is to be used as a feed for a paraboloidal reflector. Describe the principal plane patterns of a paraboloid with a half-wave dipole feed.
        What are the advantages and disadvantages of this feed compared with a circular waveguide?
        Describe some extensions of the basic half-wave dipole structure that are better feeds, giving reasons for the improved performance.

P6.13   A rectangular waveguide is chosen as a feed for a paraboloidal reflector antenna with diameter $D = 3$ m and $f/D = 0.35$. At the design frequency of 12.5 GHz, an edge taper of -12 dB is needed to satisfy sidelobe requirements. Assuming the reflector is in the far-zone region of the waveguide:
a. Calculate the reflector half-cone and the desired edge illumination;
b. Calculate approximately the waveguide dimension needed in the E-plane to produce a far-zone pattern with the attributes calculated in (a) and, hence, satisfy the design edge illumination.

P6.14   Use field correlation at the surface of a parabolic reflector between an incident linearly polarized plane wave and an axisymmetric feed to determine the aperture efficiency as a function of incident angle. The reflector has a diameter D and focal length f.

P6.15   Approximate the effect of the blockage of quad-struts supporting the feed in a reflector of diameter D by approximating the blockage at 90° apart by segments with an internal angle $\theta = \theta_s$. The field is uniform and polarized parallel to the x-direction and the struts are at 45° to this direction. Determine the loss of gain and the change in level of the first sidelobe in the two principal planes.

P6.16   For an offset reflector antenna, what type of aperture field is required to cancel out the cross-polarization in the far-field.

P6.17   Determine the efficiency of a parabolic reflector with surface errors, which have arisen in manufacturing the profile template. The surface error is circularly symmetric and sinusoidal in the radial direction with amplitude $\varepsilon \ll \lambda$ and with a period $2\pi/p$ which is comparable to the wavelength.

P6.18   For a paraboloidal reflector of diameter, what proportion of power is in the sidelobes compared with that in the main beam. As an example, consider a reflector of diameter $D = 100\lambda$, and focal length given by $f/D = 0.4$ with a feed having a Gaussian pattern function to provide an edge taper of -10 dB.

P6.19   Apply the field correlation theorem to determine an expression for the contribution to the input reflection coefficient of a feed due to reflection from the surface of a paraboloid.

Hint: $\Gamma = \dfrac{1}{2} \dfrac{\left| \displaystyle\iint_S \mathbf{E}_T \times \mathbf{H}_T \cdot \hat{n} \, dS \right|^2}{P_T^2}$ where $\mathbf{E}_T$ and $\mathbf{H}_T$ are the incident fields on S due to the feed and P_T is the total radiated power.

P6.20   For a symmetrical parabolic reflector with a diameter of $100\,\lambda$, and focal length $f/D = 0.4$, calculate the diameter of a corrugated waveguide operating at the balanced hybrid condition to obtain the best match to the focal region. What is the maximum aperture efficiency of the resulting antenna?

P6.21   Solve the equations for single reflector shaping described in Section 6.9.1, assuming that the incident and exit angles θ_1 and θ_2 are identical over the full range of angles available. Describe the solution that results for the reflector surface.

P6.22   Using the approach described in Section 6.9.1, obtain the differential equations describing the shaping of dual axisymmetric reflector antennas.

P6.23   Describe the aperture field of symmetrical reflector antenna due to a circular cup feed as shown in Figure P.6.1. The circular cup is excited from a circular waveguide of diameter h in a TE_{11} mode and the location of the phase centre is as indicated.

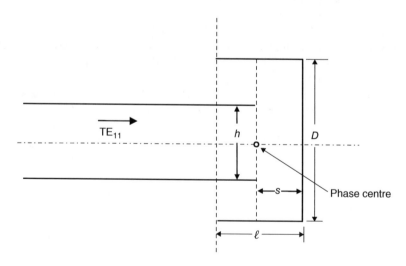

Figure P6.1   Cup feed junction

References

Bandler, J.W. and Charalambous, C. (1972), 'Practical least p-th optimization of networks', IEEE Trans. Microwave Theory Tech., Vol. **MTT-20**, pp. 834–840.

Bem, D.J. (1969): 'Electric-field distribution in the focal region of an offset paraboloid', Proc. IEEE, Vol. **116**, pp. 679–684.

Bird, T.S. and Boomars, J.L. (1980): 'Evaluation of focal fields and radiation characteristics of a dual-offset reflector antenna', Proc. IEE, Part H, Vol. **127**, pp. 209–218. Erratum: IEE Proc. (Part H), **128**, 1981, p. 68.

Bird, T.S. (1981): 'Investigation of cross-polarization in offset Cassegrain antennas, Electron.', Lett., Vol. **17**, pp. 585–586.

Brigham, E.O. (1974): 'The fast Fourier transform', Prentice-Hall Inc., Eaglewood Cliffs, New Jersey.

de Boor, C. (1978): 'A practical guide to splines', Springer-Verlag, New York.

Dixon, L.C.W. (1972): 'Nonlinear optimization', English Universities Press Ltd, London, UK.

Fletcher, R. (1972): 'Fortran subroutines for minimization by quasi-Newton methods'. A.E.R.E., Harwell, UK. AERE report 7125.

Franceschetti, G. and Mohsen, A. (1986): 'Recent developments in the analysis of reflector antennas. A review', Proc. IEE, Part H, Vol. **133**, pp. 65–76.

Fröberg, C.-F. (1974): 'Introduction to numerical analysis', Addison-Wesley, London, UK.

Galindo, V. (1964): 'Design of dual-reflector antennas with arbitrary phase and amplitude distributions', IEEE Trans. Antennas Propag., Vol. **AP-4**, No. 3, pp. 403–408.

Goldberg, D. (1989): 'Genetic algorithms in search, optimization and machine learning', Addison-Wesley Longman Publishing Company Inc., Boston, MA.

Granet, C., James, G.L. and Pezzani, J. (1997): 'A new dual-reflector feed system for the Nançay radio telescope', IEEE Trans. Antennas Propag., Vol. **45**, pp. 1366–1373.

Hay, S.G. (1999): 'Dual-shaped-reflector directivity pattern synthesis using the successive projections method', IET Microwaves Antennas Propag., Vol. **146**, pp. 119–124.

James, G.L. (1979): 'Cross-polarization performance of flanged cylindrical and conical waveguides', Proc. IREE Aust., Vol. **40**, pp. 180–184.

James, G.L. (1986): 'Geometrical theory of diffraction for electromagnetic waves', 3rd ed., Peter Peregrinus Ltd., London, UK.

Jones, E.M.T. (1954): 'Paraboloid reflector and hyperboloid lens antenna', IRE Trans. Anntenas Propag., Vol. **AP-2**, pp. 119–127.

Kennedy, J. and Eberhart, R. (1995): 'Particle swarm optimization', Proceedings of the IEEE Conference on Neural Networks, Piscataway, NJ, November–December, pp. 1942–1948.

Li, T. (1959): 'A study of spherical reflectors as wide angle scanning antennas', IRE Trans. Anntenas Propag., Vol. **AP-7**, pp. 223–226.

Minnett, H.C. and Thomas, B.M.A. (1968): 'Fields in the image space of symmetrical focussing reflectors', Proc. IEEE, Vol. **115**, pp. 1419–1430.

Poulton, G.T. (1986): 'Antenna power pattern synthesis using the method of successive projections', Electron. Lett., Vol. **22**, no. 20, pp. 1042–1043.

Poulton, G.T. & Bird, T.S. (1988): 'Earth station antennas for multiple satellite access', J. Electr. Electron. Eng. Aust., Vol. **8**, pp. 168–176.

Poulton, G.T. and Hay, S.G. (1991): 'Efficient design of shaped reflectors using successive projections', Electron. Lett., Vol. **27**, No. 23, pp. 2156–2158.

Rudge, A.W. and Adatia, N.A. (1978): 'Offset-parabolic reflector antennas: a review', Proc. IEEE, Vol. **66**, pp. 1592–1618.

Rudge, A.W., Milne, K., Olver, A.D. and Knight, P. (1983): 'The handbook of antenna design', Peter Peregrinus, London, UK, Vol. 2, Chapters 10 & 11.

Ruze, J. (1966): 'Antenna tolerance theory – a review', Proc. IEEE, Vol. **54**, pp. 633–640.

Silver, S. (1946): 'Microwave antenna theory and design', first published by McGraw-Hill Book Company, New York. Reprint published by Peter Peregrinus Ltd., London, UK, 1984.

Weinstein, L.A. (1969): 'The theory of diffraction and the factorization method', The Golem Press, Boulder, CO.

Wood, P.J. (1980): 'Reflector antenna analysis and design', Peter Peregrinus Press, London, UK.

7

Arrays of Aperture Antennas

7.1 Introduction

Arrays of aperture antennas find wide application because of their flexibility and their ability to provide shaped patterns with low sidelobes. They can be power efficient and provide significant gain. They can be used to scan the beam in almost any direction of three-dimensional space. They can be placed conformal to surfaces to provide gain over wide scan angles. In two- or three-dimensional grid arrangements, a wider variety of radiation patterns can be obtained ranging from hemispherical to a full spherical coverage. Examples include the shaped beam patterns employed on satellites to cover specific regions or countries on the earth's surface or providing selected 360 degree coverage as a wireless access antenna. Often in these applications, high-performance aperture elements are the basis of the array.

The topics in this chapter include the basic radiation patterns of arrays with aperture antennas. One of the implications in the use of an array is the mutual coupling that occurs between elements, and the analysis of this effect in aperture antennas is covered in some detail because of its importance. The physical aspects of mutual coupling are described by means of an asymptotic approach, and also a general approach for arbitrary shapes is outlined. Radiation in the presence of mutual coupling is described as are some mitigation measures if its effect must be avoided. Examples are given throughout to illustrate the techniques.

7.2 Two-Dimensional Planar Arrays

Suppose initially a finite array of aperture antennas is located in a plane. It is assumed that there is no coupling between the elements in the array. The $M \times N$ elements of the array are arranged on a regular grid in the x–y plane with a constant excitation amplitude for each element and a

Fundamentals of Aperture Antennas and Arrays: From Theory to Design, Fabrication and Testing,
First Edition. Trevor S. Bird.
© 2016 John Wiley & Sons, Ltd. Published 2016 by John Wiley & Sons, Ltd.
Companion website: www.wiley.com/go/bird448

steering phase that allows a beam to be formed in the hemisphere above the plane. Consider a single element of this array and obtain its far-field. Based on Section 3.4, the mnth element $(m, n = 0, 1, 2, \ldots, M, N)$ radiates an electric field given by

$$\mathbf{E}_{mn}(r,\theta,\phi) = \mathbf{E}_{\mathrm{o}} \frac{\exp(-jkR)}{R} F(\theta,\phi)$$

where $F(\theta, \phi)$ is the element pattern, $\mathbf{E}_{\mathrm{o}}$ is the polarization of the electric field, and A_{mn} is the excitation and from Eq. 3.12 $R \approx r - (\mathbf{r}'_{mn} \cdot \hat{r})$ where $\mathbf{r}'_{mn}$ is a vector in the x–y from the origin to the mnth element. The total field is obtained by summing the contributions from all elements. Thus

$$\mathbf{E}(r,\theta,\phi) \approx \mathbf{E}_{\mathrm{o}} \frac{\exp(-jkr)}{r} F(\theta,\phi) \sum_{m,n=0}^{M,N} A_{mn} \exp\left(jk\left(\mathbf{r}'_{mn} \cdot \hat{r}\right)\right). \tag{7.1}$$

The summation over m and n is called the array factor (AF). Its form depends on the layout of the elements in the x–y plane as will be shown in the next sections for the special cases of rectangular and hexagonal grids.

Now define

$$\mathrm{AF}(\theta,\phi) = \sum_{m,n=0}^{M,N} A_{mn} \exp\left(jk\left(\mathbf{r}'_{mn} \cdot \hat{r}\right)\right) \tag{7.2}$$

so that

$$\mathbf{E}(r,\theta,\phi) \approx \mathbf{E}_{\mathrm{o}} \frac{\exp(-jkr)}{r} F(\theta,\phi) \mathrm{AF}(\theta,\phi) \tag{7.3}$$

The gain function is given by Eq. 3.48, where

$$G(\theta,\phi) = 4\pi \frac{\left|\mathbf{E}_{\mathrm{o}} F(\theta,\phi) \mathrm{AF}(\theta,\phi)\right|^2}{P_{\mathrm{T}}} \tag{7.4}$$

where P_{T} is the total power input to the array. As a special case, suppose the apertures are fed by identical rectangular waveguides that operate only in the TE_{10} mode at frequencies well above cut-off. In that case,

$$P_{\mathrm{T}} \approx \left|\mathbf{E}_{\mathrm{o}}\right|^2 \frac{ab}{4} \sum_{m,n=0}^{M,N} \left|A_{mn}\right|^2. \tag{7.5}$$

If the array excitation is uniform so that $\left|A_{mn}\right| = 1$, the summation in Eq. 7.5 equals $(M+1)$ $(N+1)$ and, therefore,

$$P_{\mathrm{T}} \approx \left|\mathbf{E}_{\mathrm{o}}\right|^2 \frac{ab}{4} (M+1)(N+1).$$

The gain is

$$G(\theta,\phi) = \frac{16\pi}{ab(M+1)(N+1)} |F(\theta,\phi)\,\mathrm{AF}(\theta,\phi)|^2 \tag{7.6}$$

The AF dominates the directivity of an array. It is useful to calculate the directivity of the array using the AF alone. By means of Eq. 3.50, the directivity is given by

$$D = 4\pi \frac{|\mathrm{AF}_{\mathrm{peak}}|^2}{\displaystyle\int_0^{2\pi} d\phi \int_0^{\pi} d\theta \sin\theta |\mathrm{AF}(\theta,\phi)|^2} \tag{7.7}$$

where $\mathrm{AF}_{\mathrm{peak}}$ is the maximum value of the AF.

7.2.1 Rectangular Planar Array

A particular case of the array lattice is a planar rectangular grid as shown in Figure 7.1. Suppose the spacing of the $M+1$ elements in the x-direction is Δ_x, while in the y-direction the $N+1$ elements are spaced Δ_y apart. It is assumed that the origin of the co-ordinate systems occurs at the centre of the array (Figure 7.1). Let the vector to the mnth element in the x–y plane from the origin be $\mathbf{r}'_{mn} = \hat{x}m\Delta_x + \hat{y}n\Delta_y$. Therefore, from Eq. 7.1, the AF contribution comes from $\hat{r}\cdot\mathbf{r}'_{mn} = m\Delta_x u + n\Delta_y v$ where $u = \sin\theta\cos\phi$ and $v = \sin\theta\sin\phi$. In addition, suppose the element excitation be given by $A_{mn} = \exp(jm\psi_x + jn\psi_y)$. Therefore, the AF for a rectangular array is

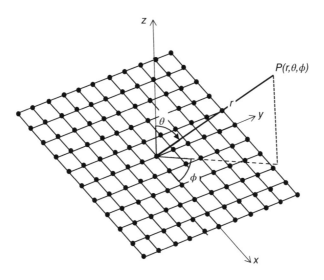

Figure 7.1 Regular planar rectangular array geometry

$$\text{AF}(\theta,\phi) = \sum_{m,n=0}^{M,N} \exp\left[jm(k\Delta_x u + \psi_x) + jn(k\Delta_y v + \psi_y)\right] \tag{7.8}$$

$$= P(M,T_x).P(N,T_y)$$

where $P(M,T_x) = \sum_{m=0}^{M} \exp[jmT_x]$, $T_x = k\Delta_x u + \psi_x$, $T_y = k\Delta_y v + \psi_y$, and as defined previously $u = \sin\theta\cos\phi$ and $v = \sin\theta\sin\phi$.

It is noted that the phases ψ_x and ψ_y steer the pointing direction of the radiation pattern and are, therefore, referred to as steering angles. Thus, the direction of the beam is usually given at

$$\theta_b = \sin^{-1}\left(\frac{1}{k}\sqrt{\left(\frac{\psi_x}{\Delta_x}\right)^2 + \left(\frac{\psi_y}{\Delta_y}\right)^2}\right) \quad \text{and} \quad \phi_b = \tan^{-1}\left(\frac{\psi_y}{\psi_x}\right) \tag{7.9}$$

because the AF pattern is usually significantly narrower than the element pattern.

The two exponential series in Eq. 7.8 are geometric series and can be summed easily. Thus, if x^n is the nth power in a series of $N+1$ terms, their sum is $S = (x^{N+1}-1)/(x-1)$. Then

$$P(M,T_x) = \sum_{m=0}^{M} \exp[jmT_x]$$

$$= \left[\frac{\exp(j(M+1)T_x)-1}{\exp(jT_x)-1}\right] \tag{7.10}$$

$$= \exp\left(\frac{jMT_x}{2}\right)\left\{\frac{\sin\left[((M+1)/2)T_x\right]}{\sin(T_x/2)}\right\}$$

$$= (M+1)\exp\left(\frac{jMT_x}{2}\right)\left\{\frac{S[((M+1)/2)T_x]}{S(T_x/2)}\right\}$$

where S is the sinc function. And similarly for the series in n where N and T_y in Eq. 7.10, replace M and T_x, respectively. Therefore, the AF in Eq. 7.8 is expressed as

$$\text{AF}(\theta,\phi) = (M+1)(N+1)\exp\left[\frac{j(MT_x+NT_y)}{2}\right]\left\{\frac{S[((M+1)/2)T_x]}{S(T_x/2)}\right\}\left\{\frac{S[((N+1)/2)T_y]}{S(T_y/2)}\right\}. \tag{7.11}$$

It is emphasized that this AF is for a co-ordinate system that is located at the centre of the rectangular lattice (see Figure 7.1). Otherwise, an additional phase factor would be present for each summation giving the grid centre relative to the origin. Thus, if the origin is located at (x_0, y_0), the additional phase factor applied to Eq. 7.11 would be $\exp[j(x_0T_x + y_0T_y)]$.

The periodicity of the AF means that images of the main beam and its associated sidelobes are repeated at intervals λ/Δ_x and λ/Δ_y in u–v space. A repetition of the main beam in this way creates grating lobes. As the beam is steered, in directions determined by ψ_x and ψ_y, the main beam lies in the visible range, that is, real space. Real space corresponds to the interior of the unit circle given by $u^2 + v^2 = 1$. Values of u and v outside of real space, that is, $u^2 + v^2 > 1$, lie in imaginary space. As the beam is steered, parts of the beam lying in imaginary space can transfer into real space, and this includes any grating lobes, which are related to the lattice geometry. To avoid this happening, the element spacing should be chosen such that

$$\frac{\Delta_x, \Delta_y}{\lambda} < \frac{1}{1 + |\sin \theta(\max)|} \tag{7.12}$$

where $\theta(\max)$ is the maximum scan angle. Eq. 7.12 shows that under all conditions grating lobes are avoided if the spacing between array elements is less than half a wavelength.

Equation 7.10 predicts that the major beam and grating lobes are located at $k\Delta_x \sin \theta \cos \phi + \psi_x = \pm 2p\pi$ and similarly $k\Delta_y \sin \theta \sin \phi + \psi_y = \pm 2q\pi$ where $p,q = 0, 1, \dots$. Eqs. 7.10 and 7.11 can be used to determine θ and ϕ at the beam maxima. Examples of AF given by Eq. 7.11 are shown in Figure 7.2a is $\Delta_x = \lambda/2 = \Delta_y$ and $M + 1 = N + 1 = 8$. Also in Figure 7.2b is the case when the spacing has been increased to $\Delta_x = 0.7\lambda = \Delta_y$ and $M + 1 = N + 1 = 11$. In this latter array, the first set of sidelobes are about 17.5 dB below the peak, while the second set are about 4 dB lower.

The directivity of the AF for the rectangular array is given by Eq. 7.7 as

$$D = \frac{4\pi}{\displaystyle\int_0^{2\pi} d\phi \int_0^{\pi} d\theta \sin \theta \left| \frac{S[((M+1)/2)T_x]}{S(T_x/2)} \frac{S[((N+1)/2)T_y]}{S(T_y/2)} \right|^2}. \tag{7.13}$$

The directivity predicted by Eq. 7.13 for $\Delta_x = \Delta_y = \lambda/2$ when $M = N$ is plotted in Figure 7.3. The directivity increases smoothly with increasing M as the area occupied by the array increases.

7.2.2 Hexagonal Array

A hexagonal array geometry is illustrated in Figure 7.4. This lattice structure, which is also called an equiangular triangle array, along with the rectangular array considered in the previous section, are the most commonly used types of planar lattice geometries. The hexagonal array is often preferred with circular elements as they can be packed most efficiently in this layout. In principle, the hexagonal array could be decomposed into two overlapping rectangular grids, which are then considered as the superposition of these grids. However, here this array geometry is considered from first principles.

In the nth ring from centre of the hexagonal lattice, there are six n elements. As a result, the AF for the hexagonal array consists of summations of lattice points at azimuth angles $\Delta_\phi = \pi/3n$ apart and at radial increments spaced Δ_ρ apart. The perpendicular to the mth side is at an angle $\pi(2m-1)/6$ to the initial line as shown in Figure 7.5. The vector to the lth lattice point on the mth side of the nth ring $(n = 1, 2, \dots, N_r)$ is

$$\mathbf{r}'_{mn} = n\Delta_\rho \sec\left(\frac{(2l/n - 1)\pi}{6}\right)\left\{\hat{x} \cos\left[\frac{(m-1)\pi}{3} + \frac{l\pi}{3n}\right] + \hat{y} \sin\left[\frac{(m-1)\pi}{3} + \frac{l\pi}{3n}\right]\right\}; \tag{7.14}$$

$$n = 1, \dots, N_r; \quad m = 1, \dots, 6; \quad l = 1, \dots, n.$$

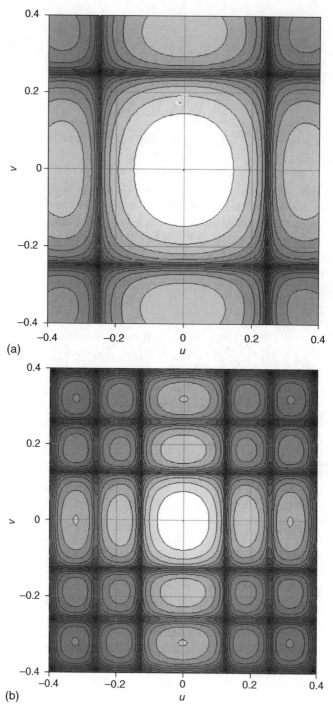

(a)

(b)

Figure 7.2 Array factor for a rectangular array with uniform excitation at 10 GHz plotted in the $u-v$ plane from the beam maximum. (a) 8×8 elements with spacing $\Delta_x = \lambda/2 = \Delta_y$. Contours are in 5.5 dB decrements below the peak value. (b) 11×11 elements with $\Delta_x = 0.7\lambda = \Delta_y$. Contours are in 5.7 dB decrements below the peak

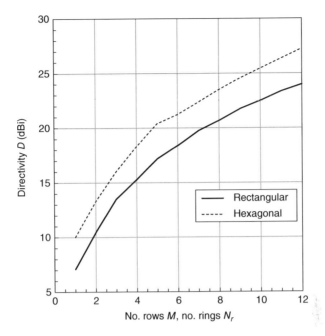

Figure 7.3 Directivity versus size of planar array. Solid curve, rectangular array versus number of elements per side $M = N$ with spacing $\Delta_x = \Delta_y = \lambda/2$ and dashed curve, hexagonal array versus ring number N_r with ring spacing $\Delta_\rho = \lambda/2$

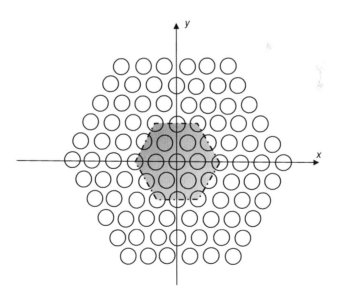

Figure 7.4 Example of a hexagonal array with 91 elements. A 19-element sub-array is shown shaded. The number of elements in this sub-array is $(1 + 3 N(N + 1))$ where N is the number of rings around the central element

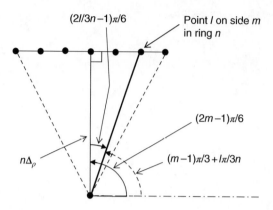

Figure 7.5 Geometry for definition of hexagonal lattice array factor

As $\hat{r} = \hat{x} \sin\theta \cos\phi + \hat{y} \sin\theta \sin\phi + \hat{z} \cos\theta$, then

$$\mathbf{r}'_{mn} \cdot \hat{r} = n\Delta_\rho \sec\left(\frac{(2l/n-1)\pi}{6}\right) \sin\theta \left[\cos\left(\phi - \frac{(m-1)\pi}{3} - \frac{l\pi}{3n}\right)\right].$$

The AF for a hexagonal lattice is therefore

$$\mathrm{AF}(\theta,\phi) = 1 + \sum_{n=1}^{N_r}\sum_{m=1}^{6}\sum_{l=1}^{n} A_{mn} \exp\left\{jnk\Delta_\rho \sin\theta \sec\left[\frac{(2l/n-1)\pi}{6}\right]\cos\left[\phi - \frac{(m-1)\pi}{3} - \frac{l\pi}{3n}\right]\right\}.$$

$$(7.15)$$

As a check on the validity of Eq. 7.15, let $\theta = 0 = \phi$, and with $A_{mn} = 1$, it is found that $\mathrm{AF}(0,0) = 1 + 3N_r(N_r+1)$. This is the number of point radiators in N rings of a hexagonal array. Contour plots of Eq. 7.15 are shown in Figure 7.6 with uniform excitation for $\Delta_\rho = \lambda/2$ and $N_r = 2$ (i.e. 19 elements) and also $N_r = 8$ (217 elements). The AF patterns shown in Figure 7.6 are typical of a hexagonal array. The pattern consists of a central beam with six radial lobes $60°$ apart. The height of the lobes decrease in amplitude with increasing $nk\Delta_\rho$. In Figure 7.6a, the first six sidelobes surrounding the main beam are about 4.5 dB below the peak value, while in Figure 7.6b they are about 17.6 dB down.

The gain can be calculated from Eq. 7.4. Suppose p_o is the power radiated by one waveguide. In the absence of mutual coupling and $|A_{mn}| = 1$, the total radiated power from a hexagonal array is

$$P_T \approx |\mathbf{E}_o|^2 \frac{ab}{4}(1 + 3N_r(N_r+1)).$$

resulting in the gain function

$$G(\theta,\phi) = \frac{16\pi}{ab(1 + 3N_r(N_r+1))}|F(\theta,\phi)\mathrm{AF}(\theta,\phi)|^2.$$

$$(7.16)$$

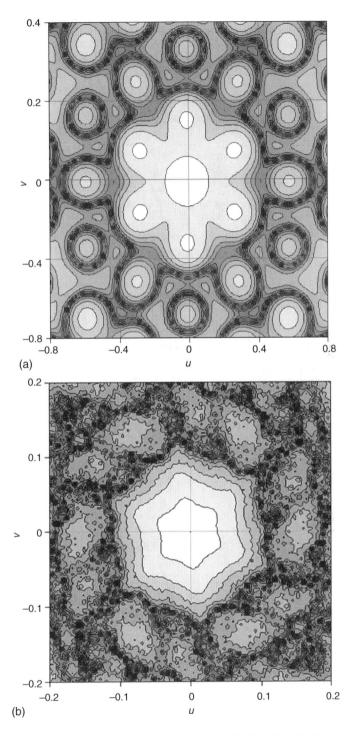

Figure 7.6 Array factor of a hexagonal array with uniform excitation plotted in the $u-v$ plane from the beam maximum. (a) 19 elements in two rings contour at 5.7 dB decrements below the peak, and (b) 217 elements in eight rings contour at 5.2 dB decrements below the peak. Frequency is 10 GHz and spacing $\Delta_\rho = \lambda/2$

The AF directivity is obtained from Eq. 7.7 as

$$D = \frac{2\pi}{3} \frac{|(1 + 3N_r(N_r + 1))|^2}{\int_0^{\pi/3} d\phi \int_0^{\pi} d\theta \sin\theta |AF(\theta, \phi)|^2} \tag{7.17}$$

where $AF(\theta, \phi)$ is given by Eq. 7.15 and in Eq. 7.17 the hexagonal lattice's sixfold symmetry has also been used. The directivity for the hexagonal array given by Eq. 7.17 as a function of number of rings up to $N_r = 12$ is plotted in Figure 7.3 for closer comparison with the directivity of a square array. It is seen that the directivity is not strongly dependent on the type of lattice although the number of elements in a hexagonal array increases more rapidly with ring number N_r compared with row number $N = M$ in the square array and the directivity is higher. However, in particular cases, the reverse occurs. For example, a rectangular 8×8 array with $\lambda/2$ element spacing has an area of 12.25 λ^2 for a directivity of 19.7 dBi, while a hexagonal lattice of four rings spaced $\lambda/2$ apart has 61 elements occupies a slightly larger area of 12.57 λ^2, has a lower directivity of 18.31 dBi.

7.3 Mutual Coupling in Aperture Antennas

The mutual coupling between small dipoles has been described in Section 2.3.3. Its prediction for an array of aperture antennas is important for accurate design and performance. The interaction between antenna elements modifies the overall radiation pattern as well as the individual aperture reflection characteristics. On some occasions, mutual coupling can reduce the overall system performance, for example, by enhancing the generation of grating lobes (Hansen, 1998). However, unraveling the details behind its effect can help to understand mutual coupling and how it may even be used to improve overall performance. Because of the importance of mutual coupling, it is valuable to review its cause, physical properties and some of the techniques used to predict it.

Mutual coupling was recognized by the early pioneers in antennas as important for design. Brillouin (1922) was probably one of the earliest to detail a method of analysis, and like many early workers, he was concerned with calculating radiation resistance rather than the complex impedance at the input of array elements. A systematized approach that developed, called the emf method, was applied by Pistolkors (1929) to find the radiation resistance of various dipole array configurations. Another approach adopted by the pioneer antenna designers was the Poynting vector method, which is the present-day standard method of calculating radiation resistance from the integrated normal energy flow through a surface surrounding the antenna. This second approach was used, for example, by Bontsch-Bruewitsch (1926) and Knudsen (1952) to analyse coupling in dipoles. Both approaches mentioned above are equivalent and can be converted into the other by means of Gauss' law, as shown by Bechmann (1931). The emf method gained wide acceptance after the work of Carter (1932), who used reciprocity and the emf method to determine expressions for self and mutual impedances for a variety of two-dipole arrangements. Carter's paper profoundly influenced much of the subsequent literature on antenna coupling because for the first time the coupling problem was expressed as an equivalent circuit. All the aforementioned papers were concerned with dipole elements.

Aperture antenna coupling such as between waveguides and slots were investigated initially in detail by Booker (1946), who derived the admittance of a slot from the impedance of its dual,

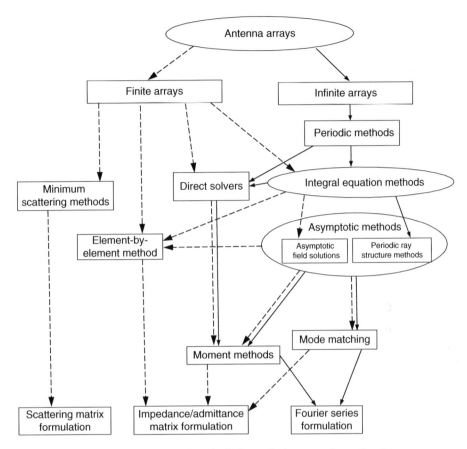

Figure 7.7   Relationships of methods for analysing mutual coupling in arrays

the dipole. The variational nature of the impedance formula was established by Miles (1949) and Storer (1952). General methods of obtaining admittances in ground planes were described by Harrington (1961). An early impetus for analysing mutual coupling effects in aperture antennas came from the need to counter blind spots in arrays, which subsequently was found to be due to surface waves on the array excited by coupling. Other more detailed methods came from the extension of methods for solving integral equations as well as others based on modal methods or periodic structures with Floquet modes. The range of methods used for analysing mutual coupling and their relationships are summarized in Figure 7.7.

The methods of periodic structures such as applied in solid-state physics (e.g. Floquet method) were not used in antennas until the 1960s. The Floquet approach found favour for a range of array structures and provided an early means of understanding coupling effects in large arrays. Array edge effects were seen to be important for small- and medium-sized arrays where the infinite array solution became a poor approximation to reality. This led in the 1960s and 1970s to the development of methods of analysing mutual coupling in finite arrays. These methods were improved throughout the 1980s and 1990s as computer technology became more capable. At the same time, the development of numerical methods and computer codes allowed the analysis of more complicated antennas such as arrays and their supporting structures. The computer codes available today allow mutual coupling to be included in the design as a matter of course. Nevertheless, the physical aspects of mutual coupling need to be understood to minimize its disadvantages and maximize the advantages. Mutual coupling can even provide benefits in some designs. For example, in an array used to feed a reflector antenna used to create a shaped beam, the gain achievable across the beam can be 0.2–0.3 dB higher if mutual coupling is properly incorporated in the initial design. This increase appears relatively modest except when you realize in some applications, such as in satellite communications, small gain increases of 0.1 dB are generally achieved at great expense (Bird & Sroka, 1992).

There are two basic approaches for analysing mutual coupling in arrays are as follows: through analysis methods for finite arrays and those for infinite arrays. Some techniques are common to both, as indicated in Figure 7.7. However, the end solution is usually different because the impedance and pattern characteristics of infinite periodic arrays are identical from one unit cell to the next. In the next section, mutual coupling in an infinite array is described and in the following section the element-by-element approach is detailed for finite arrays.

7.3.1   Infinite Periodic Arrays

When the array contains identical elements and the element spacing is regular, a solution can be expressed in terms of Floquet modes. Since the array geometry is a periodic function of the geometry, the field representation is periodic also except for a phase function which varies linearly across each cell of the array. This periodic function with phase progression is called a Floquet mode. Periodicity allows considerable simplification of the computing problem because once a solution is obtained for one cell, the solution for other cells is the same except for a progressive phase factor. By means of this representation, large arrays have been analysed with some success. Historically, in the analysis of mutual coupling in antennas, infinite periodic arrays were analysed first in detail. The literature on infinite arrays is substantial, and for details, the reader should consult the references (Hansen, 1966, 1998; Diamond, 1968; Farrell & Kuhn, 1968; Amitay et al., 1972; Rudge et al., 1983).

Consider an infinite array of identical aperture as shown in Figure 7.8. The location of an element in the transverse plane is specified by the vector

$$\boldsymbol{\rho}_{tmn} = m\mathbf{d}_1 + n\mathbf{d}_2; \quad |m| \leq M \text{ and } |n| \leq N$$

where $\mathbf{d}_1$ and $\mathbf{d}_2$ are vectors, which define the array lattice. In general, these vectors are not orthogonal. The electric field in the mnth unit cell is

$$\mathbf{E}_{mn}(\boldsymbol{\rho}_{tmn}) = \mathbf{E}_o(\boldsymbol{\rho} - \boldsymbol{\rho}_{tmn})e^{-jk\mathbf{W}\cdot\boldsymbol{\rho}_{tmn}}$$

where $\mathbf{W} = \hat{x}u + \hat{y}v$ with $u = \sin\theta\cos\phi$ and $u = \sin\theta\sin\phi$. The total electric field in the transverse aperture plane $z = 0$ is found by summing these individual contributions:

$$\mathbf{E}_t(\boldsymbol{\rho}_{tmn}) = \sum_{m=-M}^{M}\sum_{n=-N}^{N} \mathbf{E}_o(\boldsymbol{\rho} - \boldsymbol{\rho}_{tmn})e^{-jk\mathbf{W}\cdot\boldsymbol{\rho}_{tmn}}.$$

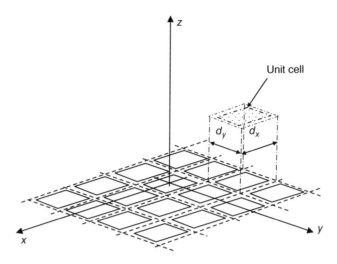

Figure 7.8 Infinite array of apertures and identification of a unit cell

In turn, the total radiated field is obtained by integrating over the infinite aperture. That is,

$$\mathbf{E}(r,u,v) = \frac{1}{4\pi^2} \int_{-\infty}^{\infty} \int_{-\infty}^{\infty} \sum_{m=-M}^{M} \sum_{n=-N}^{N} \mathbf{E}_0(\boldsymbol{\rho} - \boldsymbol{\rho}_{tmn}) e^{-jk\mathbf{W}\cdot\boldsymbol{\rho}_{tmn}} dxdy. \tag{7.18}$$

The summations can be completed in closed form, and the Fourier transform of the aperture field of the unit cell identified. This results in

$$\mathbf{E}(r,u,v) = \mathbf{E}(r,u,v) \left[\frac{\sin\left((M+1/2)(\boldsymbol{\kappa}_t \cdot \mathbf{d}_1)\right)}{\sin\left(1/2\boldsymbol{\kappa}_t \cdot \mathbf{d}_1\right)} \right] \left[\frac{\sin\left((N+1/2)(\boldsymbol{\kappa}_t \cdot \mathbf{d}_2)\right)}{\sin\left(1/2\boldsymbol{\kappa}_t \cdot \mathbf{d}_2\right)} \right] \tag{7.19}$$

where

$$\mathbf{E}(r,u,v) = \frac{1}{4\pi^2} \int\int_{\text{unit cell}} \mathbf{E}_0(x,y) e^{-jk(ux+vy)} dxdy$$

is the transform of the aperture field in the unit cell and $\boldsymbol{\kappa}_t = k(\hat{x}(1-u) + \hat{y}(1-v))$.

The assumption of an infinite periodic array simplifies the analysis of mutual coupling. Such an approach is useful for analysing the performance of large arrays of identical elements. In the following, it is assumed that the elements are identical, the spacing is regular, and there is a uniform phase shift across the array, which is applied with a linear phase shift from one element to the next. The analysis of mutual coupling in a periodic array is facilitated by means of the unit cell. A close view of a single unit cell of a planar array is shown in Figure 7.9.

Consider a field that is TE to the z-direction in the region above the aperture shown in Figure 7.9. The z-directed magnetic field is a solution of Helmholtz's equation of the form

$$\left[\nabla_t^2 + \left(k^2 + \gamma^2\right)\right] h_z(x,y) = 0$$

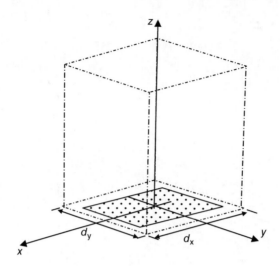

Figure 7.9   Unit cell of periodic planar array

where $H_z(x,y,z) = h_z(x,y)\exp(-j\gamma z)$ and $\nabla_t^2 = \partial^2/\partial^2 x + \partial^2/\partial^2 y$ is the Laplacian in the transverse components. The array is periodic in the x- and y-directions with spacing d_x in the x-direction and d_y in the y-direction. If $h_z(x, y)$ is a periodic function in x and y, a general solution is of the form

$$f\left(x+d_x, y+d_y\right) = h_z(x,y)\,\exp\left(-j\left(k_x d_x + k_y d_y\right)\right)$$

where k_x and k_y are the propagation constants in the x- and y-directions. This statement is a particular form of Floquet's theorem. Now let $h_z(x,y) = X(x)Y(y)$ where X and Y satisfy scalar wave equations with separation constants κ_x^2 and κ_y^2 such that

$$\gamma^2 = k^2 - \kappa_x^2 - \kappa_y^2.$$

As well, assume that the scan phase is a progressive on the excitation from cell to cell, which for element mn is of the form $V_{mn}\exp\left(j2\pi\left(m\psi_x + n\psi_y\right)\right)$ where steering phases are given by

$$\psi_x = \frac{d_x}{\lambda}\sin\theta_d\cos\phi_d \quad \text{and} \quad \psi_y = \frac{d_y}{\lambda}\sin\theta_d\sin\phi_d$$

where θ_d, ϕ_d are the drive angles. Thus, a general displacement Δ in the x-direction is $X(x+\Delta) = X(x)\exp(j2\pi m\psi_x\Delta/d_x)$, and similarly for $Y(y)$ in the y-direction. Furthermore, $X(x)$ can be represented by a Fourier series of the form

$$X(x) = \sum_{m=-\infty}^{\infty} A_m \exp\left(\frac{j2\pi mx}{d_x}\right)\exp\left(\frac{j2\pi m\psi_x x}{d_x}\right) = \sum_{n=-\infty}^{\infty} A_n \exp\left(\frac{j2\pi m(1+\psi_x)x}{d_x}\right).$$

Similarly in the y-direction, it is shown that

$$Y(y) = \sum_{n=-\infty}^{\infty} B_n \exp\left(\frac{j2\pi n(1+\psi_y)y}{d_y}\right).$$

Thus, the axial magnetic field can be expressed as a double summation of plane waves in the form

$$H_z = \sum_{m=-\infty}^{\infty} \sum_{n=-\infty}^{\infty} D_{mn} \exp(-j\gamma_{mn}z) \exp(j\kappa_{mx}x) \exp(j\kappa_{ny}y) \tag{7.20}$$

where $\kappa_{mx} = k_x = 2m\pi(1+\psi_x)/d_x$, $\kappa_{ny} = k_y = 2n\pi(1+\psi_y)/d_y$, and $\gamma_{mn}^2 = k^2 - \kappa_{mx}^2 - \kappa_{ny}^2$.

D_{mn} are currently unknown coefficients. The transverse field components can be obtained from Eq. 7.20 and Maxwell's equations in their reduced form for transverse electric (TE) fields as given by Eqs. 2.6. In the present instance of a rectangular geometry, let $\boldsymbol{\kappa}_t = \hat{x}k_x + \hat{y}k_y$ and therefore $\kappa_t^2 = \boldsymbol{\kappa}_t \cdot \boldsymbol{\kappa}_t = k_x^2 + k_y^2$. From Maxwell's equations, Eq. 7.20 results in

$$\mathbf{E}_t = -k\eta_0 \sum_{m=-\infty}^{\infty} \sum_{n=-\infty}^{\infty} D_{mn} \exp(-j\gamma_{mn}z) \left(\frac{\hat{z}\times\boldsymbol{\kappa}_t}{\kappa_t^2}\right) \exp(j\kappa_{mx}x) \exp(j\kappa_{ny}y) \tag{7.21a}$$

$$\mathbf{H}_t = \sum_{m=-\infty}^{\infty} \sum_{n=-\infty}^{\infty} \gamma_{mn} \left\{ D_{mn}\left(\frac{\boldsymbol{\kappa}_t}{\kappa_t^2}\right) \right\} \exp(-j\gamma_{mn}z) \exp(j\kappa_{mx}x) \exp(j\kappa_{ny}y) \tag{7.21b}$$

Suppose the aperture at $z=0$ is located in a ground plane and the electric field in the aperture is given by $\mathbf{E}_t(x,y,0) = \mathbf{E}_A$ over the domain of the unit cell denoted by D. With this boundary constraint, and after crossmultiplying both sides with $\hat{z}$, Eq. 7.21a becomes

$$\hat{z}\times\mathbf{E}_A = k\eta_0 \sum_{m=-\infty}^{\infty} \sum_{n=-\infty}^{\infty} D_{mn}\left(\frac{\boldsymbol{\kappa}_t}{\kappa_t^2}\right) \exp(j\kappa_{mx}x) \exp(j\kappa_{ny}y)$$

Taking the inverse transform of this complex series, then

$$D_{mn}\left(\frac{\boldsymbol{\kappa}_t}{\kappa_t^2}\right) = \frac{1}{k\eta_0 d_x d_y} \iint_{D'} (\hat{z}\times\mathbf{E}_A(x',y')) \exp(-j\kappa_{mx}x') \exp(-j\kappa_{ny}y') dx' dy'. \tag{7.22}$$

Let the aperture field components be given by $\mathbf{E}_A = \sum_{p=1}^{N} V_p \mathbf{e}_p$ and $\mathbf{H}_A = \mathbf{H}_t(x,y,0) = \sum_{p=1}^{N} I_p(\hat{z}\times\mathbf{e}_p)$ where $\mathbf{e}_p(x,y)$ are shape functions that satisfy the boundary conditions as well as orthogonality. These are often chosen to be the modes of the domain D, and N is the number of shape functions used. The term in the braces in Eq. 7.21b is replaced by Eq. 7.22. The expansions of the aperture fields are substituted into the result. With these substitutions, Eq. 7.21b now becomes

$$\sum_{p=1}^{N} I_p\left(\hat{z}\times\mathbf{e}_p\right) = \frac{1}{k\eta_o d_x d_y}\sum_{m=-\infty}^{\infty}\sum_{n=-\infty}^{\infty}\gamma_{mn}\exp(j\kappa_{mx}x)\exp(j\kappa_{ny}y)$$

$$\times\left(\sum_{q=1}^{N}V_q F_q\left(m,n,-\kappa_{mx},-\kappa_{ny}\right)\right)$$

where $F_p\left(m,n,\kappa_{mx},\kappa_{ny}\right) = \iint_{D'}\left(\hat{z}\times\mathbf{e}_p(x',y')\right)\exp(j\kappa_{mx}x')\exp(j\kappa_{ny}y')dx'dy'$.

The vector $\left(\hat{z}\times\mathbf{e}_p\right)(p=1,\dots N)$ is projected onto both sides of the linear equation and the result integrated in the unprimed co-ordinates across the domain D. Because of orthogonality, the equation simplifies to

$$I_p N_p = \frac{Y_o}{d_x d_y k}\sum_{q=1}^{N}\gamma_{mn}F_p\left(m,n,\kappa_{mx},\kappa_{ny}\right)F_q\left(m,n,-\kappa_{mx},-\kappa_{ny}\right)V_q$$

where $Y_o = 1/\eta_o$ and $N_p = \iint_D \mathbf{e}_p\cdot\mathbf{e}_p dx'dy'$. That is,

$$I_p = \sum_{q=1}^{N}Y_{pq}\left(\psi_x,\psi_y\right)V_q$$

or, alternatively, in matrix form $\mathbf{I} = \mathbf{Y}\,\mathbf{V}$ where $\mathbf{Y}$ is the admittance matrix with coefficients

$$Y_{pq}\left(\psi_x,\psi_y\right) = \frac{2Y_o}{d_x d_y k N_p}\sum_{m=-\infty}^{\infty}\sum_{n=-\infty}^{\infty}\gamma_{mn}F_p\left(m,n,\kappa_{mx},\kappa_{ny}\right)F_q\left(m,n,-\kappa_{mx},-\kappa_{ny}\right) \quad (7.23)$$

for a rectangular periodic array. The series in Eq. 7.23 converges rapidly with increasing indices m and n. As a first approximation, consider an infinite periodic array of rectangular apertures supporting only the TE$_{10}$ mode. When only one mode is used in Eq. 7.23, the result is called a 'grating lobe' series (Amitay et al., 1972; Hansen, 1998). This can be a useful approach for understanding coupling problems as well as the impact of mutual coupling on the grating lobes although care is required and also sometimes the single mode approximation is not very accurate. Continuing, let $N=1$ and $e_1 = \hat{y}\,\cos(\pi x/a)$ over part of the domain given by $|a/2|\geq x$, $|b/2|\geq y$ and be zero elsewhere. It can be shown that

$$F_1\left(\psi_x,\psi_y\right) = -\left(\frac{2ab}{\pi}\right)C\left(\frac{\kappa_{mx}a}{2}\right)S\left(\frac{\kappa_{ny}b}{2}\right)$$

where the functions $C(x)$ and $S(x)$ are defined in Appendix A. Also, the normalization is $N_1 = (ab)/2$.

Therefore,

$$Y_{11}\left(\psi_x,\psi_y\right) = \frac{Y_o 8ab}{\pi^2 d_x d_y k}\sum_{m=-\infty}^{\infty}\sum_{n=-\infty}^{\infty}\gamma_{mn}C\left(\frac{\kappa_{mx}a}{2}\right)C\left(\frac{-\kappa_x a}{2}\right)S\left(\frac{\kappa_y b}{2}\right)S\left(\frac{-\kappa_y b}{2}\right). \quad (7.24)$$

The series converges reasonably quickly if the summation in m has limits $M \geq \pm 4(1 + 1.5d_x/a)$ and similarly for n. Eq. 7.24 is similar to a result given by Diamond (1968) for the element driving admittance, which can be also written as

$$Y_{11}\left(\psi_x, \psi_y\right) = \sum_{m=-\infty}^{\infty} \sum_{n=-\infty}^{\infty} \bar{Y}(m,n)$$

where

$$\bar{Y}(m,n) = \frac{Y_o 8ab}{\pi^2 d_x d_y k} \sqrt{k^2 - \kappa_{mx}^2 - \kappa_{ny}^2} \, C\left(\frac{\kappa_{mx}a}{2}\right) C\left(\frac{-\kappa_x a}{2}\right) S\left(\frac{\kappa_y b}{2}\right) S\left(\frac{-\kappa_y b}{2}\right).$$

When all sources are turned on with the selected steering angles, an active reflection coefficient can be defined, which in the presence of only one mode is

$$\Gamma_a\left(\psi_x, \psi_y\right) = \left[\frac{1 - Y_{11}\left(\psi_x, \psi_y\right)}{1 + Y_{11}\left(\psi_x, \psi_y\right)}\right].$$

The coupling coefficients are given by the following complex Fourier series:

$$\Gamma_a\left(\psi_x, \psi_y\right) = \sum_{m=-\infty}^{\infty} \sum_{n=-\infty}^{\infty} S_{mn}^0 \exp(jm\psi_x) \exp(jn\psi_y). \tag{7.25}$$

where S_{mn}^0 is the coupling coefficient from the mnth element to the driven element. When the Fourier series is inverted, it is found that

$$S_{mn}^0 = \frac{1}{4\pi^2} \int_{-\pi}^{\pi} \int_{-\pi}^{\pi} \Gamma_a\left(\psi_x, \psi_y\right) \exp(-jm\psi_x) \exp(-jn\psi_y) \, d\psi_x d\psi_y.$$

As an example, a single-mode estimate of the active reflection coefficient of a periodic array of square waveguides on a square grid with dimensions $d_x = d_y = 0.5714\lambda$ is shown in Figure 7.10. Also shown is the magnitude from a multimode solution (Galindo & Wu, 1966). The calculated active reflection coefficient with a single mode at a scan angle of 20° is found to be $\Gamma_a = 0.30\angle-80°$. This compares with the value of $\Gamma_a = 0.32\angle149.4°$ quoted by Amitay et al. (1972), which was obtained with a basis of 30 pulse functions.

7.3.2   Finite Arrays

Ideally, the designer of arrays would wish to analyse a finite array as it can take into account factors such as edge elements and an irregular element arrangement and have different types of elements. Most arrays can be analysed as finite arrays although the computational effort can be considerable. The most common way of analysing mutual coupling in arrays of a finite number of elements is the element-by-element method (as indicated in Figure 7.7). This technique involves constructing an impedance or admittance matrix by considering each element of

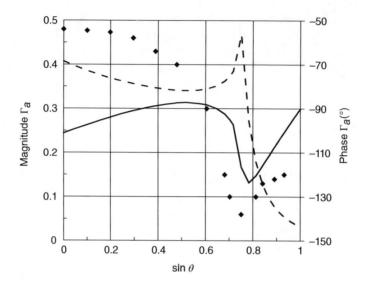

Figure 7.10 Input reflection coefficient in the *H*-plane scan of a periodic array of square waveguides obtained from a single-mode approximation. Dimensions $a = b = 0.5354\lambda$ and $d_x = d_y = 0.5714\lambda$. Solid line, magnitude; dashed line, phase; dot, Galindo and Wu (1966)

the array in turn in the presence of another element, with all other elements of the array removed either by physical means or by means of open- or short-circuiting the terminals. In the case of wire antennas, all except the two elements under consideration are open-circuited, when the mutual impedance is calculated. For slot and waveguide antennas, the mutual admittance of these two elements is obtained with the remaining elements short-circuited. (The special case of an element in isolation yields the diagonal or self-terms of the matrices.) How the mutual impedance or admittance is calculated depends on the type of antenna elements, and several formulae are in common use. To outline some of these formulae, initially consider arrays of wire antennas. Expressions for the admittance of arrays of slots are easily obtained by replacing field quantities in the usual way (see Table 2.2).

The theory given in Section 2.2 is readily extended to antennas with distributed currents from the representation given in Section 2.3.3. The extension requires taking account the phase delay between the current at the terminals of antenna 2, namely, I_2, and the current on the surface, S_2; the phase delay on antenna 1 is accounted for in the same way. Consider an infinitesimal element of antenna 2, which is illuminated by an incident electric field from antenna 1 given by $\mathbf{E_{21}}$. Suppose an electric current, $\mathbf{J_2}\,dS$, is induced on the element of antenna 2. At the terminals of antenna 2, an emf induced at the terminals is dV_{21}. For continuity of complex power, it is required that

$$I_2 dV_{21} = -\mathbf{J_2} \cdot \mathbf{E_{21}}(\mathbf{R})\,dS \qquad (7.26)$$

where I_2 is the current at the terminals of antenna 2 and $\mathbf{R}$ is the vector from antenna 1 to antenna 2. Consider initially the infinitesimal mutual impedance defined in Eq. 2.32. Now sum all such current elements over the surface of antenna 2 to give:

$$Z_{21} = \frac{V_2}{I_1} = -\frac{1}{I_1 I_2} \iint_{S_2} \mathbf{J_2 \cdot E_{21}}(\mathbf{R}) \, dS. \tag{7.27}$$

Equation 7.27 is a generalization of a formula first derived by Carter (1932). The self-impedance of the antenna, due to its own current, is given by Eq. 7.27 when the two antennas are coincident. An identical result was derived by Richmond (1961) by means of the Lorentz reciprocity theorem. Expressions for mutual impedance other than Eq. 7.27 are possible. For example, an expression that uses the complex conjugate instead of the current density arises from a derivation based on the complex Poynting vector (King, 1956). Another expression derived by Richmond (1961) uses total fields; that is, fields due to antenna 1 transmitting in the presence of antenna 2 and vice versa. It is shown in Appendix C that Eq. 7.27 is stationary with respect to the electric current and the incident field.

Expressions for the mutual impedance of two dipoles in various dispositions have been derived from Eq. 7.27 for assumed current distributions, and some of these are given by Brown (1937), Barzilai (1948) and Hansen (1966, 1998). For the case of two thin dipoles with lengths L_1 and L_2 supporting sinusoidal currents $\sin(k(L_{1,2} - |t|))$ (where $L_{1,2} > |t|$), the mutual impedance is approximately (Schelkunoff & Friis, 1952)

$$Z_{21} = \frac{j\eta_o}{4\pi \sin(kL_1/2) \sin(kL_2/2)} \int_{-L_2/2}^{L_2/2} \left(\frac{e^{-jkr'_{21}}}{r'_{21}} + \frac{e^{-jkr''_{21}}}{r''_{21}} - 2\cos\left(\frac{kL_1}{2}\right) \frac{e^{-jkr_{21}}}{r_{21}} \right) \sin\left(k\left(\frac{L_2}{2} - |t|\right) \right) dt \tag{7.28}$$

where $\mathbf{r}_{21}$, r'_{21} and r''_{21} are defined in Figure 7.11. Approximations given by Eq. 7.28 when the spacing between the dipoles, s, is greater than the length of the dipoles can be obtained (e.g. Gera, 1988). For half-wave dipoles arranged at broadside (i.e. H-plane coupling), the mutual impedance is

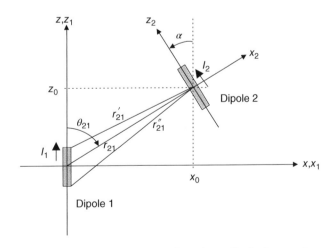

Figure 7.11   Geometry of coupling of two dipoles in x–z plane

$$Z_{21} \approx \frac{j\eta_o}{2\pi^2} \frac{e^{-jks}}{(s/\lambda)} \left[1 + \frac{((\pi/8)-(1/2\pi))}{(js/\lambda)} \right] \tag{7.29}$$

The above is accurate for $s > 0.75\lambda$. In the case of two collinear dipoles (i.e. E-plane coupling), then

$$Z_{21} \approx -\frac{\eta_o}{32\pi} \frac{e^{-jks}}{(s/\lambda)^2} \tag{7.30}$$

This equation is accurate for $s > \lambda$.

The above results are for the mutual impedance. From duality, as summarized in Table 2.2, a similar description is possible for mutual admittance. In particular, the mutual admittance obtained from the emf method is

$$Y_{21} = -\frac{1}{V_1 V_2} \iint_{S_2} \mathbf{M_2} \cdot \mathbf{H_{21}}(\mathbf{R}) \, dS \tag{7.31}$$

where V_1 and V_2 are the applied voltages, $\mathbf{M_2}$ is the magnetic current on antenna 2, and $\mathbf{H_{21}}$ is the magnetic field at antenna 2 from antenna 1. Another version of this formula that is applicable to planar waveguide and slot arrays is given by Borgiotti (1968). This expresses the mutual admittance in the form of a Fourier transform of a function that is related to the radiation pattern of the elements.

In the most general application of Eqs. 7.27 or other similar impedance formulae, the electric field at antenna 2 can be expressed in terms of a dyadic Green's function $\underset{=}{\mathbf{G}}^{(e)}$ for the electric field given as follows by

$$\mathbf{E_{21}}(\mathbf{R}) = \iint_{S_1} \underset{=}{\mathbf{G}}^{(e)}(\mathbf{R}|\mathbf{R'}) \cdot \mathbf{J_1}(\mathbf{R'}) \, dS' \tag{7.32}$$

where $\mathbf{J_1}$ is the current density on the surface of antenna 1 while $\mathbf{R}$ and $\mathbf{R'}$ are vectors from the origin to the field and the source points, respectively. Knowledge of the currents on antennas 1 and 2 is required to determine mutual impedance. Before the widespread use of digital computers, the practice was to assume a current based on approximations from on physical reasoning and to evaluate the integrals as best one could with hand calculators. Nevertheless, excellent results were achieved for dipole arrays, for example. However, improved results and mutual coupling between wider classes of antennas became possible with numerical methods, such as the method of moments. This approach for wire antennas has been described in detail in several texts including those by Harrington (1961) and also Stutzman & Thiele (1981) and, therefore, shall not be repeated here.

As mentioned earlier, one of the reasons good results are obtained with relatively simple current approximations is that Eqs. 7.27 and 7.31 are stationary with respect to small variations in the assumed currents. A proof of this stationary or variational behaviour of the mutual impedance expression Eq. 7.27 is given in Appendix C. Similar conclusions apply to the magnetic currents and the corresponding mutual admittance. This time the magnetic field is expressed in terms of the dyadic Green's function $\underset{=}{\mathbf{G}}^{(h)}$ for the magnetic field as follows (Collin, 1960):

$$\mathbf{H_{21}}(\mathbf{R}) = \iint_{S_1} \underset{=}{\mathbf{G}^{(h)}}(\mathbf{R}|\mathbf{R}') \cdot \mathbf{M_1}(\mathbf{R}') \, dS' \tag{7.33}$$

where $\mathbf{M_1}(\mathbf{R}')$ is the magnetic current on antenna 1. Therefore, the mutual admittance is expressed as

$$Y_{21} = -\frac{1}{V_1 V_2} \iint_{S_2} dS_2 \, \mathbf{M_2}(\mathbf{R}) \cdot \iint_{S_1} \underset{=}{\mathbf{G}^{(h)}}(\mathbf{R}|\mathbf{R}') \cdot \mathbf{M_1}(\mathbf{R}') \, dS_1. \tag{7.34}$$

As a particular case, an expression is obtained for the mutual admittance of two identical rectangular waveguides that terminate in a ground plane. The aperture width and height dimensions, respectively, are a and b. The free-space dyadic magnetic Green's function can be shown to be (see Appendix D)

$$\underset{=}{\mathbf{G}^{(h)}} = \frac{-jk}{2\pi\eta_0} \left(\mathbf{I} + \frac{1}{k^2} \nabla_t \nabla_t \right) G_0(|\mathbf{R} - \mathbf{R}'|) \tag{7.35}$$

where

$$|\mathbf{R} - \mathbf{R}'| = \sqrt{(x-x')^2 + (y-y')^2 + (z-z')^2}$$

$G_0(R) = \exp(-jkR)/R$ is the scalar free-space Green's function, $\mathbf{I}$ is the unit dyadic, and ∇_t is the gradient operator in the transverse plane. Let us approximate the magnetic current in the apertures using the TE field of the fundamental mode of rectangular waveguide, that is, $\hat{y} \cos(\pi x/a)$ and $\mathbf{M_1} = \mathbf{E_1} \times \hat{z} = \hat{x} V_1 \cos(\pi x/a)$. Similarly, $\mathbf{M_2} = \hat{x} V_2 \cos(\pi x'/a)$.

Therefore,

$$Y_{21} = \frac{-jk}{2\pi\eta_0} \iint_{S_2} dx' dy' \hat{x} \cos\left(\frac{\pi x'}{a}\right) \cdot \iint_{S_1} \left(\mathbf{I} + \frac{1}{k^2} \nabla_t \nabla_t \right) G_0(|\mathbf{R} - \mathbf{R}'|) \cdot \hat{x} \cos\left(\frac{\pi x}{a}\right) dx dy$$

or

$$Y_{21} = \frac{-jk}{2\pi\eta_0} \iint_{S_2} dx' dy' \cos\left(\frac{\pi x'}{a}\right) \iint_{S_1} \left(1 + \frac{1}{k^2} \frac{\partial}{\partial x'} \frac{\partial}{\partial x} \right) G_0(|\mathbf{R} - \mathbf{R}'|) \cos\left(\frac{\pi x}{a}\right) dx dy. \tag{7.36}$$

This expression can be integrated numerically with some difficulty due to the singularity in the Green's function when the field point approaches the source point. As is shown in Section 7.3.5, Eq. 7.36 can be expressed in closed form using a transformation of variables.

7.3.3   Mutual Impedance and Scattering Matrix Representation

In the element-by-element approach, finite arrays of antennas can be represented in terms of network parameters that depend on the type of array element. Consider initially an N element array of electric dipole antennas. Suppose that the dipoles have external voltages

V_p^{in} ($p = 1, \ldots, N$) applied to the input terminals. Application of the emf method to all dipoles requires that

$$\mathbf{V}^{in} = \mathbf{Z}\mathbf{I} \tag{7.37}$$

where $\mathbf{Z}$ is the impedance matrix for the mutual coupling in the array

$$\mathbf{Z} = \begin{bmatrix} Z_{11} & Z_{12} & \cdots & Z_{1N} \\ Z_{21} & Z_{22} & \cdots & Z_{2N} \\ \cdots & \cdots & \cdots & \cdots \\ Z_{N1} & \cdots & \cdots & Z_{NN} \end{bmatrix}. \tag{7.38}$$

The currents $\mathbf{I}$ at the terminals of the dipoles are obtained from the inverse of $\mathbf{Z}$ as follows:

$$\mathbf{I} = \mathbf{Z}^{-1}\mathbf{V}^{in}. \tag{7.39}$$

For arrays of magnetic dipoles, apertures and slots, the expressions are the dual of those for the electric dipole. In matrix notation, the relation between the externally applied current vector $\mathbf{I}^{in}$ and the induced voltage is obtained from

$$\mathbf{V} = \mathbf{Y}^{-1}\mathbf{I}^{in} \tag{7.40}$$

where $\mathbf{Y}$ is the admittance matrix.

It is often convenient to use a scattering matrix formulation because feeding networks can be fully included in the design. To relate this formulation to the impedance and admittance network representations, voltage and current need to be defined in terms of the complex amplitudes of forward and backward travelling waves on these lines. Let

a_k = complex amplitude of a wave travelling towards port k
b_k = complex amplitude of a wave travelling away from port k
Z_{ok} = characteristic wave impedance at port k

The associated wave admittance is $Y_{ok} = 1/Z_{ok}$. The voltage and current at port k are defined as follows:

$$V_k = Z_{ok}^{1/2}(a_k + b_k) \tag{7.41}$$

$$I_k = Z_{ok}^{-1/2}(a_k - b_k).$$

After substituting Eq. 7.41 into Eq. 7.40 and rearranging, it is found that

$$\mathbf{b} = \mathbf{S}\,\mathbf{a} \tag{7.42}$$

where

$$S = (z - I)/(z + I) = I - 2(I + z)^{-1}$$

is the scattering matrix of the array, **a** and **b** are column vectors of forward and backward travelling waves, **I** is the unit matrix, and **z** is a matrix of normalized impedances with elements

$$z_{pq} = Z_{pq} Z_{op}^{-1/2} Z_{oq}^{-1/2}$$

The normalization is usually relative to the free-space wave impedance η_o and the normalized impedance is indicated by the lower case letter z.

For an array of only two elements, the scattering matrix is

$$S = \frac{1}{\Delta} \begin{bmatrix} (z_{11} - 1)(z_{22} + 1) - z_{12}z_{21} & 2z_{12} \\ 2z_{21} & (z_{11} + 1)(z_{22} - 1) - z_{21}z_{12} \end{bmatrix}$$

where $\Delta = (z_{11} + 1)(z_{22} + 1) - z_{12}z_{21}$. Note that if the elements are identical and their inputs are well matched $z_{11} = z_{22} \approx 1$, then the coupling coefficient between the antennas is $S_{12} = S_{21} \approx z_{12}/2$. That is, as a general rule in a well-matched array of identical elements, the coupling coefficient is approximately half or 3 dB below the magnitude of the mutual impedance (or admittance), In the case of magnetic dipoles, Eq. 7.42 remains the same except that the scattering matrix is expressed as

$$S = \frac{(I - y)}{(I + y)} = 2(I + y)^{-1} - I \tag{7.43}$$

where **y** is the normalized admittance matrix with elements

$$y_{pq} = Y_{pq} Y_{op}^{-1/2} Y_{oq}^{-1/2}.$$

The normalization is relative to the free-space wave admittance and is denoted by a lower case letter y.

The scattering matrix **S** represents the coupling between the elements due to the external radiating region only. As a result, it is convenient to now replace the external region scattering matrix by $S^{(0)}$ that is fed by forward and reverse waves $a^{(0)}$ and $b^{(0)}$. Generally, each array feed element has its own individual scattering matrix, which connects to $S^{(0)}$, as illustrated in Figure 7.12. This shows that at the output of element p feeding into part of the external network, there are forward and reverse travelling waves with amplitudes $a_O^{(p)}$ and $b_O^{(p)}$. In turn, this network has input travelling waves with amplitudes $a_I^{(p)}$ and $b_I^{(p)}$, which are related through a scattering matrix $S_{ij}^{(p)}$. Suppose there are a total of N elements. These are related to the forward and backward waves at the aperture through

$$\begin{bmatrix} b_I \\ a_O \end{bmatrix} = \begin{bmatrix} S_{11} & S_{12} \\ S_{21} & S_{22} \end{bmatrix} \begin{bmatrix} a_I \\ b_O \end{bmatrix} \tag{7.44}$$

in which the elements of the scattering matrix are partitioned as shown:

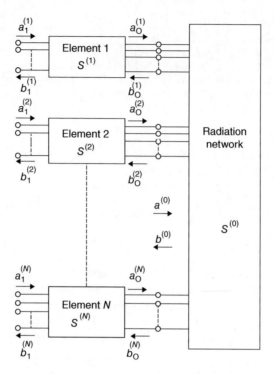

Figure 7.12   Network representation of an array

$$
\mathbf{S_{ij}} = \begin{bmatrix} S_{ij}^{(1)} & 0 & \dots & 0 \\ 0 & S_{ij}^{(2)} & \dots & 0 \\ \dots & \dots & \dots & \dots \\ 0 & \dots & \dots & S_{ij}^{(P)} \end{bmatrix}.
\tag{7.45}
$$

As the forward and reverse amplitudes $\mathbf{a_O} = \mathbf{a}^{(0)}$ and $\mathbf{b_O} = \mathbf{b}^{(0)}$ are related via Eq. 7.42, the reflected waves can be expressed in terms of the input wave amplitudes by

$$
\mathbf{b}_I = \left[\mathbf{S}_{11} + \mathbf{S}_{12}\mathbf{S}^{(0)} \left(\mathbf{U} - \mathbf{S}_{22}\mathbf{S}^{(0)} \right)^{-1} \mathbf{S}_{21} \right] \mathbf{a_I}
\tag{7.46}
$$

Equation 7.46 shows that mutual coupling, here given by $\mathbf{S}^{(0)}$, can have a significant influence on the input reflection coefficient if this coupling is strong and also the input network is not well matched.

7.3.4   Analysis of Arrays of Aperture Antennas by Integral Equation Methods

As Figure 7.7 shows both infinite and finite arrays can be analyzed with integral equation methods, which for arrays is a class of methods for solving problems where the unknowns are inside integral expressions. This commonly occurs for arrays of apertures on surfaces including curved ones. Such a method is useful as these types of arrays occur in a wide range of applications in communications and radar, especially where accuracy, efficiency and low sidelobes are required. The apertures may be physical apertures, such as arrays of waveguides (Figure 1.1a–c, 1) or two-dimensional microstrip patches (Figure 1.1k) mounted on other surfaces. Suppose the array is located on a surface that is designated by S_o. When the array is operating, let the fields radiated externally to this surface be $\mathbf{E}_1$, $\mathbf{H}_1$ and the fields inside the surface $\mathbf{E}_2$, $\mathbf{H}_2$. It is convenient to use the field equivalence principle to replace the field inside S_o by the null field. To do this and maintain continuity of the field, electric and magnetic currents $\mathbf{J}_o$, $\mathbf{M}_o$ are introduced on the surface. At this point, two types of arrays of practical interest can be identified which permit further simplification: (i) the array which comprises separate isolated electric conductors separated by homogeneous materials and (ii) the complementary situation when S_o is an electric conductor and the array comprises apertures in the conductor. Both of these antennas can be analysed rigorously by the integral equation method, and in this section, a solution of the latter problem is considered.

For an array of apertures in a conductor, the original equivalent sources are the electric current, $\mathbf{J}_o$, over the entire surface and a magnetic current, $\mathbf{M}_o$ which is non-zero in the apertures only. To remove the need to determine both sources, it is convenient to introduce an electric conductor into the null field region inside S_o and place it adjacent to the surface so that it completely shorts out $\mathbf{J}_o$. The result is the equivalent problem of an array of patches of magnetic current that are backed by an electric conductor. To solve this problem, one needs to know the Green's function for a magnetic dipole radiating in the presence of an electric conductor of shape given by S_o. Closed-form solutions may be found to this problem for some elementary surfaces, for example, the plane, the cylinder or the sphere (e.g. Felsen & Marcuvitz (1973)). The Green's function for the plane is derived in Appendix D. One may also consider the external region to comprise several homogeneous layers or shells (Galejs, 1969), but here for simplicity's sake, only a homogeneous external region is considered. When the surfaces have large curvatures, asymptotic solutions may be usefully employed. Further, by following the general principles of the geometrical theory of diffraction (GTD), asymptotic solutions for elementary surfaces may be generalized to approximate fields on arbitrary surfaces (e.g. Shapira et al., 1974; Pathak et al. 1981).

Let $\hat{n}$ be the unit outward normal to the surface S_o. If $\mathbf{E}_t$ is the tangential component of the electric field in the apertures, the magnetic current is given by

$$\mathbf{M}_o = \mathbf{E}_t \times \hat{n} \qquad (7.47a)$$

In the region outside S_o, the tangential component of the magnetic field is given by

$$\mathbf{H}^{\text{ext}}(\mathbf{R}) = \iint_{S_1} \underline{\underline{\mathbf{G}}}^{(h)}(\mathbf{R}|\mathbf{R}') \cdot \mathbf{M}_o(\mathbf{R}') dS \qquad (7.47b)$$

where $\mathbf{R}$ and $\mathbf{R}'$ are vectors from the origin to the field and source points respectively. In the special case when S_o is a plane, the appropriate Green's dyadic is best obtained by the method of images. Equations 7.47 form the basis of the present analysis of apertures on a plane. In the next chapter, a formulation will be described for curved surfaces.

Since the remaining patches of magnetic current are backed by a plane, electric conductor image theory says the field due to this arrangement is identical to a magnetic current of twice the strength that operates in free space. As shown in Appendix D, the dyadic Green's function for a planar source is given by

$$\underline{\mathbf{G}}^{(h)} = \frac{-jk}{2\pi\eta_o}\left(\mathbf{I} + \frac{1}{k^2}\nabla_t\nabla_t\right)G_o(|\mathbf{R}-\mathbf{R}'|) \tag{7.48}$$

where

$$|\mathbf{R}-\mathbf{R}'| = \sqrt{(x-x')^2 + (y-y')^2 + (z-z')^2},$$

$G_o(R) = \exp(-jkR)/R$ is the scalar free-space Greens function, $\mathbf{I}$ is the unit dyadic, and ∇_t is the gradient operator in the transverse plane. Now consider the region inside the surface S_o. The magnetic field therein consists of an incident field $\mathbf{H}^{\text{inc}}$ and a reverse travelling field due to the discontinuity at the apertures, $\mathbf{H}^{\text{rev}}$. For continuity of the field, the sum of the magnetic fields tangential to the inside surface should equal the tangential component of the field just outside. That is, at the surface,

$$\mathbf{H}_t^{\text{inc}} + \mathbf{H}_t^{\text{rev}} = \mathbf{H}_t^{\text{ext}} \tag{7.49}$$

The fields inside the waveguides leading to the apertures can be usefully approximated by an expansion of waveguide modes. Modal field solutions may be obtained by analytical methods for many common types of waveguide cross sections represented by separable co-ordinates such as circular, elliptical and rectangular geometries, while structures with general cross sections can be analysed by numerical means, such as the finite element method. Whatever method is adopted, a set of known modes with transverse fields is assumed to be available in the form $(\mathbf{e}_{pi}, \mathbf{h}_{pi})$. In waveguide, i the total transverse field $\left(\mathbf{E}_t^{(i)}, \mathbf{H}_t^{(i)}\right)$ consisting of both forward and reverse travelling waves is approximated as a finite sum of $M(i)$ modes as follows:

$$\mathbf{E}_t^{(i)} = \sum_{p=1}^{M(i)} \left(a_{pi}e^{-j\gamma_{pi}z} + b_{pi}e^{+j\gamma_{pi}z}\right)\mathbf{e}_{pi}(x,y)Y_{pi}^{-1/2} \tag{7.50a}$$

$$\mathbf{H}_t^{(i)} = \sum_{p=1}^{M(i)} \left(a_{pi}e^{-j\gamma_{pi}z} - b_{pi}e^{+j\gamma_{pi}z}\right)\mathbf{h}_{pi}(x,y)Y_{pi}^{+1/2} \tag{7.50b}$$

where

$$\gamma_{pi} = \beta_{pi} - j\alpha_{pi} = \text{propagation constant of mode } p \text{ in region } I,$$

$$\mathbf{h}_{pi} = \hat{z} \times \mathbf{e}_{pi}$$

and

$$\int_{S_i} dS\mathbf{e}_{pi} \times \mathbf{h}_{pi}\cdot\hat{z} = 2\delta_{pq} \tag{7.51}$$

where δ_{pq} is the Kronecker delta and Y_{pi} is the mode admittance of mode p in waveguide I, given by

$$Y_{pi} = \frac{1}{\eta_o} \begin{cases} \gamma_{pi}/k & \text{TE modes} \\ \varepsilon_r^{(i)} k/\gamma_{pi} & \text{TM modes} \end{cases} \tag{7.52}$$

where $\varepsilon_r^{(i)}$ is the relative permittivity of the filling in aperture i.

At this stage, the fields in the waveguides have been detailed, particularly the magnetic field, leading to the apertures and exterior regions, and these may be employed to satisfy, approximately, continuity of the magnetic field at the apertures (Eq. 7.49). It is the choice of the type of expansion functions in these approximate representations that has resulted in the coining of several different solution methods. Two of the many approaches possible are described in the following sections.

7.3.4.1 Moment Method Approach

In this approach, the field travelling away from the apertures is expressed also in terms of the magnetic current on the aperture. This was first detailed by Harrington and Mautz (1976) and since then has been used by others and to analyse arrays of rectangular waveguides (Fenn et al., 1982; Luzwick & Harrington, 1982), and also circular horns (Silvestro & Collin, 1989). Since the tangential component of the electric field should be continuous across the aperture, the magnetic current inside S_o must be $-\mathbf{M}_o$ (Eq. 7.47a). This is substituted into Eq. 7.49 to give

$$\mathbf{H}_1^{\text{inc}} + \mathbf{H}_t^{\text{rev}}(-\mathbf{M}_o) = \mathbf{H}_t^{\text{ext}}(\mathbf{M}_o). \tag{7.53}$$

$\mathbf{H}_1^{\text{rev}}$ may be expressed in the form of Eq. 7.47b, where $\underline{\underline{\mathbf{G}}}^{(h)}$ is replaced by an appropriate Green's dyadic for a magnetic source inside the conductor, although it is sometimes preferable to use a simpler representation. Further, one usually requires the amplitude of the modes in the waveguides to be an outcome of the analysis.

To satisfy Eq. 7.53 and solve for the magnetic current, the magnetic current is expressed in terms of N_e general expansion functions, here denoted by $\{\boldsymbol{\alpha}_n | n = 1, ..., N_e\}$, with unknown coefficients V_n as follows:

$$\mathbf{M}_o = \sum_{n=1}^{N_e} V_n \boldsymbol{\alpha}_n. \tag{7.54}$$

This approximation is now substituted into Eq. 7.53, and use is made of the linearity of the fields. However, since it cannot be assumed that continuity is satisfied exactly, define the residual of Eq. 7.53 as follows:

$$\mathbf{R} = \sum_{n=1}^{N_e} \left(\mathbf{H}_t^{\text{rev}}(\boldsymbol{\alpha}_i) + \mathbf{H}_t^{\text{ext}}(\boldsymbol{\alpha}_i) \right) V_i - \mathbf{H}_t^{\text{inc}}. \tag{7.55}$$

The approach adopted now is the method of weighted residuals (Finlayson, 1972). For the purpose of making $\mathbf{R}$ as small as possible, define a set of weighting functions $\{\boldsymbol{\beta}_n | n = 1, \ldots, N_e\}$ that are valid over S_o. In weighted residuals, the inner product (i.e. the integral over S_o) of the weighting functions and the residual is set to be zero; thus

$$(\boldsymbol{\beta}_i, \mathbf{R}) = \iint_{S_o} dS \; \boldsymbol{\beta}_i \cdot \mathbf{R} = 0 \tag{7.56}$$

The simplest set of weighting functions is constants, and in that case, Eq. 7.56 requires that the average residual is zero. Another simple set of weighting functions is delta functions, and use of these results in a point matching solution. The method of moments is another special case. One particular version that uses sub-sectional basis functions is popular for field calculations (Harrington, 1968). The most common approach is to use identical weighting and expansion functions, that is, $\{\boldsymbol{\alpha}_n = \boldsymbol{\beta}_n | n = 1, \ldots, N_e\}$. This approach is known as the Galerkin method, which is used here, and results in

$$(\boldsymbol{\beta}_i, \mathbf{R}) = \iint_{S_o} dS \; \boldsymbol{\alpha}_i \cdot \mathbf{R}$$
$$= \left[\left(\left(\boldsymbol{\alpha}_i, \mathbf{H}_t^{\mathrm{rev}}(\boldsymbol{\alpha}_j)\right) + \left(\left(\boldsymbol{\alpha}_i, \mathbf{H}_t^{\mathrm{ext}}(\boldsymbol{\alpha}_j)\right)\right)\right] \mathbf{V}_j - \left(\boldsymbol{\alpha}_i, \mathbf{H}_t^{\mathrm{inc}}\right) = 0. \tag{7.57}$$

The first two terms in the last line are admittance matrices for the reflected field and the exterior region given by

$$\mathbf{Y}^{\mathrm{rev}} = \left[Y_{ij}^{\mathrm{rev}}\right] = \left(\boldsymbol{\alpha}_i, \mathbf{H}_t^{\mathrm{rev}}(\boldsymbol{\alpha}_j)\right)$$

and

$$\mathbf{Y}^{\mathrm{ext}} = \left[Y_{ij}^{\mathrm{ext}}\right] = \left(\boldsymbol{\alpha}_i, \mathbf{H}_t^{\mathrm{ext}}(\boldsymbol{\alpha}_j)\right)$$

In addition, define a current source vector

$$\mathbf{I}^{\mathrm{inc}} = \left[\left(\boldsymbol{\alpha}_i, \mathbf{H}_t^{\mathrm{inc}}(\boldsymbol{\alpha}_j)\right)\right]$$

Therefore, Galerkin's method results in the matrix equation

$$\left(\mathbf{Y}^{\mathrm{rev}} + \mathbf{Y}^{\mathrm{ext}}\right) \mathbf{V} = \mathbf{I}^{\mathrm{inc}} \tag{7.58}$$

The physical picture provided by Eq. 7.58 is of two general admittances in parallel with a current source. By inverting the sum of the admittance matrices, the coefficients of magnetic current expansion are obtained from

$$\mathbf{V} = \left(\mathbf{Y}^{\mathrm{rev}} + \mathbf{Y}^{\mathrm{ext}}\right)^{-1} \mathbf{I}^{\mathrm{inc}}. \tag{7.59}$$

Once $\mathbf{V}$ has been found, the complex amplitudes of the reverse travelling fields in the wave-guides can be obtained. When Eqs. 7.50a and 7.54 are substituted into Eq. 7.47a, the result is

$$\sum_{i=1}^{N_e} \alpha_i V_i = \sum_{m=1}^{M_o} (a_m + b_m) Y_m^{-1/2} \mathbf{e}_m(x,y) \times \hat{z}. \tag{7.60}$$

Next, taking the inner product with $\mathbf{h}_n$ on both sides of this equation for the region S_o and using mode orthogonality through Eq. 7.51 results in

$$(a_m + b_m) = -\frac{Y_m^{1/2}}{2} \sum_{n,i=1}^{N_e} (\alpha_i, \mathbf{h}_m) V_i \tag{7.61}$$

As mentioned previously, several workers have used the moment method to analyse mutual coupling in rectangular apertures. All employ Galerkin's method, but differences occur in the choice of expansion functions. One approach (Luzwick & Harrington, 1982) uses linearly polarized expansion functions that are co-sinusoidal over the apertures, while another (Arndt et al., 1989) adopt a combination of pyramidal and triangular functions. In a further approach, Fenn et al. (1982) use two orthogonally polarized sets of overlapping piecewise sinusoidal-uniform surface patches as expansion and weighting functions. This leads to mutual coupling calculations involving only a single integration. Booker's relation Eq. 2.41 can be used to convert the mutual admittance formula to determine the mutual impedance of two parallel-staggered electric surface sources. Arrays of rectangular waveguides have been studied (Fenn et al., 1982), during which the method of images was used to determine the generalized admittance for the internal waveguide region.

7.3.4.2 Mode Matching in Arrays

A special case of the method of weighted residuals arises when the expansion functions are also the mode functions of the waveguides feeding the apertures. This time Eq. 7.50b is used in the equation of continuity of the magnetic field (Eq. 7.49). Further, Eq. 7.50a is used in Eq. 7.47b to determine the magnetic current. With these approximations, the residual of Eq. 7.47b at the ith aperture is

$$\mathbf{R}_i = \sum_{p=1}^{M(i)} (a_{pi} - b_{pi}) \mathbf{h}_{pi} Y_{pi}^{1/2} - \sum_{j=1}^{N} \sum_{q}^{M(j)} (a_{qj} + b_{qj}) Y_{qj}^{1/2} \int_{S_j} dS' \underline{\underline{\mathbf{G}}}^{(h)} (\mathbf{R}|\mathbf{R}') \cdot [\mathbf{e}_{qj} \times \hat{n}] \tag{7.62}$$

where without loss of generality it is assumed that $z = 0$. Using the $\mathbf{h}_{pi}$ as testing functions, the inner product is taken of both sides of Eq. 7.62 over S_i domain to obtain

$$(\mathbf{h}_{pi}, \mathbf{R}_i) = (a_{pi} - b_{pi}) 2 Y_{pi}^{1/2} - \sum_{j=1}^{N} \sum_{q}^{M(j)} (a_{qj} + b_{qj}) Y_{qj}^{1/2} \left(\mathbf{h}_{pi}, \int_{S_j} dS' \underline{\underline{\mathbf{G}}}^{(h)} (\mathbf{R}|\mathbf{R}') \cdot [\mathbf{e}_{qj} \times \hat{n}] \right) = 0$$

$$\tag{7.63}$$

where in addition use has been made of mode orthogonality given by Eq. 7.51. The series in the middle of Eq. 7.63 involves the normalized mutual admittance between modes p and q in apertures i and j which is given by

$$y_{ij}(p|q) = \frac{1}{2\sqrt{Y_{pi}Y_{pj}}} \int_{S_i} dS\, \mathbf{h}_{pi}(u_1,u_2) \cdot \int_{S_j} dS'\, \underset{=}{\mathbf{G}}^{(h)}(\mathbf{R}-\mathbf{R}') \cdot \left[\mathbf{e}_{qj}(u_1',u_2') \times \hat{n} \right] \qquad (7.64)$$

where (u_1, u_2) are co-ordinates on aperture i and similarly the primed co-ordinates refer to aperture j. It is seen that the mutual admittance is symmetric with respect to interchange of modes and apertures; thus

$$y_{ji}(q|p) = y_{ij}(p|q). \qquad (7.65)$$

With the weighted residual set the zero, Eq. 7.63 becomes

$$\mathbf{a} - \mathbf{b} = \mathbf{y}(\mathbf{a} + \mathbf{b})$$

and

$$\mathbf{b} = (1+\mathbf{y})^{-1}(1-\mathbf{y})\mathbf{a}$$
$$= \mathbf{S}\mathbf{a}$$

where $\mathbf{S} = (\mathbf{I}+\mathbf{y})^{-1}(\mathbf{I}-\mathbf{y})$ is the scattering matrix of the apertures. The scattering matrix enables the amplitudes of the reverse travelling mode amplitudes to be obtained. In the special case of apertures in a conducting ground plane, Eq. 7.64 may be simplified using the fact that the mode functions satisfy the boundary conditions on the waveguide walls. It is found that the elements of the normalized admittance matrix can be further expressed (Bird, 1979) as

$$y_{ij}(p|q) = \frac{jk}{4\pi\eta_0\sqrt{Y_{pi}Y_{qj}}} \int_{S_i} dS\, \boldsymbol{\Psi}_{pi}(u_1,u_2) \cdot \int_{S_j} dS'\, \boldsymbol{\Psi}_{qj}(u_1',u_2') G(|\mathbf{R}-\mathbf{R}'|) \qquad (7.66)$$

where a vector mode function for the admittance is defined by

$$\boldsymbol{\Psi}_{pi}(x,y) = \mathbf{h}_{pi} + \hat{z}\frac{1}{jk}\nabla_t \cdot \mathbf{h}_{pi}$$
$$= \mathbf{h}_{pi} + \hat{z}\frac{\beta}{k}h_{z,pi}. \qquad (7.67)$$

It is observed that Eq. 7.66 is valid for arbitrarily shaped apertures, which are homogeneously filled, provided that the fields are solutions of Maxwell's equations and that they satisfy the boundary conditions for the given geometry.

The method described above can be extended to the modes of partially dielectric loaded waveguides (Bird & Hay, 1990), and although the admittance matrix elements are given by Eq. 7.66, this time the mode function is expressed in terms of longitudinal section electric (LSE) or longitudinal section magnetic (LSM) modes. In this case,

$$\mathbf{\Psi}_{pi}(x,y) = \hat{z} \times \mathbf{e}_{pi} + \hat{z}\eta_o h_{z,pi} \tag{7.68}$$

where $\mathbf{e}_{pi}$ and $h_{z,pi}$ are the TE field and axial magnetic field components, respectively, of LSE or LSM mode p in waveguide i.

The number of modes needed for an accurate representation of the aperture field depends on the operating frequency. If the waveguide or horn operates in the fundamental mode and all other modes are well below cut-off, a good estimate of reflection is obtained by considering the fundamental mode only in the scattering matrix. Use of several high-order modes is recommended, however, for accurate predictions. Satisfaction of the edge condition is not critical except when there is a thin iris at the aperture, although Hockham (1975) showed that for rectangular waveguide, inclusion of TE_{m0} ($m = 3, 5, \ldots$) and TE_{0n} ($n = 2, 4, \ldots$) modes improve solution convergence.

The mode matching approach has been applied to study mutual coupling and radiation in a variety at waveguides terminated in a ground plane. These have closed-form solutions, which are described in the next section.

7.3.5   Mutual Coupling Analysis in Waveguide Apertures

The approach so far has been developing a general formalism for planar arrays. In this section, results for a number of practical finite planar arrays are obtained. The methods described earlier have been applied to specific aperture geometries, namely, rectangular, circular coaxial and elliptical, where there are closed-form modal solutions in a range of waveguides (circular (Bird, 1979, 1996), coaxial (Bird, 2004), rectangular (Bird, 1990a) and elliptical (Bird, 1990b)) that terminate in a ground plane. For more general waveguide geometries, a numerical method is required, and a simple approach based on the results of a finite element method is also described.

7.3.5.1   Rectangular Waveguide Arrays

Arrays of rectangular waveguides or horns are used in many applications that require good polarization characteristics and simplicity. Horns can be analysed accurately through more sophisticated techniques such as mode matching (refer to Section 4.5.2). The advantage of the latter approach is that the mode representation is consistent with the radiating aperture case. Consider a suitable vector for use with the mutual admittance formula (Eq. 7.66). In this section, mutual coupling is described for a finite array of different-sized rectangular waveguides. Expressions for the mutual admittance of all possible combinations of mode coupling are provided. A formula is given for reducing the order of integration in these expressions, allowing the quadruple integral for mutual admittance to be expressed as several double integrals instead. Results are provided that show excellent agreement between theory and experiment.

The transverse field vectors for rectangular waveguides $\mathbf{e}_p$ and $\mathbf{h}_p$ (the single integer subscript p is interchanged on occasions with the dual indices mn to simplify notation) that are normalized for unit power transfer are summarized as follows:

TE$_{mn}$ Modes
The TE field is given by

$$\mathbf{e}_{mni}(x,y) = \sqrt{\frac{2\varepsilon_{om}\varepsilon_{on}}{a_i b_i k_{c,mn}^2}}\left[\hat{x}\frac{n\pi}{b_i}\cos\left(\frac{m\pi}{a_i}x\right)\sin\left(\frac{n\pi}{b_i}y\right) - \hat{y}\frac{m\pi}{a_i}\sin\left(\frac{m\pi}{a_i}x\right)\cos\left(\frac{n\pi}{b_i}y\right)\right]; \quad m,n \geq 0 \tag{7.69}$$

where $k_{c,mn} = \sqrt{((m\pi)/a_i)^2 + (n\pi/b_i)^2}$ and $\varepsilon_{ov} = \begin{cases} 1 & v=0 \\ 2 & v>0 \end{cases}$.

The corresponding transverse magnetic field is

$$\mathbf{h}_{pi} = \mathbf{h}_{mni} = \hat{z} \times \mathbf{e}_{mni}$$

and the axial magnetic field component is

$$h_{zpi} = \frac{1}{j\beta}k_{c,mn}^2 \cos\left(\frac{m\pi}{a_i}x\right)\cos\left(\frac{n\pi}{b_i}y\right).$$

Therefore, the vector field function for admittance calculations (Eq. 7.67) in rectangular apertures is

$$\mathbf{\Psi}_{mni}(x,y) = \sqrt{\frac{2\varepsilon_{om}\varepsilon_{on}}{a_i b_i k_{c,mn}^2}}\left[\hat{x}\frac{m\pi}{a_i}\sin\left(\frac{m\pi}{a_i}x\right)\cos\left(\frac{n\pi}{b_i}y\right) + \hat{y}\frac{n\pi}{b_i}\cos\left(\frac{m\pi}{a_i}x\right)\sin\left(\frac{n\pi}{b_i}y\right)\right.$$
$$\left. + \hat{z}\frac{k_{c,mn}^2}{jk}\cos\left(\frac{m\pi}{a_i}x\right)\cos\left(\frac{n\pi}{b_i}y\right)\right] \tag{7.70}$$

TM$_{mn}$ Modes
The vector field function for these modes are obtained from

$$\left(\mathbf{\Psi}_{mn}\right)\big|_{\text{TM}} = \hat{z} \times \left(\mathbf{\Psi}_{mn}\right)\big|_{\text{TE}}. \tag{7.71}$$

Substitution of Eqs. 7.70 and 7.71, respectively, into Eq. 7.66 for each case, results in a mutual admittance expression for rectangular apertures given by

$$y_{ij}(m,n|m'n') = \frac{jk\pi Y_o}{8}\alpha_{mn}\alpha_{m'n'}\left(c_x I_x + c_y I_y + c_z I_z\right) \tag{7.72}$$

where

$$\alpha_{mn} = \sqrt{\frac{\varepsilon_{om}\varepsilon_{on}}{a_i b_i Y_{mn} k_{c,mn}^2}}$$

$$\begin{cases} I_x \\ I_y \\ I_z \end{cases} = \int\limits_{D_i} dS \int\limits_{D_j} dS' \, G(x-x',y-y') \begin{cases} \sin \\ \cos \\ \cos \end{cases} \left(\frac{m\pi x}{a_i}\right) \begin{cases} \cos \\ \sin \\ \cos \end{cases} \left(\frac{n\pi y}{b_i}\right) \times \begin{cases} \sin \\ \cos \\ \cos \end{cases} \left(\frac{m'\pi x'}{a_j}\right) \begin{cases} \cos \\ \sin \\ \cos \end{cases} \left(\frac{n'\pi y'}{b_j}\right)$$

$$(7.73)$$

The coefficients c_x, c_y, c_z in Eq. 7.72 are listed in Table 7.1 for the four combinations of mode coupling.

The integrals over the domains D_i and D_j can be completed in closed form leaving only a double integral in the transform domain. The most direct way of doing this is through numerical evaluation of the double integrals by means of either conventional numerical quadrature (Fröberg, 1974) or the fast Fourier transform (FFT) algorithm (Brigham, 1974; Oppenheim & Shafer, 1975). In practice, both approaches are similar; the main difficulty being to ensure that the integral converges in the region $k \leq |\xi|$, $|\eta| < \infty$ in particular. A way of accelerating the convergence of integrals in this region has been described by Kitchiner et al. (1987).

An analytical approach to evaluating the admittance formula and reducing the computation time is to employ Lewin's method (Lewin, 1951). This method can be extended to handle mode coupling between separate, different-sized rectangular apertures. In this approach, the order of integration in the four-dimensional integrals is changed to an integration over the (x, x') domain and separately the (y, y') domain as illustrated in Figure 7.13., which leaves two surface integrals having the same form. Considering the (x, x') domain, the integral is

$$I_{xx'} = \int_0^{a_i} dx \int_0^{a_j} dx' \, B(x-x') \frac{\cos}{\sin}\left(\frac{m\pi x}{a_i}\right) \frac{\cos}{\sin}\left(\frac{m'\pi x'}{a_j}\right) \qquad (7.74)$$

where $B(x-x')$ is a function with integrable singularities due to the scalar Green's function. Initially, let $\sigma = x - x'$ and $\lambda = y - y'$. As two other variables are required for this four-dimensional integration, let $\nu = x + x' - a_i$ and $\mu = y + y' - b_i$. The original and new regions of integration are

Table 7.1 Coefficients of admittance formula

Mode coupling	c_x	c_y	c_z
$TE_{mn} \leftrightarrow TE_{m'n'}$	$\dfrac{mm'}{a_i a_j}$	$\dfrac{nn'}{b_i b_j}$	$\left(\dfrac{k_{c,mn} k_{c,m'n'}}{\pi k}\right)^2$
$TM_{mn} \leftrightarrow TM_{m'n'}$	$\dfrac{nn'}{b_i b_j}$	$\dfrac{mm'}{a_i a_j}$	0
$TE_{mn} \leftrightarrow TM_{m'n'}$	$-\dfrac{mn'}{a_i b_j}$	$\dfrac{nm'}{b_i a_j}$	0
$TM_{mn} \leftrightarrow TE_{m'n'}$	$-\dfrac{nm'}{b_i a_j}$	$\dfrac{mn'}{a_i b_j}$	0

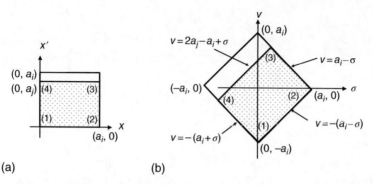

Figure 7.13 Regions of integration in analysis of mutual coupling in different-sized apertures. (a) Original domain and (b) domain after replacement of variables

shown in Figure 7.13. Now introduce the change of variable into the admittance formula and in Eq. 7.74. Note that $dxdx' = d\sigma d\nu/2$. With these substitutions, the integral over μ can be evaluated in closed form so that $I_{xx'}$ becomes

$$I_{xx'} = \pm \frac{1}{4} \left\{ \int_0^{a_i} d\sigma \left[\begin{array}{l} B(-\sigma)M_\pm\left(-\sigma, \sigma - a_i, 2a_j - a_i - \sigma, s_m, d_m\right) + \\ B(\sigma + a_i - a_j)M_\pm\left(\sigma + a_i - a_j, \sigma - a_j, a_j - \sigma, s_m, d_m\right) \end{array} \right] \right.$$
$$\left. - \int_0^{a_i - a_j} d\sigma B(\sigma)M_\pm\left(\sigma, \sigma - a_i, 2a_j - a_i + \sigma, s_m, d_m\right) \right\}$$

(7.75)

where $s_m, d_m = 1/2\left(m \pm m'\left(a_i/a_j\right)\right)$ and

$$M_\pm\left(\sigma, \nu_1, \nu_2, s, d\right) = \frac{2a_i}{\pi} \left\{ \frac{1}{s} \sin\left[\frac{s\pi}{2a_i}(\nu_1 - \nu_2)\right] \cdot \cos\left[s\pi\left(1 + \frac{(\nu_1 + \nu_2)}{2a_i}\right) + \frac{\sigma d\pi}{a_i}\right] \right.$$
$$\left. \pm \frac{1}{d} \sin\left[\frac{d\pi}{2a_i}(\nu_1 - \nu_2)\right] \cdot \cos\left[d\pi\left(1 + \frac{(\nu_1 + \nu_2)}{2a_i}\right) + \frac{\sigma s\pi}{a_i}\right] \right\}.$$

(7.76)

The above integral reduction formula (Eq. 7.75), is also applicable to the (y, y') domain. With both integrals simplified as in Eq. 7.75, the mutual admittance in Eq. 7.72 becomes

$$y_{ij}(m,n,m',n') = -\frac{jk\pi Y_o}{64}\alpha_{mn}\alpha_{m'n'}\left[\int_0^{a_j} d\sigma \left\{ \int_0^{b_j} d\lambda[T(-\sigma, \sigma_1, \sigma_2 | -\lambda, \lambda_1, \lambda_2)] \right. \right.$$
$$+ T(-\sigma, \sigma_1, \sigma_2 | \lambda_3, \lambda_4, -\lambda_4) + T(\sigma_3, \sigma_4, -\sigma_4 | -\lambda, \lambda_1, \lambda_2) + T(\sigma_3, \sigma_4, -\sigma_4 | \lambda_3, \lambda_4, -\lambda_4)]$$
$$\left. + \int_0^{b_i - b_j} d\lambda[T(-\sigma, \sigma_1, \sigma_2 | \lambda, \lambda_1, \lambda_2) + T(\sigma_3, \sigma_4, -\sigma_4 | \lambda, \lambda_1, \lambda_2)] \right\}$$
$$+ \int_0^{a_i - a_j} d\sigma \left\{ \int_0^{b_j} d\lambda[T(\sigma, \sigma_1, \sigma_2 | -\lambda, \lambda_1, \lambda_2) + T(-\sigma, \sigma_1, \sigma_2 | \lambda_3, \lambda_4, -\lambda_4)] \right.$$
$$\left. \left. + \int_0^{b_i - b_j} d\lambda T(\sigma, \sigma_1, \sigma_2 | \lambda, \lambda_1, \lambda_2) \right\} \right]$$

(7.77)

where

$$\sigma_1 = \sigma - a_i$$
$$\sigma_2 = 2a_j - a_i - \sigma$$
$$\sigma_3 = \sigma + a_i - a_j$$
$$\sigma_4 = \sigma - a_j$$
$$\lambda_1 = \lambda - b_i$$
$$\lambda_2 = 2b_j - b_i - \lambda$$
$$\lambda_3 = \lambda + b_i - b_j$$
$$\lambda_4 = \lambda - b$$

and

$$T(x,\nu_1,\nu_2|y,\mu_1,\mu_2) = G(X_{0i} - X_{0j} + x, Y_{0i} - Y_{0j} + y)$$
$$\times [c_x M_-(x,\nu_1,\nu_2,s_m,d_m) M_+(y,\mu_1,\mu_2,s_n,d_n)$$
$$+ c_y M_+(x,\nu_1,\nu_2,s_m,d_m) M_-(y,\mu_1,\mu_2,s_n,d_n)$$
$$+ c_z M_+(x,\nu_1,\nu_2,s_m,d_m) M_+(y,\mu_1,\mu_2,s_n,d_n)].$$

In the special case of identical apertures, that is, $a_i = a_j = a$ and $b_i = b_j = b$, the integrand simplifies to the single double integral, which is due to the first integral on the right side of Eq. 7.77 as there are no contributions from the other integrals. The term in parentheses in Eq. 7.72 becomes

$$c_x I_x + c_y I_y + c_z I_z = \left[\int_0^a d\sigma \left\{ \int_0^b d\lambda F(\sigma,\lambda,m,n,m',n') \right. \right.$$
$$\times [G(\sigma,\lambda) + \cos m\pi \cos m'\pi G(-\sigma,\lambda) + \cos n\pi \cos n'\pi G(\sigma,-\lambda)$$
$$+ \cos m\pi \cos m'\pi \cos n\pi \cos n'\pi G(-\sigma,-\lambda)]$$

The function in the integrand $F(\sigma, \lambda, m, n, m', n')$ is summarized in Table 7.2 for each case. Further simplification is possible for modes coupling in the same aperture. As will now be described in the next section.

7.3.5.2 Self-Admittance of TE$_{10}$ Mode

The self-admittance of the TE$_{10}$ mode at the aperture of the open rectangular waveguide is required in most calculations of mutual coupling and can be a good first approximation to the practical value. Now consider the fundamental mode only in the same aperture by letting $m = m' = 1$ and $n = n' = 0$. As a result of the simplifications obtained for same-sized apertures in the previous section, the self-admittance of the fundamental mode is given by

Table 7.2　Function in integrand of mutual admittance of modes in isolated rectangular apertures

Mode coupling	$F(\sigma, \lambda, m, n, m'n')$
$TE_{mn} \leftrightarrow TE_{m'n'}$	$\dfrac{mm'}{a^2}L_-(\sigma,a,s_m,d_m)L_+(\lambda,b,s_n,d_n) + \dfrac{nn'}{b^2}L_+(\sigma,a,s_m,d_m)L_-(\lambda,b,s_n,d_n)$
	$+ \left(\dfrac{k_{c,mn}k_{c,m'n'}}{\pi k}\right)^2 L_+(\sigma,a,s_m,d_m)L_+(\lambda,b,s_n,d_n)$
$TM_{mn} \leftrightarrow TM_{m'n'}$	$\dfrac{nn'}{b^2}L_-(\sigma,a,s_m,d_m)L_+(\lambda,b,s_n,d_n) + \dfrac{nn'}{a^2}L_+(\sigma,a,s_m,d_m)L_-(\lambda,b,s_n,d_n)$
$TE_{mn} \leftrightarrow TM_{m'n'}$	$\dfrac{nm'}{ab}L_+(\sigma,a,s_m,d_m)L_-(\lambda,b,s_n,d_n) - \dfrac{mn'}{ab}L_-(\sigma,a,s_m,d_m)L_+(\lambda,b,s_n,d_n)$
$TM_{mn} \leftrightarrow TE_{m'n'}$	$\dfrac{mn'}{ab}L_+(\sigma,a,s_m,d_m)L_-(\lambda,b,s_n,d_n) - \dfrac{nm'}{ab}L_-(\sigma,a,s_m,d_m)L_+(\lambda,b,s_n,d_n)$

$L_\pm(x,a,p,q) = \cos\left[\pi\left(p+\frac{qx}{a}\right)\right]K(-x,a,p) \pm \cos\left[\pi\left(q+\frac{px}{a}\right)\right]K(-x,a,q)K(-x,a,p) = (a-x)S\left[p\left(1-\frac{x}{a}\right)\right]$

$$y_{11}(1,0) = \left(\frac{2j\pi k^2}{a^3 b \beta_{10} k_{c,10}^2}\right)\left[\int_0^a d\sigma \int_0^b d\lambda\, G(\sigma,\lambda)F(\sigma,\lambda,1,0,1,0)\right] \tag{7.78}$$

where

$$F(\sigma,\lambda,1,0,1,0) = (b-\lambda)\left[\left(1+\left(\frac{\pi}{ka}\right)^2\right)\left(\frac{a}{\pi}\right)\sin\left(\frac{\pi\sigma}{a}\right) + \left(1-\left(\frac{\pi}{ka}\right)^2(a-\sigma)\cos\left(\frac{\pi\sigma}{a}\right)\right)\right]$$

$G(\sigma,\lambda) = \exp\left(-jk\sqrt{\sigma^2+\lambda^2}\right)/\sqrt{\sigma^2+\lambda^2}$, $k_{c,10} = \pi/a$, and β_{10} is the TE_{10} propagation constant. The reflection coefficient is approximately obtained from Eq. 7.78 in the usual way through $\Gamma \approx (1-y_{11}(1,0))/(1+y_{11}(1,0))$. Eq. 7.78 agrees with the result obtained by Lewin (1951, p. 126) for a waveguide with no dielectric filling. There is a singularity at the origin, and this can be eliminated by the use of polar co-ordinates as follows. Let $\sigma = t\cos\alpha$ and $\lambda = t\sin\alpha$ where $d\sigma d\lambda = tdt\,d\alpha$. As a consequence, Eq. 7.78 is expressed alternatively as

$$y_{11}(1,0) = \left(\frac{2j\pi k^2}{a^3 b \beta_{10} k_{c,10}^2}\right)\left\{\left(\int_0^{\alpha_o} d\alpha \int_0^{a\sec\alpha} tdt + \int_{\alpha_o}^{\pi/2} d\alpha \int_0^{b\cosec\alpha} tdt\right)\right.$$

$$\left. \exp(-jkt)F(t\cos\alpha, t\sin\alpha, 1,0,1,0)\right\} \tag{7.79}$$

where $\tan\alpha_o = b/a$. The inner integral with respect to t can be integrated out in terms of elementary functions, but it is generally simpler and cleaner to integrate Eq. 7.79 numerically.

　　A computer program (called RECAR) based on the analysis described in this and the previous section has been implemented to allow for all possible combinations of mode coupling in the calculation of radiation from a finite array of different-sized rectangular

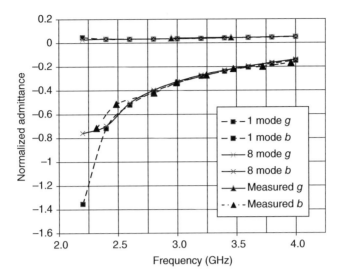

Figure 7.14 Computed and measured (Adams, 1966) conductance and susceptance of open-ended rectangular waveguide with $a = 2.186$ cm and $b = 1.016$ cm loaded with dielectric $\varepsilon_r = 9.68$

apertures. Further, provision has been made for slightly flared sections, thus allowing pyramidal horns to be analysed, and also for parasitic elements. Whatever the array configuration, by far, the greatest computational effort and time goes into determining the admittance matrix. As mentioned earlier, the admittance of modes coupling in the same waveguide is calculated using one-dimensional numerical integration. For all other mode admittances, integration in two dimensions is needed. The program calculates both single and double integrals by means of Simpson's rule algorithm with interval bisection until the error in the approximation of the integral is $<10^{-3}$.

As an example, Figure 7.14 shows the reflection coefficient of an open-ended WG-16 waveguide that has been computed from Eq. 7.79. Also shown are the results of a more accurate calculation with eight modes and also some experimental results (Adams, 1966). The waveguide dimensions are $a = 2.286$ cm and $b = 1.016$ cm, and it is loaded with Stycast HI-K material with dielectric constant 9.68, which reduces the cut-off frequency to 2.12 GHz.

The mutual coupling analysis has been verified for different-sized waveguides by comparing computed results with measurements made with a three-element test array. A square waveguide (element #1) with side length $a_1 = b_1 = 22.8$ mm is located at the centre of a ground plane, and two rectangular waveguides with dimensions $a_{2,3} = 15.7$ mm and $b_{2,3} = 7.7$ mm are located at $(0, -30.0)$ mm and $(30.0, 0)$ mm relative to the central of element 1. All three elements were connected to waveguide-to-coaxial adaptors for measurement with a network analyser. For the computations, seven modes were used in each waveguide – namely, TE_{10}, TE_{11}, TM_{11}, TE_{02}, TE_{20}, TE_{12} and TM_{12} modes. The reflection coefficient in the central square waveguide due to transition to free space is given in Figure 7.15a, and the coupling coefficients for $(TE_{10})_1 \leftrightarrow (TE_{10})_2$ in apertures 1 and 2 and $(TE_{10})_1 \leftrightarrow (TE_{10})_3$ in apertures 1 and 3 are shown in Figure 7.15b and c, respectively. The oscillation in the measured data is mainly due to diffraction from the edge of the ground plane, which is not included in the present

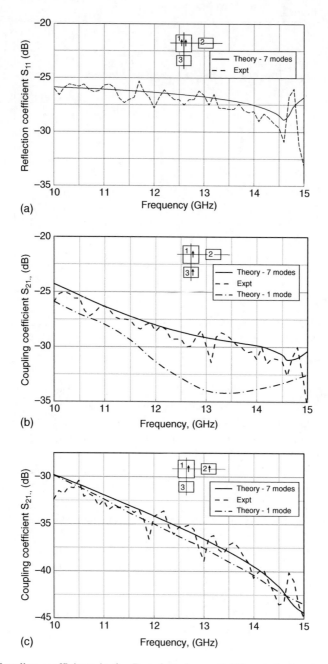

Figure 7.15 Coupling coefficients in the *E*- and *H*-planes of different-sized rectangular waveguide. Element 1 is square with side length 22.8 mm, and elements 2 and 3 have width 15.7 mm and height 7.7 mm. The centre-to-centre spacing from the square element is 30 mm in both directions. (a) Reflection coefficient at central square aperture. (b) *E*-plane coupling (apertures 1 and 3) and (c) *H*-plane coupling (apertures 1 and 2) (Bird, 1990a)

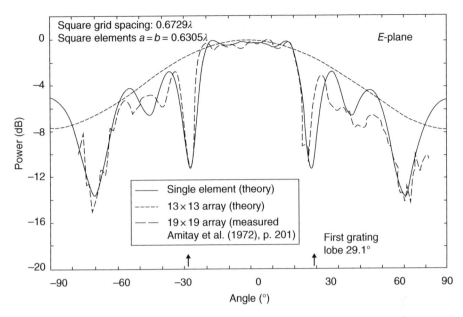

Figure 7.16   Radiation pattern of an element at the centre of a square lattice array of square waveguides $a = b = 0.6305\lambda$ with lattice grid spacing 0.6729λ.

analysis described here. Figure 7.15 shows that a seven-mode solution is in good agreement with experiment.

The importance of including modes other than the fundamental TE_{10} mode in the calculations has been observed, particularly when these modes are above cut-off. Calculations for reflection coefficient, E-plane coupling coefficients and cross-polar radiated fields are especially sensitive.

The impact of mutual coupling on the radiation pattern is demonstrated in Figure 7.16, for a finite rectangular array of square elements with side length $a = b = 0.6305\lambda$. It shows results for the single isolated waveguide and a 13×13 array with element spacing 0.6729λ, which were obtained with the computer program RECAR in which five modes were specified in each aperture, namely, TE_{10}, TE_{12}, TM_{12}, TE_{11} and TM_{11}. Also shown in the figure are measured results for a 19×19 array (Amitay et al., 1972, p. 201).

The approach for reducing the order of integration, namely, Lewin's method, has wider application than the one described in this section. With minor modification, it has been used in coupling problems involving different-sized rectangular subregions of bigger apertures such as can occur in single or multiple apertures partially loaded with different dielectric regions.

7.3.5.3   Arrays of Circular and Coaxial Waveguides

The geometry for the coupling between two coaxial waveguides is shown in Figure 7.17. The case when $b_i = 0$ corresponds to circular waveguide. The mode functions are summarized below for waveguide i:

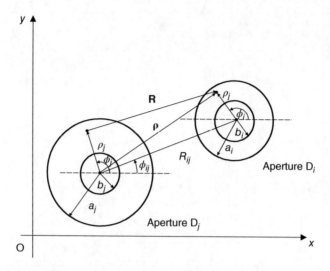

Figure 7.17   Coupling between two different coaxial waveguides

$(TE)_m \equiv TE_{pq}$ Modes

$$\Psi_{mix} = \frac{\chi_{mi}}{2}\left[-Z_{p-1}(k_{mi}\rho,\beta_{mi})\cos\left[(p-1)\phi-\psi_{mi}\right] + Z_{p+1}(k_{mi}\rho,\beta_{mi})\cos\left[(p+1)\phi-\psi_{mi}\right]\right]$$

$$\Psi_{miy} = \frac{\chi_{mi}}{2}\left[Z_{p-1}(k_{mi}\rho,\beta_{mi})\sin\left[(p-1)\phi-\psi_{mi}\right] + Z_{p+1}(k_{mi}\rho,\beta_{mi})\sin\left[(p+1)\phi-\psi_{mi}\right]\right]$$

$$\Psi_{miz} = -j\frac{\chi_{mi}k_{mi}}{k_o}Z_p(k_{mi}\rho,\beta_{mi})\cos\left[p\phi-\psi_{mi}\right] \qquad (7.80)$$

$(TM)_m \equiv TM_{pq}$ Modes

$$\Psi_{mix} = -\frac{\chi_{mi}}{2}\left[\Lambda_{p-1}(k_{mi}\rho,\beta_{mi})\sin\left[(p-1)\phi-\psi_{mi}\right] + \Lambda_{p+1}(k_{mi}\rho,\beta_{mi})\sin\left[(p+1)\phi-\psi_{mi}\right]\right]$$

$$\Psi_{miy} = \frac{\chi_{mi}}{2}\left[-\Lambda_{p-1}(k_{mi}\rho,\beta_{mi})\cos\left[(p-1)\varphi-\psi_{mi}\right] + \Lambda_{p+1}(k_{mi}\rho,\beta_{mi})\cos\left[(p+1)\varphi-\psi_{mi}\right]\right] \quad (7.81)$$

$TEM \equiv TM_{00}$ Modes

$$\Psi_{mix} = -\frac{\chi_{mi}}{\rho}\sin\phi, \quad \Psi_{miy} = \frac{\chi_{mi}}{\rho}\cos\phi \qquad (7.82)$$

where by definition $\Psi_{miz}=0$ for the TM and TEM modes. The functions Z_p and Λ_p are combinations of ordinary Bessel functions of order p as described in Appendix B, k_{mi} is the cut-off wavenumber of mode m in aperture i, $\alpha_{mi}=k_{mi}a_i$, $\beta_{mi}=k_{mi}b_i$, a_i and b_i are the outer and inner

Table 7.3 Mutual admittance $y_{ij}(m, n)$ of modes $m \equiv (ps)$ and $n \equiv (qt)$ in separate coaxial apertures i and j. Self-admittance is recovered when $R_{ij} \to 0$

	$(TE_{qt})_j$	$(TM_{qt})_j$	$(TEM)_j$
$(TE_{ps})_i$	$\kappa_{mn}\left[\dfrac{G^x_{mn}}{k^2 a_i a_j} + \dfrac{pq}{(\alpha_{mi}\alpha_{nj})^2} F^x_{mn} \right]$	$\kappa_{mn} p N^x_{mn}$	$\kappa_{mn}(-1)^p p C^{(1)}_{p,0,p}$ $(d_{mi}, 1; 0, 0; R_{ij})$ $\times \sin(p\phi_{ij} - \psi_{mi})$
$(TM_{ps})_i$	$\kappa_{mn} q M^x_{mn}$	$\kappa_{mn} L^x_{mn}$	$\kappa_{mn}(-1)^p C^{(1)}_{p,0,p}$ $(c_{mi}, 1; u_m, 0; R_{ij})$ $\times \cos(p\phi_{ij} - \psi_{mi})$
$(TEM)_i$	$\kappa_{mn}(-1)^q q C^{(1)}_{0,q,q}$ $(1, d_{nj}; 0, 0; R_{ij})$ $\times \sin(q\phi_{ij} - \psi_{nj})$	$\kappa_{mn}(-1)^q C^{(1)}_{0,q,q}$ $(1, c_{nj}; 0, u_n; R_{ij})$ $\times \cos(q\phi_{ij} - \psi_{nj})$	$\kappa_{mn} C^{(1)}_{0,0,0}(1, 1; 0, 0; R_{ij})$

conductor radii of aperture i, ψ_{mi} is the polarization angle of mode m and χ_{mi} is a scale factor in each case as follows:

$$\chi_{mi} = \sqrt{\frac{2\varepsilon_{0p}}{\pi}} \begin{cases} \left[a_i^2 Z_p^2(\alpha_{mi}, \beta_{mi})(1 - p^2/\alpha_{mi}^2) - b_i^2 Z_p^2(\beta_{mi}, \beta_{mi})(1 - p^2/\beta_{mi}^2) \right]^{-1/2} & \text{TE} \\[2ex] \left[a_i^2 \Lambda_p'^2(\alpha_{mi}, \beta_{mi}) - b_i^2 \Lambda_p'^2(\beta_{mi}, \beta_{mi}) \right]^{-1/2} & \text{TM} \\[2ex] \left[2\ln(a_i/b_i) \right]^{-1/2} & \text{TEM} \end{cases}$$

in which ε_{0p} is Neumann's number that has a value 1 if $p = 0$ or 2 if $p > 0$. A prime on Λ_p designates the first derivative with respect to the first argument.

The mutual coupling between coaxial elements is found by substituting the mode functions for the three mode types given by Eqs. 7.80–7.82 into the mutual admittance formula Eq. 7.66. The results are summarized in Table 7.3.

The functions used in Table 7.3 are defined hereunder:

$$F^x_{mn} = -S^{(1)}_{p,q} C^{(1)}_{p,q,p+q}(d_{mi}, d_{nj}; 0, 0; R_{ij}) + D^{(1)}_{p,q} C^{(1)}_{p,q,q-p}(d_{mi}, d_{nj}; 0, 0; R_{ij}) \tag{7.83a}$$

$$G^x_{mn} = S^{(1)}_{p,q} C^{(2)}_{p,q,p+q}(d'_{mi}, d'_{nj}; u_m, u_n; R_{ij}) + D^{(1)}_{p,q} C^{(2)}_{p,q,q-p}(d'_{mi}, d'_{nj}; u_m, u_n; R_{ij}) \tag{7.83b}$$

$$L^x_{mn} = S^{(1)}_{p,q} C^{(1)}_{p,q,p+q}(c_{mi}, c_{nj}; u_m, u_n; R_{ij}) + D^{(1)}_{p,q} C^{(1)}_{p,q,q-p}(c_{mi}, c_{nj}; u_m, u_n; R_{ij}) \tag{7.83c}$$

$$M^x_{mn} = S^{(2)}_{p,q} C^{(1)}_{p,q,p+q}(c_{mi}, d_{nj}; u_m, 0; R_{ij}) - D^{(2)}_{p,q} C^{(1)}_{p,q,q-p}(c_{mi}, d_{nj}; u_m, 0; R_{ij}) \tag{7.83d}$$

$$N^x_{mn} = S^{(2)}_{p,q} C^{(1)}_{p,q,p+q}(d_{mi}, c_{nj}; 0, u_n; R_{ij}) + D^{(2)}_{p,q} C^{(1)}_{p,q,q-p}(d_{mi}, c_{nj}; 0, u_n; R_{ij}) \tag{7.83e}$$

where

$$S_{p,q}^{(1,2)} = (-1)^p \begin{Bmatrix} \cos \\ \sin \end{Bmatrix} [(p+q)\phi_{ij} - \psi_{mi} - \psi_{nj}],$$

$$D_{p,q}^{(1,2)} = \begin{Bmatrix} \cos \\ \sin \end{Bmatrix} [(p-q)\phi_{ij} - \psi_{mi} + \psi_{nj}],$$

$$c_{mi} = \frac{b_i \Lambda_p'(\beta_{mi}, \beta_{mi})}{a_i \Lambda_p'(\alpha_{mi}, \beta_{mi})},$$

$$d_{mi} = \frac{Z_p(\beta_{mi}, \beta_{mi})}{Z_p(\alpha_{mi}, \beta_{mi})},$$

$$d_{mi}' = \frac{b_i Z_p(\beta_{mi}, \beta_{mi})}{a_i Z_p(\alpha_{mi}, \beta_{mi})}.$$

The remaining functions in Table 7.3 are Hankel transforms involving products of Bessel functions:

$$C_{p,q,\nu}^{(1)}(\alpha, \beta; u_m, u_n; s) = \int_0^\infty dw \frac{w^3}{\sqrt{1-w^2}} \frac{[J_p(ka_iw) - \alpha J_p(kb_iw)]}{(w^2 - u_m^2)}$$

$$\times \frac{[J_q(ka_jw) - \beta J_q(kb_jw)]}{(w^2 - u_n^2)} J_\nu(ksw) \tag{7.84a}$$

and

$$C_{p,q,\nu}^{(2)}(\alpha, \beta; u_m, u_n; s) = \int_0^\infty dw w \sqrt{1-w^2} \frac{[J_p'(ka_iw) - \alpha J_p'(kb_iw)]}{(w^2 - u_m^2)}$$

$$\times \frac{[J_q'(ka_jw) - \beta J_q'(kb_jw)]}{(w^2 - u_n^2)} J_\nu(ksw) \tag{7.84b}$$

where $u_m = \alpha_{mi}/ka_i$, $y_m = Z_p(\alpha_{mi}, \beta_{mi})$, $y_m' = \Lambda_p'(\alpha_{mi}, \beta_{mi})$, $\alpha_{mi} = k_{mi}a_i$, and the scale factor of the admittance, κ_{mn}, is listed in Table 7.4. The properies of the Hankel transforms are described in more detail in Appendix F. The mutual admittances for the cases given in Table 7.3 also apply for self-admittance in which case $R_{ij} \to 0$. Also the functions in Table 7.3 are replaced as follows:

$$F_{mn}^x \Rightarrow F_{mn} = C_{p,q,0}^{(1)}(d_{mi}, d_{nj}; 0,0; 0) \cos(\psi_{nj} - \psi_{nj}) \tag{7.85a}$$

$$G_{mn}^x \Rightarrow \frac{2G_{mn}}{\varepsilon_{om}} = 2C_{p,q,0}^{(2)}(d_{mi}', d_{nj}'; u_m, u_n; 0) \frac{\cos(\psi_{nj} - \psi_{nj})}{\varepsilon_{0m}} \tag{7.85b}$$

Table 7.4　Scale factors κ_{mn} in mutual admittances for coaxial waveguide modes

	$(TE_{qt})_j$	$(TM_{qt})_j$	$(TEM)_j$
$(TE_{ps})_i$	$\dfrac{\pi a_i a_j y_m y_n \alpha_{mi} \alpha_{nj} \chi_{mi} \chi_{nj} Y_o}{2\sqrt{Y_{mi}Y_{nj}}}$	$\dfrac{-\pi a_i a_j y_m y'_n \chi_{mi} \chi_{nj} Y_o}{2\alpha_{mi}\sqrt{Y_{mi}Y_{nj}}}$	$\dfrac{\pi a_i y_m \chi_{mi} \chi_{nj} Y_o}{\alpha_{mi}\sqrt{Y_{mi}Y_{nj}}}$
$(TM_{ps})_i$	$\dfrac{-\pi a_i a_j y'_m y_n \chi_{mi} \chi_{nj} Y_o}{2\alpha_{nj}\sqrt{Y_{mi}Y_{nj}}}$	$\dfrac{\pi a_i a_j y'_m y'_n \chi_{mi} \chi_{nj} Y_o}{2\sqrt{Y_{mi}Y_{nj}}}$	$\dfrac{-\pi a_i y'_m \chi_{mi} \chi_{nj} Y_o}{\sqrt{Y_{mi}Y_{nj}}}$
$(TEM)_i$	$\dfrac{\pi a_j y_n \chi_{mi} \chi_{nj} Y_o}{\alpha_{nj}\sqrt{Y_{mi}Y_{nj}}}$	$\dfrac{-\pi a_j y'_n \chi_{mi} \chi_{nj} Y_o}{\sqrt{Y_{mi}Y_{nj}}}$	$\left(\sqrt{\varepsilon_{ri}}\ln\frac{a_i}{b_i}\cdot\sqrt{\varepsilon_{rj}}\ln\frac{a_j}{b_j}\right)^{-1/2}$

Table 7.5　Mutual admittance $y_{ij}(m, n)$ of modes $m\equiv(ps)$ and $n\equiv(qt)$ in separate circular apertures i and j

	$(TE_{qt})_j$	$(TM_{qt})_j$
$(TE_{ps})_i$	$\kappa_{mn}\left[\dfrac{G^x_{mn}}{k^2 a_i a_j} + \dfrac{pq}{(\alpha_{mi}\alpha_{nj})^2}F^x_{mn}\right]$	$\kappa_{mn}p N^x_{mn}$
$(TM_{ps})_i$	$\kappa_{mn}q M^x_{mn}$	$\kappa_{mn}L^x_{mn}$

$$L^x_{mn} \Rightarrow \frac{2L_{mn}}{\varepsilon_{0m}} = 2C^{(1)}_{p,q,0}\left(c_{mi},c_{nj};u_m,u_n;0\right)\frac{\cos\left(\psi_{nj}-\psi_{nj}\right)}{\varepsilon_{0m}} \tag{7.85c}$$

$$M^x_{mn} \Rightarrow M_{mn} = C^{(1)}_{p,q,0}\left(c_{mi},d_{nj};u_m,0;0\right)\sin\left(\psi_{nj}-\psi_{nj}\right) \tag{7.85d}$$

$$N^x_{mn} \Rightarrow N_{mn} = C^{(1)}_{p,q,0}\left(d_{mi},c_{nj};0,u_n;0\right)\sin\left(\psi_{nj}-\psi_{nj}\right). \tag{7.85e}$$

When the radii of the inner conductors shrink to zero, that is, $b_{i,j}\to 0$, the mode functions for TE and TM modes reduce to those of circular waveguide. In this instance, it can be shown that $c_{mi}\to 0$, $d_{mi}\to 0$ and $d'_{mi}\to 0$. Thus, the mode coupling formulae for the TE and TM modes coupling in different circular waveguides are given in Table 7.5 where for $b_{i,j}\to 0$ it can be shown that

$$F^x_{mn} = -S^{(1)}_{p,q}C^{(1)}_{p,q,p+q}\left(0,0;0,0;R_{ij}\right) + D^{(1)}_{p,q}C^{(1)}_{p,q,q-p}\left(0,0,0;R_{ij}\right) \tag{7.86a}$$

$$G^x_{mn} = S^{(1)}_{p,q}C^{(2)}_{p,q,p+q}\left(0,0;u_m,u_n;R_{ij}\right) + D^{(1)}_{p,q}C^{(2)}_{p,q,q-p}\left(0,0;u_m,u_n;R_{ij}\right) \tag{7.86b}$$

$$L^x_{mn} = S^{(1)}_{p,q}C^{(1)}_{p,q,p+q}\left(0,0;u_m,u_n;R_{ij}\right) + D^{(1)}_{p,q}C^{(1)}_{p,q,q-p}\left(0,0;u_m,u_n;R_{ij}\right) \tag{7.86c}$$

$$M^x_{mn} = S^{(2)}_{p,q}C^{(1)}_{p,q,p+q}\left(0,0;u_m,0;R_{ij}\right) - D^{(2)}_{p,q}C^{(1)}_{p,q,q-p}\left(0,0;u_m,0;R_{ij}\right) \tag{7.86d}$$

$$N^x_{mn} = S^{(2)}_{p,q}C^{(1)}_{p,q,p+q}\left(0,0;0,u_n;R_{ij}\right) + D^{(2)}_{p,q}C^{(1)}_{p,q,q-p}\left(0,0;0,u_n;R_{ij}\right) \tag{7.86e}$$

where Eqs. 7.84 become

$$C^{(1)}_{p,q,\nu}(0,0;u_m,u_n;s) = \int_0^\infty dw \frac{w^3}{\sqrt{1-w^2}} \frac{J_p(ka_iw)\,[J_q(ka_jw)]}{(w^2-u_m^2)\,(w^2-u_n^2)} J_\nu(ksw) \tag{7.87a}$$

and

$$C^{(2)}_{p,q,\nu}(0,0;u_m,u_n;s) = \int_0^\infty dw\, w\sqrt{1-w^2} \frac{J'_p(ka_iw)\,J'_q(ka_jw)}{(w^2-u_m^2)\,(w^2-u_n^2)} J_\nu(ksw) \tag{7.87b}$$

The Hankel transforms should be integrated with care in the vicinity of the poles and the branch cut located at $w=1$. In some cases, the result can be represented by series. Appendix F should be consulted for further details of this type of Hankel transforms.

7.3.5.4 Self-Admittance of TE$_{11}$ Mode in Circular Waveguide

The self-admittance of the TE$_{11}$ mode in the aperture of an open-ended circular waveguide is an example and an important practical result both for single apertures and also for mutual coupling calculations. The mutual admittance for a single aperture is given by the top-left element of Table 7.5 with G^x_{mn} and F^x_{mn} replaced by the functions G_{mn} and F_{mn} as given by Eqs. 7.85. For the TE$_{11}$ mode $m=1=n$, it is found that the self-admittance of the TE$_{11}$ mode is

$$y_{11}(1,1) = \left(\frac{2x_1^4 k}{(x_1^2-1)\beta_{11}}\right)\left(\frac{G_{11}}{(ka)^2} + \frac{F_{11}}{(x_1)^4}\right) \tag{7.88}$$

where $x_1 = 1.841184$ is the first zero of $J'_1(x_1)=0$ corresponding to the TE$_{11}$ mode, β_{11} is the propagation constant, and

$$F_{11} = C^{(1)}_{1,1,0}(0,0;0,0;0) = \int_0^\infty dw \frac{(J_1(kaw))^2}{w\sqrt{1-w^2}}$$

$$= \int_0^1 dw \frac{(J_1(kaw))^2}{w\sqrt{1-w^2}} + j \int_1^\infty dw \frac{(J_1(kaw))^2}{w\sqrt{w^2-1}} \tag{7.89}$$

$$= \left(\frac{1}{2} - \frac{J_1(2ka)}{2ka}\right) + j\frac{\tilde{H}_1(2ka)}{(2ka)}$$

where $u_1 = x_1/ka$ and $\tilde{H}_1$ is a Struve function of order 1, which can be expressed in a power series (Abramowitz & Stegun, 1965) as follows:

$$\tilde{H}_1(x) = \frac{2}{\pi}\left(1 + \sum_{q=0}^\infty \frac{\varepsilon_{0n}J_{2q}(x)}{(4q^2-1)}\right)$$

and

$$G_{11} = C_{1,1,0}^{(2)}(0,0;u_1,u_1;0) = \int_0^\infty dw\, w\sqrt{1-w^2}\,\frac{\left[J_1'(kaw)\right]^2}{\left(w^2-u_1^2\right)^2}.$$

The functions G_{11} and also $C_{p,q,\nu}^{(1,2)}$ can be integrated directly with some care. The contours of both integrals run along the real positive w-axis, which can lead to oscillatory integral values and slow convergence. The integration path can be deformed into the complex plane, and this improves convergence. For integrals of this type, one of the Bessel functions in the integrand can be replaced by their equivalent in terms of Hankel functions of first and second kind. Thus, in G_{11}, one of the Bessel functions in the product is replaced with $J_p'(x) = \left(H_p^{(1)'} + H_p^{(2)'}\right)/2$. Therefore,

$$G_{11} = \frac{1}{2}\int_0^\infty dw\, w\sqrt{1-w^2}\,\frac{J_1'(kaw)\left(H_1^{(1)'} + H_1^{(2)'}\right)}{\left(w^2-u_1^2\right)^2}$$

$$= G_1 + G_2$$

The contour for G_1 involving $H_1^{(1)'}$ can now be deformed into the $w = u + jv$-plane along the positive jv axis, and this enables the replacement of the Hankel functions by their modified Bessel equivalents (see Appendix B), which helps to achieve better convergence. Similarly, the contour for G_2 is deformed around the branch cut and along the $-jv$ line. The details of this approach have been described by the author (Bird, 1979) and will not be repeated here except to quote the final results for both contour integrals

$$G_1(ka) = -\frac{j}{\pi}\int_0^\infty dv\,\frac{v\sqrt{1+v^2}I_1'(kav)K_1'(kav)}{\left(v^2+u_1^2\right)^2} \tag{7.90a}$$

and

$$G_2(ka) = \int_0^1 du\,\frac{u\sqrt{1-u^2}J_1'(kau)H_1^{(2)'}(kau)}{\left(u^2-u_1^2\right)^2}$$

$$-\frac{j}{\pi}\int_0^\infty dv\,\frac{v\sqrt{1+v^2}I_1'(kav)K_1'(kav)}{\left(v^2+u_1^2\right)^2} \tag{7.90b}$$

$$+\frac{\pi ka}{4u_1}\sqrt{1-u_1^2}J_1''(kau_1)Y_1'(kau_1)H(u_1-1)$$

where I_1' and K_1' are derivatives of the modified Bessel function (Appendix B) and $H()$ is the Heaviside step function. The third term is a contribution from the pole on the real u-axis when the waveguide is operating above its cut-off (i.e. $u_1 > 1$).

As an example, consider an isolated circular waveguide in a ground plane. The waveguide has a diameter of 5.7 cm, and its reflection coefficient was measured from 3.3 to 5.0 GHz.

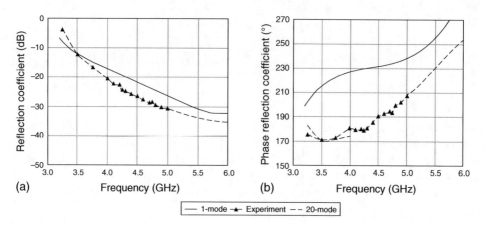

Figure 7.18 Reflection coefficient of circular waveguide with diameter 5.7 cm. (a) Magnitude in dB and (b) phase

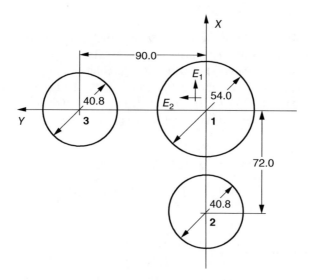

Figure 7.19 Array of three circular waveguides in an infinite ground plane. Dimensions are in mm

The experimentally obtained values are shown in Figure 7.18 along with results calculated using 20 modes. The latter result closely approximates the measured data, which becomes noisy above 4.5 GHz as the calibration of the network analyser became more difficult due to over-moding in the transition to the antenna under test. For comparison, the result given by the TE_{11} mode self-admittance (1-mode) in the top-left column of Table 7.3 is also shown in Figure 7.18.

An array example is illustrated in Figure 7.19. This array consists of three waveguides with apertures that terminate in a large metallic ground plane. A 5.4 cm-diameter waveguide is located at the centre of a large ground plane, and two 4.08 cm apertures are located nearby.

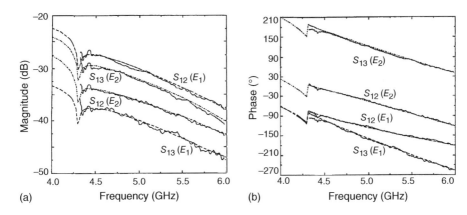

Figure 7.20   Coupling coefficients for the three-element array of circular waveguides for elements 1 and 2 (S_{12}) and elements 1 and 3 (S_{13}) corresponding to polarizations E_1 and E_2. Solid curve, experiment; dashed, theory. (a) Magnitude and (b) phase. *Source*: Reproduced with permission of IET (Bird, 1996)

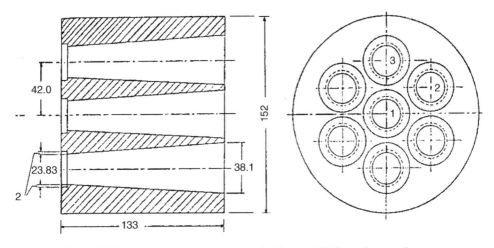

Figure 7.21   Seven-element array of conical horns. All dimensions are in mm

Measurements were taken over a 3.25–6 GHz frequency range for the two polarizations E_1 and E_2. Computations were made assuming nine modes in each waveguide, namely, TE_{1n} and TM_{1n} ($n = 1, 2$), TE_{2n} and TM_{2n} ($n = 1, 2$) and TM_{01}. The results obtained are shown in Figure 7.20 and are compared with the experimental coupling coefficients for the TE_{11} mode in each waveguide when the input to the waveguides is assumed matched. Excellent agreement is obtained between experiment and theory.

As a final example of circular aperture coupling, consider the seven-element array of conical horns illustrated in Figure 7.21. All aperture diameters are 3.81 cm.

Figures 7.22 and 7.23 show the measured and computed mutual coupling between elements 1 and 3 for cases when horn 1 is excited with vertical and horizontal polarization. These plots

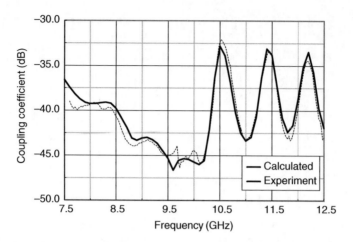

Figure 7.22   Measured and computed E-plane coupling of elements 1 and 3 in the seven-element conical horn array. Element 1 is vertically polarized

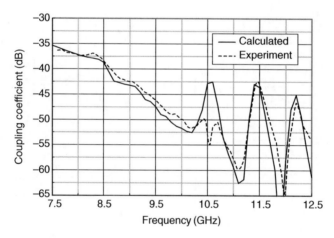

Figure 7.23   Measured and computed H-plane coupling of elements 1 and 3 in the seven-element conical horn array. Element 1 is horizontally polarized

correspond to E- and H-plane coupling in the seven-element array. The computations were performed with 16 modes in each waveguide (viz. TE_{1n} and TM_{1n} ($n = 1,\ 4$); TM_{01}, TM_{02}, TE_{2n} and TM_{2n} ($n = 1, 2$); TE_{31} and TM_{31}), and the flare was modelled using the mode matching method. Once again, the simulated results are in excellent agreement with the measured data.

7.3.5.5   Mutual Coupling in Other Geometries

Waveguides and horns that have general cross sections can be analysed for mutual coupling using the admittance formula given in Eq. 7.66 (Kuehne & Marquardt, 2001). For separable

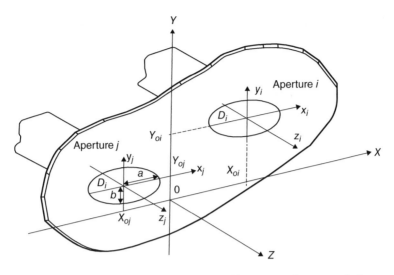

Figure 7.24 Geometry of elliptical waveguide apertures in a ground plane

co-ordinate systems, mode functions can be readily obtained in closed form and in some instances the integrals can also be evaluated in closed form. This depends on the availability of transform functions in those domains. The fourfold integral in Eq. 7.66 must be evaluated directly for general cross sections. In some separable co-ordinate systems, the number of integrals may be reduced by replacing the Green's function with an equivalent transform representation. Such a representation exists for elliptical co-ordinates, but this offers few advantages for numerical evaluation of Eq. 7.66 as the resulting integral is still difficult. One advantage though is that the transform representation is useful for identifying modes that couple.

Horns with an elliptical cross section (see Figure 7.24) find application as sources of circular polarization and as feeds for reflectors for producing shaped beams from satellites. Radiation from elliptical corrugated horns (Vokurka, 1979) has been studied in the past, but relatively little attention has been paid to the smooth-wall variety, and in particular reflection at the aperture. There are infinite sets of TE and transverse magnetic modes in elliptical waveguide, designated TE_{cmn}, TE_{smn}, TM_{cmn} and TM_{smn} where in the subscripts c and s refer to the even and odd types of Mathieu functions (McLachlan, 1964) required for each mode set. Mention is made of the work of Müller (1960) who described solutions for the radiation characteristics of smooth-wall elliptical waveguides and gave results for radiation in the fundamental TE_{c11} mode. Experience with smooth-wall horn radiators of rectangular and circular cross section has shown that significant reflection can occur from apertures of moderate size, resulting in a mismatch at the horn input. A complete solution for elliptical apertures has been given by the author (Bird, 1990b), and both theoretical and experimental results were provided. As shown in the previous section for circular apertures, the free-space Green's function can be used this time for elliptical apertures in a ground plane. The aperture field and ground plane are replaced by a magnetic current distribution as described in Section 7.3.4. Of the four basic mode sets in elliptical regions the c modes are have the electric field parallel to the minor axis (y-axis in Figure 7.24) and the s modes are polarized parallel to the major axis.

The fundamental mode of elliptical waveguide limits to the TE_{11} mode in circular waveguide as $b \to a$. The mutual admittance of the TE_{cmn} and $\mathrm{TE}_{cm'n'}$ modes in elliptical waveguide apertures D and D' is given by

$$y_{ij}(mn|m'n') = \frac{jkY_o hh' D_{mn} D_{m'n'}}{4\pi\sqrt{Y_{mn}Y_{m'n'}}} \int_0^{\xi_o} d\xi \int_0^{2\pi} d\eta \int_0^{\xi_o'} d\xi' \int_0^{2\pi} d\eta' G(X-X'|Y-Y')$$

$$\times \left[f_x(\xi,\eta,q)f_x(\xi',\eta',q') + f_y(\xi,\eta,q)f_y(\xi',\eta',q') - \left(\frac{k_{c,mn}^2 hk_{c,m'n'}^2 h'}{k^2} \right) f_z(\xi,\eta,q)f_z(\xi',\eta',q') \right]$$

(7.91)

where the functions in the integrand are as follows:
TE_{cmn} modes

$$f_x(\xi,\eta,q) = Ce_m'(\xi,q)ce_m(\eta,q)v - Ce_m(\xi,q)ce_m'(\eta,q)u$$

$$f_y(\xi,\eta,q) = Ce_m'(\xi,q)ce_m(\eta,q)u + Ce_m(\xi,q)ce_m'(\eta,q)v \qquad (7.92)$$

$$f_z(\xi,\eta,q) = \left(\cos h^2\xi - \cos^2\eta \right) Ce_m(\xi,q)ce_m(\eta,q)$$

where $u = \cos h\xi \sin \eta$ and $v = \sin h\xi \cos \eta$, (ξ, η, z) are elliptical cylinder co-ordinate, q is a separation variable, $h = \sqrt{a^2-b^2}$, $\cos h\xi_o = a/b$, and also $ce_m(\eta, q)$ and $Ce_m(\xi, q)$ are even order are ordinary and modified Mathieu functions respectively of order m. The multiplier D_{mn} is a normalization constant, which is chosen to give unit power transfer in each mode and a symmetric scattering matrix, and is given by

$$D_{mn} = \sqrt{\frac{2}{\pi}} \left[\int_0^{\xi_o} d\xi \left(Ce_m'^2(\xi,q) + \Theta_m Ce_m^2(\xi,q) \right) \right]^{-1/2} \qquad \mathrm{TE}_{c,mn}$$

for the two types of TE modes.

TE$_{smn}$ modes

Same as for Eq. 7.92 and D_{mn} with replacements, $Ce \to Se$ and $ce \to se$.

TM$_{cmn}$ modes

Same as Eq. 7.92 with replacements $(f_y)_{\mathrm{TE}} \Rightarrow (f_x)_{\mathrm{TM}}$, $(f_x)_{\mathrm{TE}} \Rightarrow (-f_y)_{\mathrm{TM}}$ and $(f_z)_{\mathrm{TM}} = 0$.

TM$_{smn}$ modes

Same as for the TM$_{cmn}$ modes except with replacements, $Ce \to Se$ and $ce \to se$.

In addition, $k_{c,mn} = 2h\sqrt{q_{mn}}$ where q_{mn} is the nth zero of the following:

$$\text{TE modes}: \begin{cases} \text{Even}; & Ce'_m(\xi_0,q) \\ \text{Odd}; & Se'_m(\xi_0,q) \end{cases} = 0$$

$$\text{TM modes}: \begin{cases} \text{Even}; & Ce_m(\xi_0,q) \\ \text{Odd}; & Se_m(\xi_0,q) \end{cases} = 0$$

It can be shown that the dominant TE_{c11} couples only to the TE_{cmn} and TM_{smn} modes when m is an odd integer. Thus, the TE_{c11} mode and its odd mode counterpart, the TE_{s11} mode, are not coupled at the aperture, which is not unexpected as the dominant field polarizations are orthogonal.

The reflection coefficient of the TE_{c11} mode in a single aperture can be calculated from Eq. 7.73 by letting $m = 1 = m'$ and $n = 1 = n'$. Convergence is a problem due to the singularity of the Green's function, but this singularity can be isolated or removed by subtraction. In the latter approach, an integral having the same type of singularity with a known integral value is added to the function, and at the same time, its integral representation is subtracted under the integral sign. Such a function is the integral for static fields, which is excited by a uniform source, given by

$$I(x,y) = \iint_{D'} \frac{dx'dy'}{\sqrt{(x-x')^2 + (y-y')^2}}$$

When this singularity is subtracted, stable solutions are obtained. Results have been obtained (Bird, 1990b) as a function of normalized frequency ka with waveguide ellipticity as a parameter. In these calculations, coupling to other modes in the aperture was neglected, as was done above for circular and rectangular apertures. The reflection coefficient in elliptical waveguide increases as a/b increases in much the same way as the reflection level does for the TE_{10} mode of rectangular waveguide. Well above cut-off, the level of reflection of the TE_{c11} mode is approximately the same as for the TE_{10} mode in open-ended rectangular waveguide with the same aspect ratio, a/b. This is demonstrated in Figure 7.25. It is seen that the coupling coefficient in either aperture type increases comparably as frequency increases until close to the cut-off frequency of the first coupled high-order mode.

7.3.5.6   Waveguide-Fed Slot Arrays

The theory outlined earlier can be used to analyse mutual coupling in waveguide-fed slot arrays. These arrays are generally of two types: standing wave fed, where adjacent slots are one-half guide wavelength apart, and travelling wave fed. There are several methods available for feeding the parallel plate region, and these include probes, waveguide or a folded parallel plate section to allow the antenna to be fed from the rear with a conventional waveguide that has a smaller width. Whatever the feeding mechanism of the slotted section, the analysis for side-fed apertures is similar and tends to have greater complexity than the analysis given earlier for apertures fed directly by the waveguide, for example. This is because coupling must be considered between the interior waveguide and the exterior region via a ground plane of finite

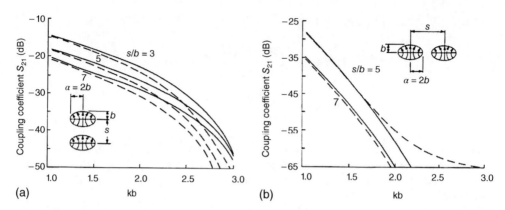

Figure 7.25   Coupling coefficient for identical (- - - -) rectangular and (-------) elliptical waveguides with a/b = 2 as a function of separation distance (s). (a) E-plane and (b) H-plane. *Source*: Reproduced with permission of IET (Bird, 1990b)

thickness. Single waveguide-fed slots have been analysed by several workers (e.g. Stevenson (1948); Oliner (1973); Vu Khac & Carson (1973); Quintez & Dudley (1976); Butler & Yung (1979); Lyon & Sangster (1981) and Stern & Elliott (1985). Vu Khac and Carson (1973) describe a moment method for solving the integral equations governing the coupling between the internal and exterior regions and determined the characteristics of a longitudinal slot in a rectangular waveguide. Lyon and Sangster (1981) extended this method to thick-walled rectangular waveguide. The case of arrays of slots has been considered by Stern and Elliott (1985), where the problem is treated by modelling the slots as centre-fed dipoles, and also by Ando, Kirokawa and co-workers (an example is Tomura et al. (2013) who have considered both centre-fed and end-fed arrays. The moment method is used to calculate the coupling between two slots in the broad wall by Shan-wei et al. (1985). They applied Galerkin's method to the equations of continuity of the tangential components of the magnetic fields, and sinusoidal basis functions were used for the magnetic currents in the slots. Excellent agreement with experiment was obtained with this approach.

In an alternative approach (Bird & Bateman, 1992), the thickness of the slots is directly accounted for by making use of the formulation in Section 7.3.5.1. The array of rectangular slots is then fed by a parallel plate waveguide as shown in Figure 7.26. Suppose there are N_s radiating slots. A flat plate antenna offers a similar gain to the reflector or lens antenna without the additional volume taken up with feeds and struts although the bandwidth may be limited if the feed mechanism is narrowband. The latter is very important for exciting the slots. In the following, it is assumed that the slots (region 2) are excited by the fundamental TM_0 mode of parallel plate waveguide (region 3). Radiation from the array (in region 1) is determined by the size of slots and their arrangement as well as the plate thickness and the termination of the mode at the end of the parallel plate waveguide. The TM_0 mode is incident from the left side of Figure 7.26b. Some of the power is coupled to the exterior region through the slots, some is reflected back into the parallel plate region, and the remainder travels towards the load at the far end. The magnetic field radiated and scattered back in regions 1 and 3 in Figure 7.26b is given by Eq. 7.33. The dyadic Green's function adopted in region 1 is the free-space dyadic given by Eq. 7.35.

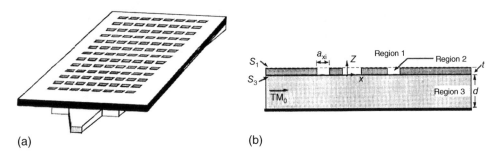

(a)   (b)

Figure 7.26 Plate array of rectangular slots. (a) Top plate and (b) excitation of slots by a TM$_0$ mode in the parallel plate region

In the parallel plate region, region 3, the dyadic is denoted by $\underline{\underline{G}}_3(x,y,z|x',y',z')$. A suitable Green's function can be found by observing that due to field equivalence a magnetic source in the top plate is equivalent to an infinite array of point sources in two planes with spacing $2d$. Thus,

$$\underline{\underline{G}}_3(x,y,z|x'y') = \sum_{n=-\infty}^{\infty} \underline{\underline{G}}_1(x,y,z|x',y',2d).$$

The summation on the right-hand side can be re-expressed in a suitable form for numerical calculation by noting that in this application $\underline{\underline{G}}_3$ has a single component, which has a scalar functional dependence given by

$$G_3(R) = \sum_{n=-\infty}^{\infty} G_0\left(\sqrt{R^2 + (z-2nd)^2}\right).$$

Use is now made of Poisson's summation formula, which is

$$\sum_{n=-\infty}^{\infty} f(n\alpha) = \frac{1}{\alpha}\sum_{n=-\infty}^{\infty} F\left(\frac{2n\pi}{\alpha}\right)$$

where the function $F(2n\pi/\alpha)$ denotes the Fourier transform of $f(n\alpha)$. As a consequence,

$$\underline{\underline{G}}_3(R) = \frac{(\hat{x}\hat{x} + \hat{y}\hat{y} + \hat{z}\hat{z})}{d}\left[\frac{-j\pi}{2}H_0^{(2)}(kR) + 2\sum_{n=1}^{\infty} K_0\left(R\sqrt{\left(\frac{n\pi}{d}\right)^2 - k^2}\right)\right] \qquad (7.93)$$

where $H_0^{(2)}$ is the Hankel function and K_0 is the modified Bessel function both of second kind and order zero. The value of G_3 converges rapidly and normally only requires two or three terms of the series for a result that is usually sufficiently accurate in practice. In the slots (region 2 in Figure 7.26b), the field is expressed as a sum of rectangular waveguide modes as in Eq. 7.50 in a short length of waveguide. By doing this, the thickness of the plate is explicitly taken into account and can be used as a design parameter.

To obtain the radiation from the slots due to the input mode, continuity is imposed on the upper and lower surfaces of the top plate. Continuity of the transverse field on the upper surface at $z = t$ and the use of Galerkin's method result in a scattering matrix formulation described in Section 7.3.3. In this case:

$$\mathbf{b}^{(1)} = \mathbf{S}^{(1)} \mathbf{b}^{(3)}$$

where

$$\mathbf{S}^{(1)} = 2\left(\mathbf{I} + \mathbf{y}^{\text{ext}}\right)^{-1} - \mathbf{I}$$

with admittance matrix $\mathbf{y}^{\text{ext}}$ for the external free-space region given in Eq. 7.66.

In the same way, continuity of the magnetic field at $z = 0$ gives $\mathbf{H}^{\text{inc}} + \mathbf{H}^{(3)} = \mathbf{H}^{(2)}$, where $\mathbf{H}^{\text{inc}}$ is the magnetic field in the incident wave, $\mathbf{H}^{(3)}$ is the scattered field in region 3, and $\mathbf{H}^{(2)}$ is the field in the slots, and then an application of Galerkin's method results in

$$\left(\mathbf{I} - \mathbf{y}^{\text{int}}\right)\mathbf{a}^{(3)} - \left(\mathbf{I} + \mathbf{y}^{\text{int}}\right)\mathbf{b}^{(3)} = \mathbf{F} \tag{7.94}$$

where $\mathbf{F}$ is a vector of forcing functions with components

$$F_{pi} = \frac{1}{2} Y_p^{-1/2} \iint_{S_i} dx' dy' H_o(y') \left(\hat{y} \cdot \underset{=}{\mathbf{G}}_3 \cdot \mathbf{h}_{pi}\right) \exp(-jkx')$$

$\mathbf{a}^{(3)}$ and $\mathbf{b}^{(3)}$ are the incident and reflected mode amplitudes in the parallel plate region. They are related to $\mathbf{a}^{(1)}$ and $\mathbf{b}^{(1)}$ by a transmission factor for the slots. $H_0^{\text{inc}}(y)$ is the transverse magnetic field of the incident TM_0 mode. The mutual admittance for the interior region $\mathbf{y}^{\text{int}}$ is given by Eq. 7.66 wherein the scalar free-space Green's function is now replaced by the scalar value of $\underset{=}{\mathbf{G}}_3 (R)$ given in Eq. 7.93.

The reflection and transmission coefficients are obtained by equating the field due to the infinite array of images to a modal expansion in the parallel plate region. For the TM_0 mode, the reflection coefficient is

$$b_0 = \frac{H_o \exp(-j\gamma_0 x_1)}{2w Y_0^{1/2}} \int_{S_{\text{pp}}} dydz \sum_{i=1}^{N_s} \int_{S_i} dx' dy' \left(\hat{y} \cdot \underset{=}{\mathbf{G}}_3 \cdot \mathbf{h}_{pi}\right) \tag{7.95}$$

where S_{pp} refers to the input cross section of the parallel plate waveguide, γ_0 and Y_0 are the propagation constant and wave admittances of the TM_0 mode, w is the width of the waveguide, x_1 is the input reference plane, and H_o is the amplitude of the incident wave.

As an example, the reflection coefficient of a single slot is given in Figure 7.27 as a function of the slot width. The calculations made with the present approach are compared with the results of Quintez and Dudley (1976) and are seen to be in good agreement.

The radiation pattern of a 15×15 element slot array has been computed by the method described later in Section 7.3.7 for an array of rectangular waveguides and the results in the two principal planes are given in Figure 7.28 at 12.5 GHz. The computed gain is 30 dBi and the radiation efficiency is 70%. The parallel plates have a spacing of $d = 4.5$ mm, and

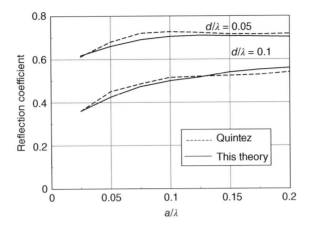

Figure 7.27 Reflection coefficient of a slot in a parallel plate waveguide with plates separated by distance d versus slot width a in the direction of incidence. This theory compared with Quintez and Dudley (1976)

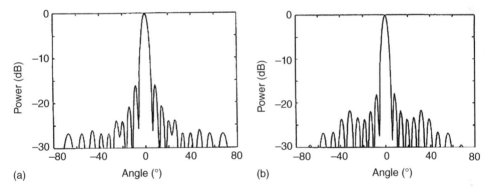

Figure 7.28 Radiation from a 15×15 element slotted planar array at 12.5 GHz. Computed co-polar patterns in (a) E-plane and (b) H-plane

the top plate has thickness $t = 0.5$ mm. The filling material in the slotted region has a dielectric constant of 2.12. In this example, the slots in the top plate are spaced one wavelength apart in the material in the direction of propagation. In the transverse direction, they are also spaced one wavelength apart. The size of the slots can be varied to control the radiation pattern, and also slot pairs can be used with advantage to control both the input reflection and the aperture distribution.

7.3.5.7 Arrays of Microstrip Patches

In the analysis of mutual coupling between microstrip antennas (see Figures 1.1k and 7.29a), two main direct approaches have been used in the literature (Pozar, 1982; Newman et al., 1983; Mohammadian et al., 1989), apart from numerical methods such as the method of

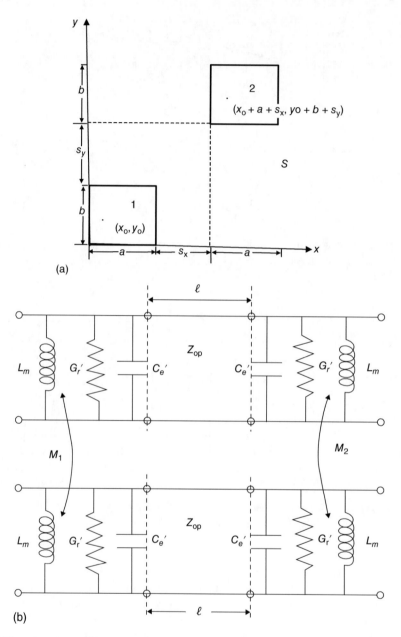

Figure 7.29   Mutual coupling between two identical microstrip patches showing the probe locations. (a) Geometry and (b) equivalent circuit representation

moments with Rao–Wilton–Glisson (RWG) (Rao et al., 1982) basis functions or the mixed potential integral equation (MPIE) method (Mosiq, 1988). An equivalent circuit representation of the coupling is shown in Figure 7.29b. In this model, mutual coupling is represented by mutual inductances across the slot apertures to show that not only does mutual coupling introduce an altered excitation, but also it loads the existing elements with an inductive element and modifies the conductance and capacitance of the single patch shown in Figure 5.5.

The first direct approach (Pozar, 1982; Newman et al., 1983) represents the patches as electric current sources and the fields are given in terms of the Green's function for the elementary electric currents on a grounded dielectric slab. The method of solution is similar to the one given earlier for apertures. Once the current on the patches is determined, the mutual impedance is calculated from Eq. 7.27. One of the complicating features of the analysis of the microstrip patch is that the probe feed to the patch is an integral part of the antenna. Therefore, for accurate results, a suitable model is required for the probe. Although in practice a coaxial transmission line-fed probe is relatively easy to implement, in theory it is quite difficult to treat accurately. The most common approach is to use an idealized feed model comprising a uniform electric current filament. Although this model gives reasonable predictions for the radiation and cross-coupling properties, it gives inaccurate results for self-impedance because the boundary conditions are not satisfied where the probe and patch connect. Aberle and Pozar (1989) describe an improved method where continuity is ensured at the connection and a moment method is used to calculate the current on the probe and patch. This has been applied successfully to both finite and infinite arrays of patches.

A second direct method (Mohammadian et al., 1989) uses the cavity model for microstrip, which was described in Chapter 5. The microstrip patch is modelled as a grounded dielectric slab with the magnetic current distributions located at the open walls formed between the edges of the patch and the ground plane. Since the electric field is zero everywhere on the surface of the cavity except on the open walls, field equivalence may be used to replace the fields on the surface by equivalent magnetic currents in front of a perfect electric conductor. Mutual coupling occurs between these sources, and the magnetic field due to these equivalent sources is needed to calculate the effect. Central to this is determining a dyadic Green's function of the magnetic type for the grounded dielectric slab. From the Green's dyadic, the magnetic field due to the magnetic currents on the open boundary may be calculated and the mutual impedance obtained from

$$Z_{21} = \frac{1}{I_1 I_2} \iint_S dS \, \mathbf{H_{21}}(S) \cdot \mathbf{M}_2 \tag{7.96}$$

where the input currents to the two patches shown in Figure 7.29a are I_1 and I_2, $\mathbf{M}_2$ is the magnetic current on the second patch, and $\mathbf{H_{21}}$ is the magnetic field at patch 2 due to an elementary source on patch 1. It is noted that the mutual impedances calculated from Eq. 7.96 can be used to contribute the values of the circuit elements due to mutual coupling in the equivalent circuit model shown in Figure 7.29b.

The magnetic current on patch 2 is given by $\mathbf{M}_2 = E_{2z} \hat{z} \times \hat{n}$ where $\hat{n}$ is the normal to the open-circuit walls on patch 2. The TM$_{01}$ mode is assumed to be generated on each patch. E_{2z} is derived from the modal expansion of the cavity model. The magnetic field on patch 2 due to patch 1 can be obtained in closed form as can the magnetic field in the dielectric slab. When these are brought together, the mutual impedance is expressed as

$$Z_{21} = \frac{-32jk\eta_o}{\pi^2}K_f \int_{-\infty}^{\infty} d\kappa_1 \int_{-\infty}^{\infty} \frac{d\kappa_2}{(\kappa_1^2+\kappa_2^2)} \sin(h_2 t)$$

$$\times \left\{ (\alpha\kappa_2^2 + \beta\kappa_1^2)S1^2 + (\alpha\kappa_1^2 + \beta\kappa_2^2)C1^2 + 2\kappa_1\kappa_2(\alpha-\beta)S1C1 \right\} \qquad (7.97)$$

$$\times \exp\left[j\kappa_2(a+s_x) + \kappa_2(b+s_y) \right]$$

where

$$K_f = \left[\frac{k_2/ab}{k_2^2-k_{10}^2} S\left(\frac{\pi y_p}{2b}\right) \cos\left(\frac{\pi y_o}{b}\right) \right]^2$$

$$\alpha = \frac{\mu_{r1}\mu_{r2}}{(h_2\mu_{r1}-h_1\mu_{r2})e^{-jh_2 t} + (h_2\mu_{r1}+h_1\mu_{r2})e^{jh_2 t}}$$

$$\beta = \frac{\mu_{r1}\mu_{r2}h_1h_2}{(k_2^2\mu_{r1}h_1-k_1^2\mu_{r2}h_2)e^{-jh_2 t} + (k_2^2\mu_{r1}h_1+k_1^2\mu_{r2}h_2)e^{jh_2 t}}$$

$$S1 = \frac{a}{2}S\left(\frac{\kappa_1 a}{2}\right)\sin\left(\frac{\kappa_2 b}{2}\right)$$

$$C1 = -\kappa_2 \left(\frac{b}{\pi}\right)^2 \cos\left(\frac{\kappa_1 a}{2}\right)C\left(\frac{\kappa_2 b}{2}\right).$$

In Eq. 7.97, t is the substrate thickness, and μ_{r1}, μ_{r2}, ε_{r1} and ε_{r2} are the relative permeabilities and permitivities of the surrounding medium (region 1) and of the substrate (region 2), $h_i = \sqrt{k_i^2-\kappa_1^2-\kappa_2^2}$ where $k_i = k\sqrt{\mu_{ri}\varepsilon_{ri}}(i=1,2)$. The spacing between the patches in the x- and y-directions are s_x and s_y as shown in Figure 7.29a. The functions α and β contain the TE and TM surface wave poles, which have singularities in the integration domain. The equations for these poles are found by setting the denominators of α and β to be zero. The multiplying factor K_f is a term contributed from the input probe where y_p is its diameter and (x_o, y_o) are the coordinates of the connection on patch 1, shown in Figure 7.29a. Typically, the probe occurs off the centre of the patch to achieve good excitation and input match. The eigenvalue k_{10} is a root of the transcendental equation

$$\tan(k_{10}b) = \frac{2k_{10}\alpha_{10}}{k_{10}^2-\alpha_{10}^2}$$

where $\alpha_{10} = jk(t/a)Y_{rw}\eta_o/\sqrt{\varepsilon_r}$. The real part of α_{10} is $\approx 0.98\pi/b$. Y_{rw} is the admittance of the radiating walls at $y=0$ and $y=b$, which are the transitions from a parallel plate waveguide filled with the substrate to a dielectric slab of thickness t. An approximate result for this admittance is (Carver & Mink, 1981)

$$Y_{rw} \approx \left(0.5 + j\left(\frac{\Delta\ell}{t}\right)\varepsilon_{eff}\right)\left(\frac{ka}{\eta_o}\right)$$

where ε_{eff} the effective dielectric constant of the patch given by

$$\varepsilon_{\text{eff}} \approx \left(\frac{\varepsilon_r + 1}{2}\right) + \left(\frac{\varepsilon_r + 1}{2}\right)\left(1 + 10\frac{t}{a}\right)^{-1/2}$$

and

$$\frac{\Delta\ell}{t} \approx 0.412\left(\frac{\varepsilon_{\text{eff}} + 0.30}{\varepsilon_{\text{eff}} - 0.26}\right)\left(\frac{a + 0.26t}{a + 0.81t}\right).$$

The double integrals in Z_{21} can be evaluated by means of numerical integration or with a FFT (Brigham, 1974). Before evaluating the integral numerically, it is convenient to change the variables of integration. To do this, let $\kappa_1 = \rho \cos \xi$ and $\kappa_2 = \rho \sin \xi$ so that $d\kappa_1 d\kappa_2 = \rho d\xi d\rho$ and

$$Z_{21} = \frac{-32jk\eta_o}{\pi^2}K_f\int_0^\infty d\rho\rho\int_0^{2\pi}d\xi\bar{Z}(\rho,\xi)\exp\left(-j\rho\left(\cos\xi(a+s_x)\right) + \sin\xi\left(b+s_y\right)\right)$$

where

$$\bar{Z}(\rho,\xi) = \sin(h_2t)\left\{\left(\alpha(\rho)\sin^2\xi + \beta(\rho)\cos^2\xi\right)S1^2 + \left(\alpha(\rho)\cos^2\xi\right.\right.$$
$$\left.\left. + \beta(\rho)\sin^2\xi\right)C1^2 + \sin 2\xi(\alpha(\rho) - \beta(\rho))S1C1\right\}.$$

When evaluating the transform integral in Z_{21}, care must be taken with the poles in the integrand. One way is described by Mohammadian et al. (1989) where the poles are subtracted and their contributions summed separately. Typical results obtained with this mutual admittance expression are shown in Figure 7.30 where good agreement is demonstrated with measured

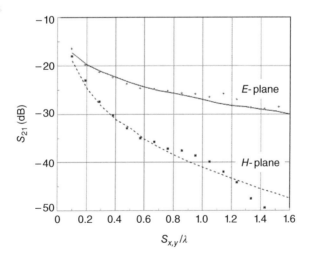

Figure 7.30 Coupling coefficients in the (a) E-plane $(s_x = -a)$ and (b) H-plane $(s_y = -b)$ of two microstrip patches at 1.405 GHz. Solid and dashed curves, theory (Mohammadian et al., 1989); points, measured (Carver & Mink, 1981). Patch dimensions $a = 10.57$ cm and $b = 6.55$ cm. Substrate thickness $t = 0.32$ cm and dielectric constant $\varepsilon_r = 2.5$. Probe position $x_o = 1.7$ cm and $y_o = 3.28$ cm

results (Mohammadian et al., 1989) for E- and H-plane coupling. These cases correspond in Figure 7.29a when to $s_x = -a$ and $s_y = -b$, respectively. When the patches are well matched, the coupling coefficient is approximately given by $S_{21} \approx Z_{21}/2\eta_o$ (see discussion under Eq. 7.42).

7.3.5.8   A Numerical Formulation of Coupling in Arbitrary Shaped Apertures*

The expression Eq. 7.66 for mutual admittance is applicable to apertures of arbitrary shape, providing that a solution can be found to Helmholtz's equation for this geometry that also satisfies the boundary conditions on the walls. An approach based on numerical field solutions has been described by Kuehne and Marquardt (2001). Alternatively, a specific numerical formulation results by assuming that the vector field function $\mathbf{\Psi}_{pi}$ for the guiding structure has been obtained by a particular method such as finite elements where the field solution is known only at a finite number of nodes N on the aperture plane. The coupled apertures are discretized with triangular elements. This discretization may be identical to the one used to obtain the field solution or a different mesh could be chosen for the mutual coupling calculation. In the latter case, interpolation may be necessary to find the magnetic field components at the nodes.

Consider the discretization of the apertures with triangular finite elements with polynomial shape functions (Silvester, 1969). These elements could be in the same aperture for self-admittance or different apertures as illustrated in Figure 7.31 for mutual admittance between two arbitrary apertures.

Consider initially the coupling between two triangular elements, namely, Δ_m and Δ_n, in aperture i and j, respectively. The contribution to the mutual admittance is

$$\left[y_{ij}(p|q)\right]_{mn} = \frac{jk}{4\pi\eta_o \sqrt{Y_{pi}Y_{qj}}} \int_{\Delta_m} dS \left[\mathbf{\Psi}_{pi}(u_1,u_2)\right]_m \cdot \int_{\Delta_n} dS' \left[\mathbf{\Psi}_{qj}(u_1',u_2')\right]_n G(|\mathbf{R}-\mathbf{R}'|).$$

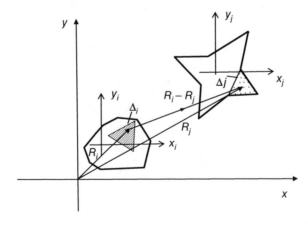

Figure 7.31   Coupling of two arbitrary waveguide apertures with fields obtained by triangular finite element method

As the fields are also known on the same triangular elements, the mode vector on triangular element m can be represented by

$$\left[\mathbf{\Psi}_{pi}(u_1,u_2)\right]_m = \sum_{\ell=1}^{N} \alpha_{\ell}^{m}(x,y)\mathbf{\Psi}_{\ell}^{m}. \tag{7.98}$$

In addition, a similar N point interpolation for the Green's function is obtained on triangle n. This is given by

$$G\left(\left|\mathbf{R}_{\mu}-\mathbf{R}_{j}\right|\right) = \sum_{\nu=1}^{N} \alpha_{\nu}^{n}(x_j,y_j)g_{\mu\nu}^{mn} \tag{7.99}$$

where $g_{\mu\nu}^{mn} = \exp\left(-jk\sqrt{\left(x_{\mu}^{m}-x_{\nu}^{n}\right)^2 + \left(y_{\mu}^{m}-y_{\nu}^{n}\right)^2}\right) \Big/ \sqrt{\left(x_{\mu}^{m}-x_{\nu}^{n}\right)^2 + \left(y_{\mu}^{m}-y_{\nu}^{n}\right)^2}$.

The co-ordinates (x_i^m,y_i^m) are located at the ith interpolation point in triangle m. The reason for interpolating $G\left(\left|\mathbf{R}_i-\mathbf{R}_j\right|\right)$ in this way is to simplify the integration and also shorten the formulation for triangular domains. A singular point in the calculation of self-admittance can be accommodated by stepping slightly away from the singularity within a circle of small radius ε and using $G(\varepsilon)$ instead. The change in the admittance value tends to be quite small.

Substituting Eqs. 7.98 and 7.99 into the mutual admittance expression (Eq. 7.66) results in

$$\left[y_{ij}(p|q)\right]_{mn} = \frac{jk}{4\pi\eta_o\sqrt{Y_{pi}Y_{qj}}} \left[\int_{\Delta_m} dS \sum_{\ell=1}^{N}\alpha_{\ell}^{m}(x,y)\mathbf{\Psi}_{\ell}^{m}\sum_{\mu=1}^{N}\alpha_{\mu}^{n}(x,y)\right.$$
$$\left.\int_{\Delta_n} dS' \sum_{t=1}^{N}\alpha_{t}^{n}(x',y')\mathbf{\Psi}_{t}^{n}\sum_{\nu=1}^{N}\alpha_{\nu}^{n}(x',y')g_{\mu\nu}^{mn}\right].$$

The integrals over Δ_m and Δ_n can be completed in closed form or replaced by the standard finite element $\mathbf{T}$ matrix (Silvester, 1969), which is defined as

$$T_{pq} = \frac{1}{\Delta}\iint_{\Delta} dS\alpha_p(x,y)\alpha_q(x,y)$$

where Δ is the area of the triangular domain. The mutual admittance is now expressed as

$$\left[y_{ij}(p|q)\right]_{mn} = \frac{jk}{4\pi\eta_o\sqrt{Y_{pi}Y_{qj}}}\Delta_m\Delta_n(\mathbf{\Psi}^{m}[\mathbf{T}][\mathbf{g}]^{mn}[\mathbf{T}]\mathbf{\Psi}^{n}) \tag{7.100}$$

where $[\mathbf{g}]^{mn}$ is an $N\times N$ matrix of the Green's function. The total admittance is found by summing all contributions such as Eq. 7.100, that is,

$$y_{ij}(p|q) = \frac{jk}{4\pi\eta_0\sqrt{Y_{pi}Y_{qj}}}\sum_m\sum_n\Delta_m\Delta_n\left((\mathbf{\Psi}^m)^T[\mathbf{T}][\mathbf{g}]^{mn}[\mathbf{T}]\mathbf{\Psi}^n\right)$$

$$= \frac{jk}{4\pi\eta_0\sqrt{Y_{pi}Y_{qj}}}\sum_m\sum_n\Delta_m\Delta_n\left((\mathbf{\Psi}_x^m)^T[\mathbf{T}][\mathbf{g}]^{mn}[\mathbf{T}]\mathbf{\Psi}_x^n\right) \qquad (7.101)$$

$$+ \left(\mathbf{\Psi}_y^m\right)^T[\mathbf{T}][\mathbf{g}]^{mn}[\mathbf{T}]\mathbf{\Psi}_y^n + \left(\mathbf{\Psi}_z^m\right)^T[\mathbf{T}][\mathbf{g}]^{mn}[\mathbf{T}]\mathbf{\Psi}_z^n\Big)$$

where the vector $\mathbf{\Psi}^m$ has been separated into its component column vectors and the superscript T denotes the transpose operation.

The result given by Eq. 7.101 has proved to be more instructive than useful in practice. It clearly shows the role of the Green's dyadic, here represented by the matrix $[\mathbf{g}]^{mn}$, of linking the field distributions $\mathbf{\Psi}^m$ and $\mathbf{\Psi}^n$ in the aperture(s). The $[\mathbf{T}]$-matrices provide the connection from the Green's dyadic and the field. As an example, consider two identical $0.3\lambda \times 0.6\lambda$ rectangular waveguides that couple in their E-planes as shown in Figure 7.32. The centre-to-centre spacing is 0.4λ.

For the purpose of this example, it is assumed that an accurate solution has been obtained for the TE_{01} mode. In this example, for simplicity, only four triangular elements with a linear shape function are used. (This is not a good approximation to the actual field as more elements are required for an accurate solution and representation of the fields.) The set of matrices for this example are easily obtained and are listed below (note that $\mathbf{\Psi}_x = 0$):

$$\mathbf{T} = \begin{bmatrix} 0.167 & 0.083 & 0.083 & 0 & 0 & 0 \\ 0.083 & 0.333 & 0.167 & 0.083 & 0 & 0 \\ 0.083 & 0.167 & 0.5 & 0.167 & 0.083 & 0 \\ 0 & 0.083 & 0.167 & 0.5 & 0.167 & 0.083 \\ 0 & 0 & 0.083 & 0.167 & 0.333 & 0.083 \\ 0 & 0 & 0 & 0.083 & 0.083 & 0.167 \end{bmatrix} \quad \mathbf{\Psi}_y = \begin{bmatrix} 0 \\ 0 \\ 10.472 \\ 10.472 \\ 0 \\ 0 \end{bmatrix} \quad \mathbf{\Psi}_z = \begin{bmatrix} -1.389j \\ -1.389j \\ 0 \\ 0 \\ -1.389j \\ -1.389j \end{bmatrix}$$

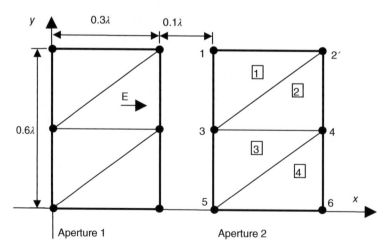

Figure 7.32 Rectangular apertures coupling in their E-planes as represented with four finite elements

and

$$
g = \begin{bmatrix}
-2.023-1.469j & -0.441+1.359j & -2 & -0.441+1.359j & -0.25+1.364j & 0.431-0.741j \\
8.09-5.878j & -2.023-1.469j & -1.278-2.892j & -2.023-1.469j & -1.278+1.034j & 0.957+0.511j \\
-2 & 0.095+1.31j & -2.023-1.469j & 0.095+1.31j & -2 & 0.921-0.262j \\
8.09-5.878j & -2.023-1.469j & -1.278-2.892j & -2.023-1.469j & -1.278+1.034j & 0.957+0.511j \\
-0.25+1.364j & 0.957+0.511j & -2 & 0.957+0.511j & -2.023-1.469j & 1 \\
-1.064+1.169j & -1.278+1.034j & -1.775-2.131j & -1.278+1.034j & 1.545-4.755j & -2.023-1.469j
\end{bmatrix}
$$

The individual [T]-matrices of the triangles have been combined into an overall [T]-matrix for the two rectangular apertures. Using Eq. 7.101 and the matrices given earlier, an estimate of mutual admittance is $0.598 - j0.782$ (-0.136 dB $\angle -52.59°$). This compares with a value of $0.45 - j0.748$ ($-1.179 \angle -58.96°$) which is obtained from Eq. 7.66 using analytical expression for the TE_{01} mode. Considering the level of approximation in the field representation as well as the integration, this result is respectable in the circumstances.

7.3.6 An Asymptotic Expression for Mutual Admittance*

Computation of mutual coupling can take a considerable amount of computation time. This can be reduced by means of an asymptotic formula to Eq. 7.66 that is accurate for large element spacings. By means of this formula, considerable savings are possible for the coupling to distant elements in very large finite arrays, for example, as well as providing helpful estimates on the level of coupling overall. This asymptotic formula expands the Green's function in a Taylor series and then evaluates the mutual admittance in terms of Fourier transforms of weighted moments of the mode function. For separable geometries such as rectangular and elliptical polar co-ordinates, these transforms can be expressed in closed form. The approach has been applied successfully to arrays with rectangular, circular and coaxial elements.

Consider the mutual admittance expression (Eq. 7.66) for modes in apertures i and j, the centres of which are separated by a distance R_{ij} that is much greater than the aperture dimensions. To a second order, the Green's function can be approximated by

$$
\begin{aligned}
G(R) \approx G(R_{ij}) \exp\left[-jk\rho_i \cos\left(\phi_i-\phi_{ij}\right) + jk\rho_j \cos\left(\phi_j-\phi_{ij}\right)\right] \\
\times \left\{ 1 - \frac{1}{R_{ij}}\left[\rho_i \cos\left(\phi_i-\phi_{ij}\right) - \rho_j \cos\left(\phi_j-\phi_{ij}\right)\right] + \frac{jk}{4R_{ij}}\left[\rho_i\rho_j \sin\left(\phi_i-\phi_{ij}\right)\sin\left(\phi_j-\phi_{ij}\right)\right] \right. \\
\left. + \rho_i^2 \cos\left[2\left(\phi_i-\phi_{ij}\right)\right] + \rho_j^2 \cos\left[2\left(\phi_j-\phi_{ij}\right)\right] - \sqrt{\rho_i^2+\rho_j^2} \right] \right\}
\end{aligned}
$$

(7.102)

where $R_{ij} = \sqrt{(x_j-x_i)^2 + (y_j-y_i)^2}$ and $G(R_{ij}) = \exp(-jkR_{ij})/R_{ij}$. The polar co-ordinates used in Eq. 7.102 can be easily transformed to other aperture co-ordinate systems such as rectangular by replacing the local coordinates (ρ_i, ϕ_i) and (ρ_j, ϕ_j) with (x_i, y_i) and (x_j, y_j) by

$x_i = \rho_i \cos \phi_i$ and $y_i = \rho_i \sin \phi_i$, similarly for the aperture j. Substitution of Eq. 7.102 into Eq. 7.66 results in the three-term approximation to admittance

$$y_{ij}(p|q) = \frac{jk}{4\pi\eta_o} G(R_{ij}) \left\{ \mathbf{S}_p^{(i)}(-k) \cdot \mathbf{S}_q^{(j)}(k) + \frac{1}{R_{ij}} \left[\mathbf{S}_p^{(i)}(-k) \cdot \mathbf{X}_q^{(j)}(k) \right. \right.$$

$$\left. \left. - \mathbf{X}_p^{(i)}(-k) \cdot \mathbf{S}_q^{(j)}(k) + jk\mathbf{T}_p^{(i)}(-k) \cdot \mathbf{T}_q^{(j)}(k) \right] \right\} \tag{7.103a}$$

which involve geometric transforms of the field functions. These transforms in Eq. 7.103 are given by:

$$\mathbf{S}_p^{(i)}(w) = \iint_{D_i} dS_i \boldsymbol{\Psi}_{pi}(\rho_i, \phi_i) \exp\left(jw\rho_i \cos\left(\phi_i - \phi_{ij}\right) \right)$$

$$\mathbf{T}_p^{(i)}(w) = \iint_{D_i} dS_i \boldsymbol{\Psi}_{pi}(\rho_i, \phi_i) \rho_i \sin\left(\phi_i - \phi_{ij}\right) \exp\left(jw\rho_i \cos\left(\phi_i - \phi_{ij}\right) \right) \tag{7.103b}$$

$$\mathbf{X}_p^{(i)}(w) = \iint_{D_i} dS_i \boldsymbol{\Psi}_{pi}(\rho_i, \phi_i) \rho_i \left[\cos\left(\phi_i - \phi_{ij}\right) + \frac{jw\rho_i}{4} \cos\left(2\left(\phi_i - \phi_{ij}\right) - 1\right) \right]$$

$$\times \exp\left(jw\rho_i \cos\left(\phi_i - \phi_{ij}\right) \right)$$

The transforms given in Eq. 7.103b can be evaluated in closed form in many cases, for example, rectangular (Bird & Bateman, 1994) and circular elements (Bird, 1979, 1996). The results can be quite accurate as will be demonstrated in the following.

An asymptotic approximation to the coupling between two rectangular waveguides can be obtained from Eq. 7.103 along with the vector field function given by Eq. 7.70. It may be shown that the first term of the asymptotic expression for the mutual admittance of TE_{10} modes in identical rectangular waveguides is

$$y_{ij}(10|10) = \frac{4jk^2 ab}{\pi^3 \beta_{10}} G(R_{ij}) \sin^2\phi_o C^2\left(\frac{ka}{2}\cos\phi_o\right) S^2\left(\frac{kb}{2}\sin\phi_o\right) \tag{7.104}$$

where ϕ_o is the angle in the x–y plane between the apertures, that is, $\tan\phi_o = (y_j - y_i)/(x_j - x_i)$, and β_{10} is the propagation constant of the TE_{10} mode. The mutual admittance in the E-plane ($\phi_o = 90°$) is

$$y_{ij}(10, 10) \approx j\frac{4k^2 ab}{\pi^3 \beta_{10}} G(R_{ij}) S^2\left(\frac{kb}{2}\right). \tag{7.105}$$

In the H-plane, it is observed that the first-order approximation predicts zero coupling when ($\phi_o = 0$). This is because the z-component is maximum at this angle and this cancels out the x-component resulting in zero for first-order coupling. However, by including second-order terms and higher in $1/R_{ij}$, it is found that H-plane coupling is not zero but is given approximately by

$$y_{ij}(10|10) \approx -\frac{8kab}{\pi^3\beta_{10}}\frac{G(R_{ij})}{R_{ij}}C^2\left(\frac{ka}{2}\right). \tag{7.106}$$

The formulation in Eq. 7.103 has a simple physical interpretation. The first term of the formula is the mutual admittance of two elemental magnetic dipoles on a ground plane weighted by the far-field patterns of the apertures in the plane of the conductor, along the trajectory between the two centres. For example, consider two identical rectangular wave-guides of width a and height b that are excited in the TE_{10} mode only. From Section 4.2, the far-field pattern in the aperture plane of the E-plane (i.e. along the x-axis) the electric field is dominated by the function $S(kb/2)$, while in the orthogonal H-plane the pattern is determined by $C(ka/2)$, where the functions S and C are defined in Appendix A. The mutual admittance of two waveguides that couple in the E-plane direction is approximately given by the product of the square of E-plane pattern times the mutual admittance of two elemental dipoles (see Problem P7.12). Similarly, for H-plane coupling, it consists of the product of the patterns in the direction between the two apertures. Both Eqs. 7.105 and 7.106 are accurate for $s>3a$ and $3b$, respectively.

To demonstrate some typical results, Figure 7.33 shows the coupling between two circular waveguides of radius 1.905 cm in a ground plane that has been computed from Eq. 7.103 assuming that only the TE_{11} mode is in both apertures. These results are compared in Figure 7.33 with measured data and computed results from the exact theory (Bailey, 1974).

A further demonstration of the accuracy of the asymptotic formula is shown in Figure 7.34 where the coupling coefficient for two circular waveguides as a function of separation distance is compared with results from the exact admittance expressions given in Section 7.3.5.3. In both analyses, five modes were used in each aperture. It has been found that the asymptotic formula is very accurate for aperture spacings $>3\lambda$.

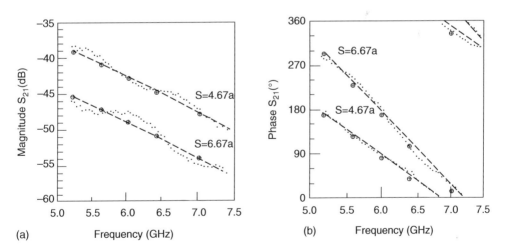

Figure 7.33 H-plane coupling coefficient versus frequency for two waveguides of radius 1.905 cm located in a ground plane and separated by $s=4.667a$ and $s=6.667a$. (a) Magnitude and (b) phase. Dashed curve, exact theory (Bailey, 1974); points, measured (Bailey, 1974); circle with cross, asymptotic 1 mode (Bird, 1979)

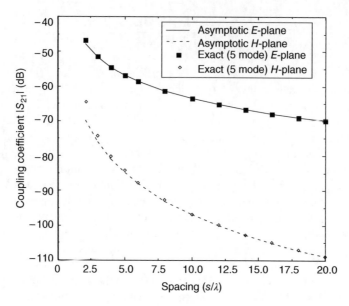

Figure 7.34 Coupling coefficient versus separation distance (s/λ) for two circular waveguides of diameter 2λ. Line, asymptotic formula; and dots, exact formula

7.3.7 Radiation from Finite Arrays with Mutual Coupling

Once the complex amplitudes of the currents have been obtained from the mutual coupling formulation, the radiation from all elements in the array can be found by superposition. In the far-field region of a single isolated antenna excited by an electric current at i, the radiated electric field is

$$\mathbf{E}_p = \mathbf{F}_p(\theta,\phi)I_i\frac{\exp(-jk|\mathbf{r}-\hat{\mathbf{r}}_i|)}{|\mathbf{r}-\hat{\mathbf{r}}_i|} \tag{7.107}$$

where I_i is the element excitation, $\mathbf{F}(\theta,\phi)$ is the element pattern, $\mathbf{r}$ is the radial vector from the origin of the element to the field point, and $\hat{\mathbf{r}}_i$ is the radial vector to the source (see Figure 7.35). The current is related to the drive voltage sources via the impedance matrix as described by Eq. 7.39. The total field is given by

$$\mathbf{E} = \sum_{i=1}^{Na}\mathbf{F}_i(\theta,\phi)I_i\frac{\exp(-jk|\mathbf{r}-\hat{\mathbf{r}}_i|)}{|\mathbf{r}-\hat{\mathbf{r}}_i|}$$

where I_i is given by Eq. 7.39 and Na is the number of apertures. Now, if the distance of the elements is small relative to the distance from the origin to the field point (i.e. $|\mathbf{r}_i| << |\mathbf{r}| = r$), then

$$\mathbf{E} \approx \frac{e^{-jkr}}{r}\sum_{i=1}^{Na}\mathbf{F}_i(\theta,\phi)I_i\exp(jk\mathbf{r}\cdot\hat{\mathbf{r}}_i)$$

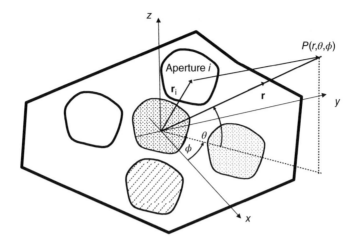

Figure 7.35   Geometry for calculation of radiation from an array of aperture antennas

Suppose all the elements are identical, that is, $\mathbf{F}_i(\theta,\phi) = \mathbf{F}_o(\theta,\phi)$, then

$$\mathbf{E} \approx \frac{e^{-jkr}}{r}\mathbf{F}_o(\theta,\phi)\sum_{i=1}^{Na} I_i \exp(jk\mathbf{r}\cdot\hat{\mathbf{r}}_i). \qquad (7.108)$$

Finally, in the ideal situation when there is no coupling (i.e. $Z_{ij}=0$ where $i \neq j$),

$$\mathbf{E} \approx \frac{e^{-jkr}}{r}\mathbf{F}_o(\theta,\phi)\frac{1}{Z_{in}}\sum_{i=1}^{Na} V_i^{in} \exp(jk\mathbf{r}\cdot\hat{\mathbf{r}}_i) \qquad (7.109)$$

where Z_{in} is the input impedance of the elements. Eq. 7.109 is the conventional form of simple array analysis and synthesis in absence of coupling fields. There are other variations of Eq. 7.108 when only one or two neighbours are included.

When the aperture array is located in a ground plane (see Figure 7.35), the radiated field is obtained by the methods of Section 3.4 and in particular Eqs. 3.26. The aperture field at $z=0$ is given by Eq. 7.50a as

$$\mathbf{E}_a^{(i)} = \sum_{m=1}^{M(i)} (a_{mi} + b_{mi})\ \mathbf{e}_{mi}(x,y)\ Y_{mi}^{-1/2}$$

and, as a consequence, the transform vector Eq. 3.24a is given by

$$\mathbf{N}(u,v) = \sum_{i=1}^{Na}\sum_{m=1}^{M(i)} Y_{mi}^{-1/2}(a_{mi}+b_{mi})\iint_{A_i} dS'\mathbf{e}_{mi}(x',y')\exp(j2\pi(ux'+vy')). \qquad (7.110)$$

Let $\mathbf{F}_{mi}(u,v) = Y_{mi}^{-1/2}\iint_{A_i} dS'\mathbf{e}_{mi}(x',y')\exp(j2\pi(ux'+vy'))$.

Then expressing Eq. 7.110 in a matrix format and introducing the scattering matrix relation (Eq. 7.42) to replacing the reverse travelling wave amplitudes, it is found that

$$\mathbf{N} = \begin{bmatrix} \mathbf{N_x} \\ \mathbf{N_y} \end{bmatrix} = \mathbf{F}(\mathbf{a}+\mathbf{b}) = \mathbf{F}(\mathbf{I}+\mathbf{S})\mathbf{a}$$

where $\mathbf{N_x}$ and $\mathbf{N_y}$ are column vectors corresponding to the x- and y-components of the transforms of the array elements. The far-fields can now be calculated from

$$\mathbf{E_\theta} = \frac{jk}{2\pi}\frac{e^{-jkr}}{r}\left(\mathbf{N_x}\cos\phi + \mathbf{N_y}\sin\phi\right)$$

$$\mathbf{E_\phi} = \frac{jk}{2\pi}\frac{e^{-jkr}}{r}\left(-\mathbf{N_x}\cos\phi + \mathbf{N_y}\sin\phi\right).$$

To obtain the fields in the near- or intermediate-field regions, it is preferable to return to the original Green's function and to calculate the magnetic field by integrating across the aperture field (magnetic current), as given by Eq. 7.47b, and then obtain the other field components by means of Maxwell's equations. Thus, the transverse field, $\mathbf{H}_T$, is found this way, while the remaining magnetic field component is obtained from $\nabla\cdot\mathbf{H}=\mathbf{0}$. Thereafter, the electric field is calculated from $\mathbf{E}=(-j\eta_o/k)\nabla\times\mathbf{H}$.

7.4 Techniques for Minimizing Effects of Mutual Coupling

The neglect of mutual coupling can have serious consequences, and it is especially important to consider its effect in the design. In some applications, it is desirable that mutual coupling is as low as possible, while in others the coupling can be included in the design as a matter of course. If mutual coupling is neglected, the expected performance can depart significantly from that achieved in practice. Therefore, what can be done about it? There are several possible approaches:

- Increase antenna element spacing.
- Reduce the near-field in the direction of adjacent elements by increasing the field taper (also not practical in some cases without compromising performance).
- Isolate elements with electromagnetic 'fences'.
- Compensate for mutual coupling through signal processing methods.
- Take account of mutual coupling at the outset, as in a normal coupled circuit, using one of the methods described in Section 7.3.

Some of these approaches will now be discussed in detail.

7.4.1 Element Spacing

A change in the element spacing is an obvious approach, but it is not always practical as the apertures may be fixed in position or the beam shape or spacing may be compromised or incorrect. Care should also be taken not to introduce grating lobes especially in the coverage area or in other directions, which may introduce interference. This may mean the element spacing should be no greater that $0.7-1\lambda$ unless other means are taken to taper the pattern.

Increasing element spacing was described in the previous section for particular aperture shapes for two-element coupling. This coupling varies with distance depending on location and polarization. As we have seen, for magnetic current dominant sources, which is often the case for apertures, coupling in the E-plane falls in the limit as $1/s$ where s is the separation distance, and in the H-plane, it limits to $1/s^2$. For electric current dominant sources the reverse occurs. Elsewhere, the decay is a combination for anywhere between these limits. The elements need to be two to three wavelengths apart for the trend to be monotonic, so care is required for smaller separations. If the coupling level is unexpectedly high, the spacing could be increased as long as the spacing is not detrimental to the performance. Otherwise, another technique should be adopted.

7.4.2 Aperture Field Taper

In electromagnetic horns, an aperture field distribution may be synthesized by means of high-order modes or inserts such as dielectric and corrugations that produce little radiated power in the direction of neighbouring array elements. For example, the aperture distribution in the E-plane of rectangular horns which are excited in the TE_{10} mode is almost uniform. Corrugations or dielectric placed on the walls parallel to the H-plane increase the field taper in the E-plane and reduce the field at the edge of the aperture. The same effect is created in circular horns by exciting the correct proportions of TE_{12} and TM_{12} modes, and this results in almost equal E- and H-plane patterns and reduced cross-polarization (Schennum et al., 1978). The latter approach is a narrower-band solution than corrugations or dielectric but it is a less expensive and more lightweight option.

7.4.3 Electromagnetic Fences

Metallic guards, fences or cups are sometimes used between elements or at the edges of large arrays as shown in Figure 7.36. The perimeter of the fence is usually perpendicular to the ground plane and is located between elements. For example, arrays of dipoles placed in circular cups generally tend to have lower coupling. The purpose of the cup, and other fences, is to reduce the radiation in the direction of the other elements. The aperture excites a field in the cup, and this creates a new aperture field that depends on the diameter of the cup. Not only is its shape important, but also its height and its effect can be direction dependent. For example, Mailloux (1971) showed that if a fence is placed between two long slots, which couple in the E-plane, coupling initially increases with the height of the fence up to a height of a quarter of a wavelength and decreases thereafter, but not monotonically. By contrast, for long slots, which couple in their H-plane, the coupling decreases monotonically with fence height. Reactively loaded elements placed between adjacent elements (see Figure 7.36a) can also be used to reduce coupling in waveguide arrays (e.g. Edelberg & Oliner, 1960; Hockham, 1974). Such indirectly excited elements provide extra degrees of freedom for radiation pattern control (e.g. Silvestro, 1989).

7.4.4 Mutual Coupling Compensation

In some array applications, it is desirable to compensate for the effect of mutual coupling through appropriate design of the beamforming network. To describe how this can be achieved,

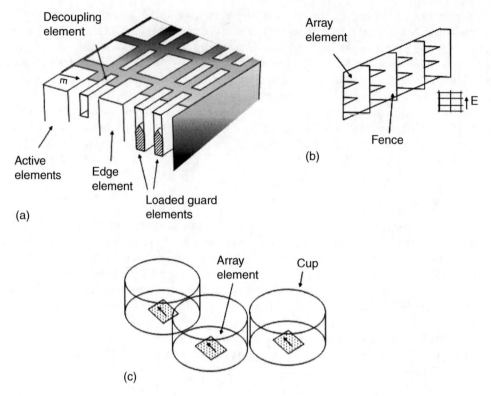

Figure 7.36 Decoupling methods for arrays. (a) Guard elements, (b) fences (sometimes called baffles) and (c) cups

consider the electric field radiated by an array of apertures with scattering matrix as shown in Figure 7.12. In matrix notation, this field is

$$\mathbf{E}(r,\theta,\phi) = \mathbf{e}(r,\theta,\phi)\underset{=}{\mathbf{W}}\,\mathbf{a_I} \tag{7.111}$$

where $\mathbf{e}(r,\theta,\phi)$ is a row vector of electric field vectors of M radiation modes,

$$\mathbf{W} = \left(\mathbf{U}+\mathbf{S}^{(0)}\right)\left(\mathbf{U}-\mathbf{S}_{22}\mathbf{S}^{(0)}\right)^{-1}\mathbf{S}_{21}, \tag{7.112}$$

which is an $M \times N$ matrix, and $\mathbf{a_I}$ is a column vector of N inputs. The other quantities were defined earlier in Section 7.3.3. It is clear that the radiated field would be uncoupled for new array coefficients $\mathbf{c}$ if

$$\mathbf{E}(r,\theta,\dot{\phi}) = \mathbf{e}(r,\theta,\phi)\mathbf{c}$$

where $\mathbf{c}$ is a column vector.

   This occurs for the array input vector given by

$$\mathbf{a_I} = \mathbf{W}^{-1}\mathbf{c} \tag{7.113}$$

Thus, if **c** specifies a given pattern distribution, such as Taylor or Chebyshev, and **W** is known, then the input excitation required to produce this uncoupled distribution is found by multiplying **c** by $\mathbf{W}^{-1}$. An example of this type of compensation is described by Steyskal and Herd (1990). They showed that the technique is most beneficial for digital beamforming where $\mathbf{W}^{-1}$ was realized by Fourier decomposition of the measured array element patterns. They presented results for a linear array of eight X-band waveguide elements that have a common *E*-plane. Excitations were synthesized for a 30 dB Chebychev pattern. However, the measured pattern for the array with coupling, but with no compensation, had sidelobes at the 20 dB level. Practical and computed results showed that following compensation the sidelobe level was reduced to the required 30 dB level. Compensation is possible also by connecting a network between the input ports and the antenna ports. Andersen and Rasmussen (1976) describe necessary conditions for such a lossless network to achieve complete decoupling and de-scattering.

7.4.5 *Power Pattern Synthesis Including the Effect of Mutual Coupling*

The obvious way to minimize the impact of mutual coupling is to take it into account in the design process. One design approach that is sensitive to neglecting mutual coupling is the design of shaped beam. The effect of this on the design of arrays has been considered by several workers. In adaptive arrays, the effect of mutual coupling was considered by Gupta and Ksienski (1983) and also Yirnin et al. (1985) to be significant even for inter-element spacing's greater than half a wavelength, and its effect was drastic when the element spacing was within this distance. Perrott (1985) showed that the accuracy of an interferometer direction finding system was improved when mutual coupling was included. A power pattern synthesis formulation (Bird, 1982) that includes mutual coupling is described in the next section.

7.5 Low-Sidelobe Arrays and Shaped Beams

Antennas with shaped beams or low sidelobes are required in many applications from on-board satellites to surveillance radar. Efficient illumination of irregularly shaped regions on the earth's surface from a geostationary satellite requires the radiation pattern to be contoured to suit the desired coverage. In addition, it is desirable for the beam to have low sidelobes outside this coverage region, and in frequency reuse applications, cross polarization should also be low. Contoured beam illumination may not be the only task of such an antenna as it may also be required to produce multiple pencil beams. A suitable antenna to perform these functions usually involves a directly radiating array or array feed cluster in combination with a reflector or lens. The antenna adopted in this section is an array-fed offset reflector antenna as illustrated in Figure 7.37 although a similar approach could also be used for a directly radiating array of horns.

Individual beams are generated by exciting clusters of elements so as to achieve high beam efficiency and low cross-polarization. Although only selected elements may be involved in producing a particular beam, mode coupling in the array ensures all elements are involved to some extent depending on location, polarization and neighbouring structures. In the presence of this coupling, the objective is to determine the required excitation to produce a particular beam. Once the excitation of each beam is known, the excitation of multiple antenna beams follows by superposition.

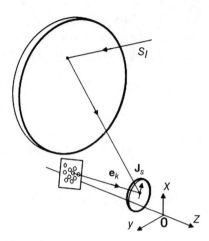

Figure 7.37   Array-fed reflector and shaped beam synthesis

In this section, a method is described for obtaining the array excitation for generating spot beams with maximum efficiency and zero cross-polarization in the beam direction. Control of the cross-polar null is important as it can be positioned to increase polarization diversity by reducing cross-coupling between adjacent beams. By adopting a similar approach to the spot beam analysis, a beam having a desired shape and cross-polar isolation can be synthesized. The formulation to achieve this is described here. The approach accounts for deficiencies of the antenna by including diffraction from a reflector and also feed element coupling.

Consider a feed cluster in the focal plane of a reflector antenna, which may consist of one or more reflecting surfaces, as shown in Figure 7.37. When this antenna is illuminated by a wave from the direction s_l in the far-field, a current $\mathbf{J}_s$ is induced on the reflector surface(s). Consider the reflector nearest to the array, in this case the subreflector. From reciprocity, a beam that is produced in the direction s_l is achieved by exciting the array with a selected driving function. The field due to an N element array is represented by M radiating modes in each aperture. In addition, suppose only the first J modes are accessible, that is, modes that can be addressed directly at the inputs. Some modes are generated electromagnetically at the same ports or through reflections at other ports. There are thus JN accessible ports at the MN ports that are available for producing a beam. Let $\mathbf{E}_k$ be the electric field radiated by port k and let its total amplitude be $A_k = a_k + b_k$ where a_k and b_k are amplitudes of the incident and reflected modes. The next step, when constraints are implemented, determines whether the antenna has low sidelobes. The constraint conditions are a special case of those for a general shaped beam, and also the array amplitudes are coupled through the scattering matrix $\mathbf{S}$. The approach is essentially the same for an array of directly radiating horns or a reflector antenna.

Under these conditions, the fraction of the power radiated by the array in the beam direction is given by the power coupling theorem Eq. 3.63 as

$$\eta(\theta,\phi,\mathbf{a}) = \frac{QQ^*}{P_r} \tag{7.114}$$

where η is the aperture efficiency and

$$Q = \sum_{k=1}^{MN} A_k t_k \qquad (7.115)$$

where

$$t_k = \frac{1}{4\sqrt{P_i}} \iint_{\Sigma} dS \ \mathbf{e}_k \cdot \mathbf{J}_s.$$

P_r and P_i are, respectively, the power radiated by the feed array and the power incident upon the reflector.

A similar expression as Eq. 7.114 results also for a directly radiating array of horns. In this case, the gain function of the antenna (Eq. 3.48) is used:

$$G(\theta, \phi, \mathbf{a}) = \frac{2\pi r^2}{\eta_o} \frac{|\mathbf{E}|^* |\mathbf{E}|}{P_r} \qquad (7.116)$$

where $\mathbf{E}$ is the electric field radiated by the array at a far-field distance of radius r. Compared with the gain of a uniformly illuminated aperture with the same dimensions, G_o, the array has an efficiency

$$\eta(\theta, \phi, \mathbf{a}) = \frac{2\pi r^2}{G_o \eta_o} \frac{\mathbf{E}^* \cdot \mathbf{E}}{P_r} = c \frac{\left(\sum_{k=1}^{MN} A_k \mathbf{e}_k \right)^* \cdot \left(\sum_{k=1}^{MN} A_k \mathbf{e}_k \right)}{P_r} \qquad (7.117)$$

where c is a constant, $\mathbf{e}_k$ is the radiated electric field at accessible port k, and A_k is its amplitude. The radiated power is:

$$P_r = \mathbf{a}_l^T \mathbf{D} \mathbf{a}_l$$

where

$$\mathbf{D} = \frac{1}{2} \left\{ \left[\left(\mathbf{I} - \mathbf{S}_{22} \mathbf{S}^{(0)} \right)^{-1} \mathbf{S}_{21} \right]^\dagger \left[\left(\mathbf{I} + \mathbf{S}^{(0)} \right)^\dagger \left(\mathbf{I} - \mathbf{S}^{(0)} \right) + \right. \right.$$
$$\left. \left. \left(\mathbf{I} - \mathbf{S}^{(0)} \right)^\dagger \left(\mathbf{I} + \mathbf{S}^{(0)} \right) \right] \left(\mathbf{I} - \mathbf{S}_{22} \mathbf{S}^{(0)} \right)^{-1} \mathbf{S}_{21} \right\}.$$

The dagger † is the Hermitian conjugate matrix operation (i.e. conjugate transpose).

Equation 7.114 (or Eq. 7.117) gives the antenna gain relative to a uniformly illuminated aperture and is suitable as a basis for a synthesis procedure. In addition, it is convenient to express the efficiency in matrix notation. Let

$$Q = \mathbf{u}^T \mathbf{v}$$

where $\mathbf{v}^T = [a_1 \ a_2 \cdots a_{JN}]$ is a subset of the excitation coefficients corresponding to accessible ports and T denotes the matrix transpose operation

$$\mathbf{u}^T = \mathbf{t}^T \mathbf{W} \tag{7.118}$$

with

$$\mathbf{t}^T = [t_1 \quad t_2 \quad \cdots \quad t_{MN}] \quad \text{and} \quad W_{ij} = \begin{cases} 1 + S_{ii}; & i = 1, 2, \ldots, JN \\ S_{ij}; & \text{otherwise} \end{cases}. \tag{7.119}$$

In this notation, the power radiated by the array is

$$P_r = \mathbf{v}^T \mathbf{D} \, \mathbf{v}$$

where $\mathbf{D} = \mathbf{P} - \mathbf{S}^T \mathbf{P} \mathbf{S}$ is a $JN \times JN$ positive-definite diagonal matrix given by

$$\mathbf{P} = \begin{bmatrix} p_1 & 0 & \cdots & 0 \\ 0 & p_2 & 0 & 0 \\ \vdots & & \ddots & \vdots \\ 0 & 0 & \cdots & p_{MN} \end{bmatrix}$$

where p_i the mode power at the accessible ports i.

Equation 7.114 can now be re-expressed as

$$\eta(\theta, \phi, \mathbf{v}) = \frac{(\mathbf{v}^T \mathbf{u})^* (\mathbf{u}^T \mathbf{v})}{(\mathbf{v}^T)^* \mathbf{D} \mathbf{v}}$$

and further refined to

$$\eta = \frac{\mathbf{x}^\dagger (\boldsymbol{\alpha}^\dagger \boldsymbol{\alpha}) \mathbf{x}}{\mathbf{x}^\dagger \mathbf{x}} \tag{7.120}$$

where $\mathbf{x} = \mathbf{D}^{1/2} \mathbf{v}$, $\boldsymbol{\alpha} = \mathbf{u}^T \mathbf{D}^{1/2}$. Eq. 7.120 is a quadratic form, which has a maximum value at the maximum eigenvalue of $(\boldsymbol{\alpha}^\dagger \boldsymbol{\alpha} - \lambda \mathbf{I}) \mathbf{x} = 0$. Since $\boldsymbol{\alpha}^\dagger \boldsymbol{\alpha}$ is a rank one matrix, the maximum eigenvalue is simply $\boldsymbol{\alpha}^\dagger \boldsymbol{\alpha}$, and the associated eigenvector is $\mathbf{x} = \boldsymbol{\alpha}$. Therefore, the beam efficiency is maximized when

$$\mathbf{v} = \mathbf{D}^{-1} \mathbf{W}^\dagger \mathbf{t}^*. \tag{7.121}$$

When there is no coupling or reflection at the ports, that is, the scattering matrix $\mathbf{S} = 0$, $\mathbf{W} = \mathbf{I}$ and $\mathbf{v} = \mathbf{D}^{-1} \mathbf{t}^*$. Further, as $\mathbf{D}$ is diagonal, the ith accessible port is proportional to the ith correlation coefficient. Mutual coupling causes the excitation coefficient to depend on all far-field vectors through the matrix $\mathbf{D}$. When there is no coupling, $\mathbf{D} = \mathbf{I} = \mathbf{W}$ and Eq. 7.121 predicts that for maximum gain the excitation coefficient of a particular mode should equal the complex conjugate of its own far-field vector.

The formulation above can be extended to include cross-polarization. In a specified direction, Eq. 7.114 will have a range of values depending on polarization. Let $\mathbf{u}_c$ the vector Eq. 7.118 that results from currents generated by co-polar illumination and $\mathbf{u}_x$ be the corresponding vector due to cross-polar illumination. Two functions in the form of Eq. 7.114 can be defined for co-polar and cross-polar illumination in terms of a common desired excitation $\mathbf{v}$ as follows:

$$\eta_c(\theta,\phi,\mathbf{v}) = \frac{Q_c^* Q_c}{(\mathbf{v}^T)^* \mathbf{D}\mathbf{v}} = \frac{(\mathbf{v}^T \mathbf{u}_c)^* (\mathbf{u}_c^T \mathbf{v})}{(\mathbf{v}^T)^* \mathbf{D}\mathbf{v}} \tag{7.122a}$$

$$\eta_x(\theta,\phi,\mathbf{v}) = \frac{Q_x^* Q_x}{(\mathbf{v}^T)^* \mathbf{D}\mathbf{v}} = \frac{(\mathbf{v}^T \mathbf{u}_x)^* (\mathbf{u}_x^T \mathbf{v})}{(\mathbf{v}^T)^* \mathbf{D}\mathbf{v}} \tag{7.122b}$$

where c and x indicate co-polar and cross-polar illumination, respectively. To meet the common objective of maximum efficiency and minimum cross-polarization, a single performance index is constructed and is

$$\xi = (1-\eta_c) + |\mu^2|\eta_x \tag{7.123}$$

where μ is an arbitrary complex constant. Once again, intermediate vectors are introduced in this case $\mathbf{x} = \mathbf{D}^{1/2}\mathbf{v}$, $\boldsymbol{\alpha}^\dagger = \mathbf{u}_c' \mathbf{D}^{-1/2}$ and $\boldsymbol{\beta}^\dagger = \mathbf{u}_x' \mathbf{D}^{-1/2}$. Substituting Eq. 7.122 with these intermediate vectors in Eq. 7.123 results in

$$\xi = \frac{\mathbf{x}^\dagger \left(\mathbf{I} - \boldsymbol{\alpha}^\dagger \boldsymbol{\alpha} + |\mu|^2 \boldsymbol{\beta}^\dagger \boldsymbol{\beta} \right) \mathbf{x}}{\mathbf{x}^\dagger \mathbf{x}} \tag{7.124}$$

As with the case of a single efficiency, Eq. 7.124 is a ratio of quadratic forms. In this case, the minimum of Eq. 7.124 is required, which occurs at the maximum eigenvalue of

$$\left(|\mu|^2 \boldsymbol{\beta}^\dagger \boldsymbol{\beta} - \boldsymbol{\alpha}^\dagger \boldsymbol{\alpha} - \lambda \mathbf{I} \right) \mathbf{x} = 0. \tag{7.125}$$

To find this, define $\boldsymbol{\tau} = \mu\boldsymbol{\beta}$. Resolve $\boldsymbol{\alpha}$ into components parallel and orthogonal to $\boldsymbol{\tau}$. Thus, $\boldsymbol{\alpha} = \boldsymbol{\alpha}_1 + \boldsymbol{\alpha}_2$ where $\boldsymbol{\alpha}_1 = k\boldsymbol{\tau}$ and $\boldsymbol{\alpha}_2^\dagger \boldsymbol{\tau} = 0$. It can be shown that $k = \boldsymbol{\alpha}^\dagger \boldsymbol{\tau}/\boldsymbol{\tau}^\dagger \boldsymbol{\tau}$. In addition, the source vector $\mathbf{x}$ is expressed as a linear combination of $\boldsymbol{\alpha}_1$ and $\boldsymbol{\alpha}_2$ as follows: $\mathbf{x} = c\boldsymbol{\alpha}_1 + d\boldsymbol{\alpha}_2$. This is permissible since if either $\boldsymbol{\alpha}$ or $\boldsymbol{\beta}$ are identically zero, Eq. 7.125 reduces to the rank one eigenvalue problem discussed previously. Making the above substitutions in Eq. 7.125 gives

$$\boldsymbol{\tau} \lfloor c(1-k^2)\boldsymbol{\alpha}^\dagger \boldsymbol{\tau} - kd\boldsymbol{\alpha}^\dagger \boldsymbol{\alpha}_2 \rfloor - \boldsymbol{\alpha}_2 \lfloor ck\boldsymbol{\alpha}^\dagger \boldsymbol{\tau} + d\boldsymbol{\alpha}^\dagger \boldsymbol{\alpha}_2 \rfloor = \lambda(ck\boldsymbol{\tau} + d\boldsymbol{\alpha}_2).$$

By equating the coefficients, homogeneous equations in c and d are obtained:

$$c\left(\lambda k + (k^2 - 1)\boldsymbol{\alpha}^\dagger \boldsymbol{\tau} \right) + \left(d\, k\boldsymbol{\alpha}^\dagger \boldsymbol{\alpha}_2 \right) = 0 \tag{7.126a}$$

$$c\left(k\, \boldsymbol{\alpha}^\dagger \boldsymbol{\tau} \right) + d\left(\lambda + \boldsymbol{\alpha}^\dagger \boldsymbol{\alpha}_2 \right) = 0. \tag{7.126b}$$

For a non-trivial solution of Eqs. 7.126, the determinant of the matrix relating the coefficients c and d must be zero. This leads to the following the characteristic equation for λ:

$$\lambda^2 + \lambda\left[\left(\boldsymbol{\alpha}^\dagger\boldsymbol{\alpha}\right) - \left(\boldsymbol{\tau}^\dagger\boldsymbol{\tau}\right)\right] + \left[\left(\boldsymbol{\alpha}^\dagger\boldsymbol{\tau}\right)^2 - \left(\boldsymbol{\alpha}^\dagger\boldsymbol{\alpha}\right)\left(\boldsymbol{\tau}^\dagger\boldsymbol{\tau}\right)\right] = 0. \tag{7.127}$$

The solutions of Eq. 7.127 are

$$\lambda = \frac{-\left(\left(\boldsymbol{\alpha}^\dagger\boldsymbol{\alpha}\right) - \left(\boldsymbol{\tau}^\dagger\boldsymbol{\tau}\right)\right) \pm \Delta}{2} \tag{7.128}$$

where the discriminant is

$$\Delta = |\boldsymbol{\alpha} + \boldsymbol{\tau}||\boldsymbol{\alpha} - \boldsymbol{\tau}|.$$

To determine which of the two solutions are relevant, define a real angle θ such that

$$\cos\theta = \frac{\left(\boldsymbol{\alpha}^\dagger\boldsymbol{\alpha}\right) - \left(\boldsymbol{\tau}^\dagger\boldsymbol{\tau}\right)}{\Delta}$$

which results in the two solutions of the characteristic equation given by $\Delta\sin^2(\theta/2)$ and $-\Delta\cos^2(\theta/2)$. The second solution is the required result as it gives an eigenvalue nearest to -1. Substituting this eigenvalue into Eqs. 7.126 results in $k(c-d) + \nu d = 0$ where

$$\nu = \frac{1}{2}\left[\frac{\left(\boldsymbol{\alpha}^\dagger\boldsymbol{\alpha}\right) + \left(\boldsymbol{\tau}^\dagger\boldsymbol{\tau}\right) - \Delta}{\left(\boldsymbol{\alpha}^\dagger\boldsymbol{\tau}\right)}\right].$$

Finally, $\mathbf{x} = (c\boldsymbol{\alpha}_1 + d\boldsymbol{\alpha}_2) = d(\boldsymbol{\alpha} - \mu\nu\boldsymbol{\beta})$. Without loss of generality, the multiplier can be set to unity, that is, $d = 1$. Therefore, in Eq. 7.122, let

$$Q_c = \boldsymbol{\alpha}^\dagger\mathbf{x} = \boldsymbol{\alpha}^\dagger\boldsymbol{\alpha} - \mu\nu\boldsymbol{\alpha}^\dagger\boldsymbol{\beta} \tag{7.129a}$$

and

$$Q_x = \boldsymbol{\beta}^\dagger\mathbf{x} = \boldsymbol{\beta}^\dagger\boldsymbol{\alpha} - \mu\nu\boldsymbol{\beta}^\dagger\boldsymbol{\beta}. \tag{7.129b}$$

Zero cross-polarization on-axis is achieved when $\eta_x(0, 0, \mathbf{v}) = 0$. It follows from Eq. 7.129b that $Q_x = 0$ when

$$\mu\nu = \frac{\boldsymbol{\beta}^\dagger\boldsymbol{\alpha}}{\boldsymbol{\beta}^\dagger\boldsymbol{\beta}}. \tag{7.130}$$

Equation 7.130 specifies the value of the weighting factor in the initial performance index (Eq. 7.123) to give the correct level of co- and cross-polar excitation to achieve zero cross-polarization. The co-polar beam efficiency achieved under condition Eq. 7.130 is

$$\eta_c(\theta, \phi, \mathbf{v}) = \boldsymbol{\alpha}^\dagger\boldsymbol{\alpha} - \frac{\left(\boldsymbol{\alpha}^\dagger\boldsymbol{\beta}\right)^*\left(\boldsymbol{\alpha}^\dagger\boldsymbol{\beta}\right)}{\boldsymbol{\beta}^\dagger\boldsymbol{\beta}}$$

and the corresponding eigenvector is

$$x = \alpha - \frac{\beta^{\dagger}\alpha}{\beta^{\dagger}\beta}\beta. \tag{7.131}$$

Replacing the intermediate quantities in the above with the matrices and vectors of the original formulation, the excitation vector given by Eq. 7.131 is

$$v = D^{-1}\left[u_c^* - u_x^* \frac{u_x^T D^{-1} u_c^*}{u_x^T D^{-1} u_x^*} \right]. \tag{7.132}$$

A physical interpretation of Eq. 7.132 is that the excitation for a general array consists of the excitation to achieve a maximum for an uncoupled array minus a component that cancels out the cross-polarization in that direction. The method described earlier could be extended to more directions, but it becomes increasingly difficult to obtain a closed-form solution as the number of unknowns increased. As the number of specified directions (or stations) increases, the order of the characteristic equation increases in step with the number of unknowns. Thus when there are three conditions are specified, the solution of a cubic is required, for four conditions the solution of a quartic polynomial is needed, etc.

A performance index for contoured beam synthesis requires η_c to be maximized across many stations (directions) in a selected coverage area or across the main beam and be minimized outside the coverage or main beam while at the same time η_x is normally minimized everywhere. An ideal index consists of constraints on both η_c and η_x at specified stations points throughout the field of view (FOV). Suppose there are N_s stations in the FOV and N_c of these are in desired coverage. At station j, let

$$\eta_{cj} = \eta_c(\theta_j, \phi_j, v)$$
$$\eta_{xj} = \eta_x(\theta_j, \phi_j, v)$$

c_{uj} = specified upper beam efficiency level
c_{Lj} = specified lower beam efficiency level
x_j = specified maximum cross-polar level relative to the peak
w_{cuj}, w_{cLj} = weighting on upper and lower beam efficiency levels
w_{xj} = weighting on the cross-polar level

The constraints can be included in a single function that is negative when all are met as described in Section 6.9.2. Define

$$f_i = \begin{cases} -w_{cLj}(\eta_{cj} - c_{uj}) & i=j; & j=1,2,\ldots,N_c \\ w_{cuj}(\eta_{cj} - c_{uj}) & i=N_c+j; & j=1,2,\ldots,N_c; \quad i=1,2,\ldots,N_c+2N_s \\ w_{xj}(\eta_{xj} - x_j) & i=N_c+N_s+j; & j=1,2,\ldots,N_s \end{cases} \tag{7.133}$$

The aim is to find a value for **v** for which all $f_i \leq 0$ ($i =$ I, 2, ..., $N_c + 2\,N$). One approach is to combine all these constraints into a single performance index such as the least pth index that is given by Eq. 6.123.

The advantage of the least pth performance index is that the minimum can be obtained using gradient search methods as well as more sophisticated optimization methods.

As an example of a shaped beam design, consider an array feed of rectangular horns for a parabolic reflector that is required to produce a beam in the shape of the outline of a country on the earth's surface. This occurs in the design of shaped beam antennas from satellites. A set of stations are chosen to cover the country and surrounding regions. The constraints described in Eqs. 7.133 and 7.134 are then used to determine the array excitation coefficients. An example of this approach is the production of a shaped beam to cover continental United States, Alaska, Hawaii and Puerto Rico, from satellite locations between 91°W and 133°W (Bird & Sroka, 1992). The required pattern was achieved with 33 rectangular horns, which have E-plane steps, for horizontal polarization and 26 horns for vertical polarization. The layout of the apertures in the focal plane of the reflector is shown in Figure 7.38. The resulting shaped beam pattern for the horizontal polarization is shown in Figure 7.39. The horns are analysed using the mode matching technique, in which each horn is represented by a series of uniform sections. Mutual coupling between the apertures was included by the methods described in Section 7.3. The penalty function method described earlier was used with the reflector and array analysis software to apply constraints to the assigned stations inside and outside the desired coverage region to obtain the excitation coefficients to achieve the shaped beam shown in Figure 7.39. For the results shown here 17 modes are used for calculating mode self-coupling, while the first four modes only are used in the cross-coupling calculation.

A second example is the design of a low sidelobe array that required a maximum gain and have sidelobes that are <-30 dB below the peak. The array chosen consisted of seven rectangular pyramidal horns that are spaced every $d = 2.5$ cm in their E-plane. The horns are identical with aperture dimensions of 6.0×2.3 cm and a length of 23 cm, and they are fed from WR-75 waveguide (inside dimensions 1.905×0.950 cm). The principal plane radiation pattern of the isolated horn at 10 GHz is shown in Figure 7.40.

The numerical method described earlier can be used to meet the requirements. A set of constraints were established and a satisfactory result was obtained. An alternative although classic approach is to use a standard polynomial approach such as binomial or Chebyshev synthesis (e.g. Balanis, 1982) for which the maximum sidelobe level can be set. The array function for a specified array given the number of elements and element spacing is then matched to the standard polynomial. The AF for this array is

$$\mathrm{AF} = \sum_{n=0}^{N} a_n \cos\left(2nu\right) \tag{7.134}$$

where $u = \left((kd)/2\right)\sin\theta$. The array is assumed symmetric with an odd number $N_e = 2N + 1$ of elements, where the coefficient

$$A_{\pm n} = \begin{cases} a_o; & n = 0 \\ \dfrac{a_n}{2}; & n = \pm 1, \pm 2, \ldots \end{cases}$$

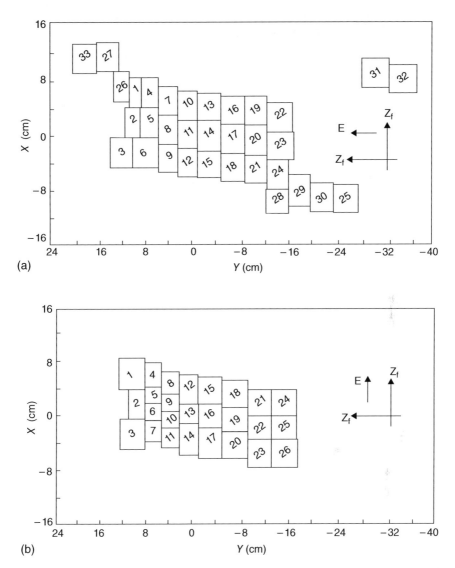

Figure 7.38 Feed horn layout in the focal plane. (a) Horizontal polarization and (b) vertical polarization (Bird & Sroka, 1992). *Source*: Reproduced by permission of The Institution of Engineers, Australia

corresponds to the excitation coefficient of horn n, which is symmetrically placed either side of the central element with excitation A_0. Eq. 7.135 is now expanded and the function $\cos(2nu)$ is expressed in terms of integer powers of $\cos(u)$. For the present example with $N = 3$, after collecting terms, it is found that

$$AF = (a_0 - a_1 + a_2 - a_3) + (2a_1 - 8a_2 + 18a_3)\cos^2 u + (8a_2 - 48a_3)\cos^4 u + 32a_3\cos^6 u.$$

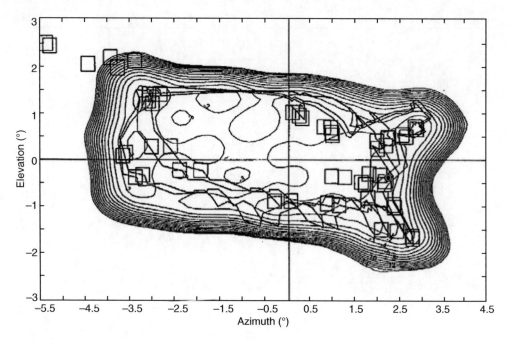

Figure 7.39 Radiation pattern of a horizontally polarized transmit beam at 11.95 GHz with mutual coupling included. Contours are in 1 dB decrements from the peak level (Bird & Sroka, 1992). *Source*: Reproduced by permission of The Institution of Engineers, Australia

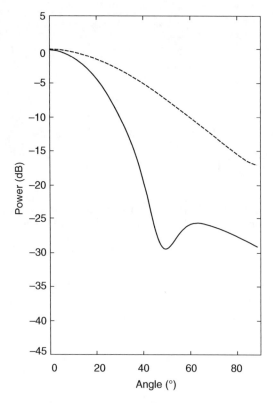

Figure 7.40 Principal plane radiation pattern of pyramidal horn with aperture dimensions 6 cm × 6 cm at 10 GHz. Solid line, *H*-plane; short dash, *E*-plane

Clearly, this step can be generalized for other values of N. Next, $\cos u$ is replaced by the substitution $\cos u = z/z_0$ to obtain polynomial powers in z. The coefficients of the polynomial that results are equated with the coefficients of a polynomial that has desirable sidelobe characteristics, for example, Chebyshev. The parameter z_0 in the expansion is defined from the final polynomial coefficient of the final polynomial of N. Note that the order of this polynomial is one less than the number of elements in the array, that is, $N_e - 1$. In this case, the order is 6. For the Chebyshev polynomial, which is adopted here, the final polynomial corresponds to $T_{N_e-1}(z_0) = \text{MSL}$ where MSL is the maximum sidelobe level relative to the peak and $T_M(z)$ is the Chebyshev polynomial order M. Formulae for the Chebyshev polynomials are given in Appendix A and in the references (Abramowitz & Stegun, 1965). It can be shown that

$$T_{N_e-1}(z_0) = \cosh\left[(N_e-1)\cosh^{-1}(z)\right] \tag{7.135}$$

for $|z_0| > 1$. Expressed alternatively,

$$z_0 = \cosh\left[\frac{1}{(N_e-1)}\cosh^{-1}(\text{MSL})\right].$$

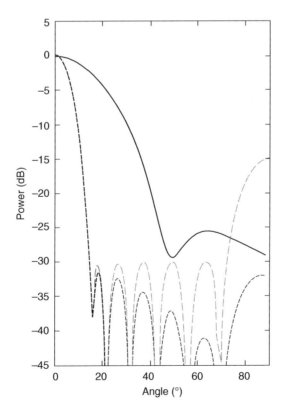

Figure 7.41 Radiation pattern of seven-element E-plane array of rectangular horns at 10 GHz. Element spacing $d = 2.5$ cm. Solid line, H-plane; short dash, E-plane (optimization); and long dash, Chebychev

In the present example of a seven-element array, $z_o = 1.248489$. The sixth-order Chebyshev polynomial is $T_6(z) = 32z^6 - 48z^4 + 18z^2 - 1$ (Abramowitz & Stegun, 1965, p. 795). The coefficients of AF are found by equating the coefficients so that $AF = T_6(z)$. Thus,

$$32 = 32 \frac{a_3}{z_o^6}$$

$$-48 = \frac{(8a_2 - 48a_3)}{z_o^4}$$

$$18 = \frac{(2a_1 - 8a_2 + 18a_3)}{z_o^2}$$

$$-1 = (a_0 - a_1 + a_2 - a_3).$$

This results in the following coefficients: $a_0 = 7.166444987$, $a_1 = 12.5242748$, $a_2 = 18.144943314$ and $a_3 = 3.787113499$. When normalized to the central horn, the coefficients are $a_0 = 1.0$, $a_1 = 1.747627286$, $a_2 = 1.136538874$ and $a_3 = 0.528450788$. The AF corresponding to these coefficients is plotted in Figure 7.41. The E-plane element that is obtained by numerical synthesis is also shown in Figure 7.41. This sidelobe level has reduced slightly with an optimization routine and with a harder sidelobe constraint of 35 dB. The gain of this design is 17.333 dBi compared with 17.319 dBi for the design with the Chebyshev coefficients. Also, the excitation that is obtained by optimization was identical to the aforementioned except for the outside horns where the coefficients are $A_{\pm 3} = a_{|3|}/2 = 0.264225394$.

7.6   Problems

P7.1 An array of microstrip patches is arranged with its elements polarized parallel to the x-axis and arranged on a regular 3×3 grid at intervals of $\lambda/2$ in the plane at $z = 0$.
   a. Neglecting mutual coupling, find the radiation pattern of the array when the elements are driven with currents of the same amplitude but with a phase-shift between adjacent elements of Δ radians.
   b. What value does the phase-shift need to be to steer the beam $\pm 20°$ either side of broadside in the E-plane?
   c. Plot the E-plane pattern of the array for patches of width $w = \lambda_g/2$ and length $\ell = 1.1\lambda_g$ on a substrate of thickness $h = 0.005\lambda_g$ and dielectric constant $\varepsilon_r = 1.5$.

P7.2 Using Lewin's formula for admittance, Eq. 7.78 and your favourite computer software, calculate the reflection coefficient in a rectangular waveguide with dimensions $a = 2.286$ cm and $b = 1.016$ cm over the frequency band 9–10 GHz.

P7.3 A uniform linear array consists of seven identical square $3\lambda/8$ waveguides spaced $\lambda/2$ apart along the y-axis in the H-plane. The waveguides are loaded with dielectric with a relative permittivity of $\varepsilon_r = 3$. The waveguides are of equal length and are fed with equal amplitude but with a uniform progressive phase shift Δ.
   a. Determine the element pattern and the AF.
   b. Plot the E-plane radiation pattern when $\Delta = -0.5$ rad.
   c. Design a feed network for the array.

d. If the input reflection coefficient of one waveguide is $-10\,\text{dB} \angle 30°$, determine the magnitude of the input reflection coefficient of the feeding network assuming all other components are matched to 50 ohms.

P7.4   Calculate the coupling coefficient, S_{12}, of two adjacent square waveguide apertures of sidelength 0.7λ as a function of separation for the two cases of E- and H-plane coupling. Assume the normalized admittance at each aperture is $y_{10} = 1.098 + j0.063$. Compare the two sets of results and discuss.

P7.5   Show that the first-order term in separation distance R_{ij} in the asymptotic formula for the mutual admittance between two circular waveguide apertures is given by

$$y_{ij}(p,q) \approx \frac{jk}{4\pi\eta_o} G(R_{ij}) \left(\frac{\chi_{mi}\chi_{nj}}{2}\right) \left(\mathbf{S}_p^{(i)}(-k) \cdot \mathbf{S}_q^{(j)}(k)\right)$$

where

$$\mathbf{S}_p^{(i)}(w) = \hat{x}\left[-Q_{p-1}(w,k_{mi},a_i)\cos \Delta_{p-1} + Q_{p+1}(w,k_{mi},a_i)\cos \Delta_{p+1}\right]$$
$$+ \hat{y}\left[Q_{p-1}(w,k_{mi},a_i)\sin \Delta_{p-1} + Q_{p+1}(w,k_{mi},a_i)\sin \Delta_{p+1}\right]$$
$$- \hat{z}\left(-2j\frac{k_{mi}}{k}\right) Q_p(w,k_{mi},a_i)\cos \Delta_p,$$

$$\Delta_p = p\phi_{ij} - \psi_m \text{ and } Q_p(w,u,\xi) = \int_0^\xi d\rho\, \rho J_p(w\rho) J_p(u\rho).$$

P7.6   Show that the self-admittance of the TEM mode in coaxial waveguide that is filled with a dielectric material with a relative dielectric constant ε_r is given by

$$y_{11}(0,0) = \frac{\pi\chi_{00}^2}{\sqrt{\varepsilon_r}} \int_0^\infty \frac{dw}{w\sqrt{1-w^2}} (J_0(ka_1 w) - J_0(kb_1 w))^2$$

where a_1 and b_1 are the outer and inner conductor radii and $\chi_{00} = 1/\sqrt{\pi \ln(a_1/b_1)}$. Assuming the dimensions a_1 and b_1 are small compared with the wavelength, show that the normalized conductance is $g_{00} \approx 2/(3\sqrt{\varepsilon_r}\ln(a_1/b_1)) \left[(\pi/\lambda)^2 (a_1^2 - b_1^2)\right]^2$ (Marcuvitz, 1986, p. 213).

P7.7   Find the location of the grating lobes in a hexagonal array of waveguide elements by examining the AF given in Eq. 7.15.

P7.8   Two identical rectangular patches are placed side by side on a substrate of thickness $0.15\,\text{cm}$ and dielectric constant 2.55. Their dimensions are $1.6 \times 1.66\,\text{cm}$. At an operating frequency of 5 GHz, calculate the E- and H-plane coupling as a function of the spacing between them from 0.1λ to 1λ. Each patch is fed with a probe located at $x_o = 0.55$ cm and $y_o = 0.8$ cm.

P7.9   Use Eq. 7.66 to obtain the reflection coefficient of the TEM mode of parallel plate waveguide of height a and width b where $a \ll b$. Assume $E_x = E_o \exp(-j\beta z)$.

P7.10 Obtain the asymptotic admittance of two slots coupling in their E-plane from the dual of Eq. 7.29. Compare this result with the mutual admittance given by Eq. 7.105 and discuss. How might the present approximation be improved?

P7.11 Suppose a cross-polarized beam is required in the same direction as the main beam with half the gain. Use the approach in Section 7.5 starting with Eq. 7.122 to obtain the array excitation to achieve the required array excitation $\mathbf{v}$.

P7.12 By means of the Chebyshev procedure described in Section 7.5 and (a) initially neglecting mutual coupling, synthesize the excitation coefficients of an 11-element linear planar array of 0.7λ long slots of width 0.1λ with spacing of 0.5λ to achieve a sidelobe level of -40 dB in the array's E-plane. (b) Introduce mutual coupling through, for example, the asymptotic admittance (Eq. 7.105), use the excitation coefficients in (a), and recalculate the E-plane radiation pattern. What is the impact on the sidelobe level? Hint: The normalized aperture admittance of the slot is $y_{10} = -0.282 - j0.321$.

References

Aberle, J.T. and Pozar, D.M. (1989): 'Analysis of infinite arrays of probe-fed rectangular microstrip patches using a rigorous feed model', IEE Proc. H, Vol. **136**, pp. 110–119.

Abramowitz, M. and Stegun, I.A. (1965): 'Handbook of mathematical functions', Dover Inc., New York.

Adams, A.T. (1966): 'Aperture admittance measurements', Proc. IEEE, Vol. **54**, No. 12, pp. 2002–2003.

Amitay, N., Galindo, V. and WU, C.P. (1972): 'Theory and analysis of phased array antennas', Wiley, New York.

Andersen, J.B. and Rasmussen, H.H. (1976): 'Decoupling and descattering networks for antennas', IEEE Trans. Antennas Propag., Vol. **AP-24**, pp. 841–846.

Arndt, F., Wolff, K.-H., Briinjes, L., Heyen, R., Siefken-Herrlich, F., Bothmer, W. and Forgber, E. (1989): 'Generalized moment method analysis of planar reactively loaded rectangular waveguide arrays', IEEE Trans. Antennas Propag., Vol. **AP-37**, pp. 329–338.

Bailey, M.C. (1974): 'Mutual coupling in a finite planar array of circular apertures', IEEE Trans. Antennas Propag., Vol. **AP-22**, No. 2, pp. 178–184.

Balanis, C.A. (1982): 'Antenna theory: analysis and design', Harper and Rowe, New York.

Barzilai, G. (1948): 'Mutual impedance of parallel aerials', Wireless Engineer, Vol. **36**, pp. 343–352.

Bechmann, R. (1931): 'On the calculation of radiation resistance of antennas and antenna combinations', Proc. IRE, Vol. **19**, pp. 1471–1480.

Bird, T.S. (1979): 'Mode coupling in a planar circular waveguide array', IEE J. Microwaves Optics Acoustics, Vol. **3**, pp. 172–180.

Bird, T.S. (1982): 'Contoured-beam synthesis for array-fed reflector antennas by field correlation', IEE Proc. H, Microwaves, Antennas Propag., Vol. **129**, pp. 293–298.

Bird, T.S. (1990a): 'Analysis of mutual coupling in finite arrays of different-sized rectangular waveguides', IEEE Trans. Antennas Propag., Vol. **AP-38**, pp. 166–172.

Bird, T.S. (1990b): 'Behaviour of multiple elliptical waveguides opening into a ground plane', IEE Proc. H, Microwaves, Antennas Propag., Vol. **137**, pp. 121–126.

Bird, T.S. (1996): 'Improved solution for mode coupling in different-sized circular apertures and its application', IEE Proc. H, Microwaves, Antennas Propag., Vol. **143**, pp. 457–464.

Bird, T.S. (2004): 'Mutual coupling in arrays of coaxial waveguides and horns', IEEE Trans. Antennas Propag., Vol. **AP-52**, pp. 821–829.

Bird, T.S. and Bateman, D.G. (1992): 'Analysis of a parallel plate waveguide-fed slot antenna including mutual coupling effects', URSI Electromagnetic Theory Symposium, Sydney, Australia, 17–20 August 1992, pp. 424–426.

Bird, T.S. and Bateman, D.G. (1994): 'Mutual coupling between rotated horns in a ground plane', IEEE Trans. Antennas Propag., Vol. **AP-42**, pp. 1000–1006.

Bird, T.S. and HAY, S.G. (1990): 'Mismatch in a dielectric-loaded rectangular waveguide antenna', Electron. Lett., Vol. **26**, pp. 59–61.

Bird, T.S. and Sroka, C. (1992): 'Design of the Ku-band antennas for the Galaxy HS601C satellites', J. Electr. Electron. Eng. Aust., Vol. **12**, pp. 267–273.

Bontsch-Bruewitsch, M.A. (1926): 'Die strahlung der komplizierten rechtwinkeligen antennen mit gleichbeschaffenen vibratoren', Ann. Phys., Vol. **81**, pp. 425–453.

Booker, H.G. (1946): 'Slot aerials and their relation to complementary wire aerials (Babinet's principle)', J. Instn. Elect. Engrs. Pt. IIIA, Vol. **93**, pp. 620–626.

Borgiotti, G.V. (1968): 'A novel expression for the mutual admittance of planar radiating elements', IEEE Trans. Antennas Propag., Vol. **AP-16**, pp. 329–333.

Brigham, E.O. (1974): 'The fast Fourier transform', Prentice-Hall Inc., Eaglewood Cliffs, New Jersey.

Brillouin, L. (1922): 'Sur l'origine de Ia resistance de rayonnement', Radioelectricite, Vol. **3**, No. 4, pp. 147–152.

Brown, G.H. (1937): 'Directional antennas', Proc. IRE, Vol. **25**, pp. 78–145.

Butler, C.M. and Yung, E.K. (1979): 'Analysis of a terminated parallel-plate waveguide with a slot in its upper plate', Ann. Telecommun., Vol. **34**, pp. 451–458.

Carter, P.S. (1932): 'Circuit relations in radiating systems and applications to antenna problems', Proc. IRE, Vol. **20**, pp. 1004–1041.

Carver, K.R. and Mink, J.W. (1981): 'Microstrip antenna technology', IEEE Trans. Antennas Propag., Vol. **AP-29**, pp. 2–24.

Collin, R.E. (1960): 'Field theory of guided waves', McGraw-Hill, New York.

Diamond, B.L. (1968): 'A generalized approach to the analysis of infinite planar array antenna', Proc. IEEE, Vol. **56**, pp. 1837–1851.

Edelberg, S. and Oliner, A.A. (1960): 'Mutual coupling effects in large arrays II: compensation effects', IRE Trans. Anntenas Propag., Vol. **AP-8**, pp. 360–367.

Farrell, G.F. and Kuhn, D.H. (1968): 'Mutual coupling in infinite planar arrays of rectangular waveguide horn', IEEE Trans. Antennas Propag., Vol. **AP-16**, pp. 405–414.

Felsen, L.B. and Marcuvitz, N. (1973): 'Radiation and scattering of waves', Prentice-Hall, Upper Saddle River, NJ.

Fenn, A.J., Thiele, G.A. and Munk, B.A. (1982): 'Moment method analysis of finite rectangular waveguide phased arrays' IEEE Trans. Antennas Propag., Vol. **AP-30**, pp. 554–564.

Finlayson, B.A. (1972): 'The method of weighted residuals and variational principles', Academic Press, New York.

Fröberg, C.-F. (1974): 'Introduction to numerical analysis', Addison-Wesley, London, UK.

Galejs, J. (1969): 'Antennas in inhomogeneous media', Pergamon Press, Oxford, UK.

Galindo, V. and Wu, C.P. (1966): 'Numerical solutions for an infinite phased array of rectangular waveguides with thick walls', IEEE Trans. Antennas Propag., Vol. **AP-14**, No. 2, pp. 149–158.

Gera, A.E. (1988): 'Simple expressions for mutual impedances (dipoles and microstrip elements)', IEE Proc. H, Microwaves, Antennas Propag., Vol. **135**, pp. 395–399.

Gupta, I.J. and Ksienski, A.A. (1983): 'Effect of mutual coupling on the performance of adaptive arrays', IEEE Trans. Antennas Propag., Vol. **AP-31**, pp. 785–791.

Hansen, R. (1966): 'Microwave and scanning antennas', Academic, New York, Vol. **3**, Chapters 2 and 3.

Hansen, R. (1998): 'Phased array antennas', Wiley, New York.

Harrington, R.F. (1961): 'Time-harmonic electromagnetic fields', McGraw-Hill, New York.

Harrington, R.F. (1968): 'Field computation by moment methods', Macmillan Co., New York.

Harrington, R.F. and Mautz, J.R. (1976): 'A generalized network formulation for aperture problems', IEEE Trans. Antennas Propag., Vol. **AP-24**, pp. 870–873.

Hockham, G.A. (1974): 'TEM waveguide coupling in presence of short-circuited slot', Electron. Lett., **10**, pp. 62–63.

Hockham, G.A. (1975): 'Use of the 'edge condition' in the numerical solution of waveguide antenna problems', Electron. Lett., **11**, pp.418–419.

King, R.W.P. (1956): 'The theory of linear antennas', Harvard University Press, Cambridge, MA.

Kitchiner, D., Raghavan, K. and Parini, C.G. (1987): 'Mutual coupling in a finite planar array of rectangular apertures', Electron. Lett., Vol. **23**, No. 11, pp. 1169–1170.

Knudsen, H.L. (1952): 'On the calculation of some definite integrals in antenna theory', Appl. Sci. Res., Sect. B, **3**, pp. 51–68.

Kuehne, R. and Marquardt, J. (2001): 'Modal analysis of waveguide antennas with arbitrary cross sections', IEEE Trans. Microwave Theory Tech., Vol. **MTT-49**, No. 11, pp. 2152–2156.

Lewin, L. (1951): 'Advanced theory of waveguides', Iliffe & Sons Ltd., London, UK, Chapter 6.

Luzwick, J. and Harrington, R.F. (1982): 'Mutual coupling analysis in a finite planar rectangular waveguide antenna array', Electromagnetics, **2**, pp. 25–42.

Lyon, R.W. and Sangster, A.J. (1981): 'Efficient moment method analysis of radiating slots in a thick-walled rectangular waveguide', IEE Proc. H, Microwaves, Antennas Propag., **128**, pp. 197–205.

Mailloux, R.J. (1971): 'Reduction of mutual coupling using perfectly conducting fences', IEEE Trans. Antennas Propag., Vol. **AP-19**, pp. 166–173.

Marcuvitz, N. (1986): 'Waveguide handbook', Peter Peregrinus, London, UK.

Mclachlan, N.W. (1964): 'Theory and application of Mathieu functions', Dover Inc., New York.

Miles, J.W. (1949): 'On the diffraction of an electromagnetic wave through a plane screen', J. Appl. Phys., Vol. **20**, pp. 760-768.

Mohammadian, A.H., Martin, N.W. and Griffin, D.W. (1989): 'A theoretical and experimental study of mutual coupling in microstrip antenna arrays', IEEE Trans. Antennas Propag., Vol. **AP-37**, pp. 1217–1223.

Mosiq, J.R. (1988): 'Arbitrarily shaped microstrip structures and their analysis with a mixed potential integral equation', IEEE Trans. Microwave Theory Tech., Vol. **MTT-36** No. 2, pp. 314–323.

Müller, K.E. (1960): 'Untersuchung des Strahlungsverhaltens ellipisher Hohlleiter sowie der Möglichkeit zur Erzeugung eines zierkular poliarisierten Strahlungsfeldes', Hochfrequenztechn. U. Elektroak., Vol. **69**, pp. 140–151.

Newman, E.H., Richmond, J.H. and Kwan, B.W. (1983): 'Mutual impedance computation between microstrip antennas', IEEE Trans. Microwave Theory Tech., Vol. **MTT-31**, pp. 941–945.

Oliner, A.A. (1973): 'The impedance properties of narrow radiating slots in the broad face of rectangular waveguide: Part I – Theory', IEEE Trans. Antennas Propag., Vol. **AP-5**, pp. 4–11.

Oppenheim, A.R. and Shafer, R.W. (1975): 'Digital signal processing', Prentice-Hall, Englewood Cliffs, NJ.

Pathak, P.H., Wang, N., Burnside, W.D. and Kouyoumjian, R.G. (1981): 'A uniform GTD solution for the radiation from sources on a convex surface', IEEE Trans. Antennas Propag., Vol. **AP-29**, pp. 609–621.

Perrott, R.A. (1985): 'The effect of mutual coupling on interferometer accuracy', IEE International Conference on Antennas & Propagation, Coventry, UK, 16–19 April, pp. 292–294.

Pistolkors, A.A. (1929): 'The radiation resistance of beam antennas', Proc. IRE, Vol. **17**, pp. 562–579.

Pozar, D.M. (1982): 'Input impedance and mutual coupling of rectangular microstrip antennas', IEEE Trans. Antennas Propag., Vol. **AP-30**, pp. 1191–1196.

Quintez, J.P. and Dudley, D.G. (1976): 'Slots in a parallel plate waveguide', Radio Sci., Vol.**11**, No. 8–9, pp. 713–724.

Rao, S.M., Wilton, D.R. and Glisson, A. (1982): 'Electromagnetic scattering by surfaces of arbitrary shape', IEEE Trans. Antennas Propag., Vol. **AP-30**, No. 5, pp. 409–418.

Richmond, J.H. (1961): 'A reaction theorem and its application to antenna impedance calculations', IRE Trans. Anntenas Propag., Vol. **AP-9**, pp. 515–520.

Rudge, A.W., Milne, K., Olver, A.D. and Knight, P. (1983): 'The handbook of antenna design', Peter Peregrinus, London, UK, Vol. **2**, Chapters 10 and 11.

Schelkunoff, S.A. and Friis, H.T. (1952): 'Antennas: theory and practice', Wiley, New York.

Schennum, G.H., Han, C.C. and Gould, H.J. (1978): 'Reduction of mutual coupling in a waveguide array', IEEE APS, Washington, DC, pp. 412–415.

Shan-Wei, L., Ji-Shi, R. and Yin, F. (1985): 'Study of coupling properties of slotted waveguide in rectangular waveguide', Acta Electron. Sinica, **13**, pp. 92–101.

Shapira, J., Felsen, L.B. and Hessel, A. (1974): 'Surface ray analysis of mutually coupled arrays on variable curvature cylindrical surfaces', Proc. IEEE, **62**, pp. 1482–1492.

Silvester, P. (1969): 'High-order polynomials triangular finite elements for potential problems', Int. J. Eng. Sci., Vol. **7**, pp. 849–861.

Silvestro, J.W. (1989): 'Mutual coupling in a finite planar array with interelement holes present', IEEE Trans. Antennas Propag., Vol. **AP-37**, pp. 791–794.

Silvestro, J.W. and Collin, R.E. (1989): 'Aperture impedance of flared horns', IEE Proc. H, Microwaves, Antennas Propag., Vol. **136**, pp.235–240.

Stern, G.J. and Elliott, R.S. (1985): 'Resonant length of longitudinal slots and validity of circuit representation: theory and experiment', IEEE Trans. Antennas Propag., Vol. **AP-33**, pp. 1264–1271.

Stevenson, A.F. (1948): 'Theory of slots in rectangular waveguides', J. Appl. Phys., **19**, pp. 24–38.

Steyskal, H. and Herd, J.S. (1990): 'Mutual coupling compensation in small array antennas', IEEE Trans. Antennas Propag., Vol. **AP-38**, pp. 1971–1975.

Storer, J. (1952): 'Variational solution to the problem of the symmetrical cylindrical antenna', Cruft Lab. Report. TR 101, Cambridge, MA, USA.

Stutzman, W.L. and Thiele, G.A. (1981): 'Antenna theory and design', Wiley, New York, Chapter 7.

Tomura, T., Hirokawa, J., Hirano, T. and Ando, M. (2013): 'A 16 × 16-element plate-laminated-waveguide slot array with 19.2% bandwidth in the 60-GHz band', IEEE International Antennas and Propagation Symposium, Orlando, FL, USA, 7–13 July, pp. 53–54.

Vokurka, V.J. (1979): 'Elliptical corrugated horn for broadcasting satellite antennas', Electron. Lett., Vol. **15**, pp. 652–654.

Vu Khac, T. and Carson, C.T. (1973): 'Impedance properties of a longitudinal slot in the broad face of rectangular waveguide', IEEE Trans. Antennas Propag., Vol. **AP-21**, pp. 708–710.

Yirnin, Z., Hirasawa, K. and Fujimoto, K. (1985): 'Performance of power inversion adaptive array with the effect of mutual coupling', Proceedings of ISAP '85, Kyoto, Japan, 20–22 August, pp. 803–806.

8

Conformal Arrays of Aperture Antennas

8.1 Introduction

A major limitation of the planar-phased array is that as the beam is scanned from broadside, the beam broadens and the pattern deteriorates. On the other hand, an array that is conformal to a convex surface, such as a circular cylinder or a cone, has a beam that can be scanned in discrete steps through an arc while maintaining a constant pattern. Antennas mounted on moving platforms such as on an aircraft are usually covered by aerodynamic radomes to reduce drag and improve performance. This can lead to wastage of space in the radome. An antenna that is flush with the surface can reduce space and avoid hindering the aerodynamics. Although there are some well-documented examples of conformal arrays, their use by and large is not widespread mainly because of the complexity of the feed network as well the array itself. Recent improvements in microwave and optical components have simplified the design of feed networks, thereby making conformal arrays an attractive alternative for applications such as wide-angle scanned beam antennas with ultra-low sidelobes. Of importance in the design of low sidelobe antenna arrays is predicting the effect of mutual coupling between the array elements.

Difficulties can arise in the design of conformal arrays because the elements are pointing in different directions and the curvature can create propagating waves on the surface, called 'creeping waves'. Their impact on performance can be significant and difficult to predict without detailed analysis. General surfaces usually involve esoteric functions and transformations that considerably complicate the field analysis. An approximate approach involves obtaining solutions for common surfaces such as planes, cylinders or spheres as canonical surfaces and to generalize these results by allowing basic quantities to vary such as a radius by introducing radius of curvature. In the following chapter, the canonical structure is a circular cylinder but the approach is equally applicable to a cone or sphere.

In the next section, the radiation is determined from some typical conformal arrays. Next, mutual coupling and radiation is described for a finite array of rectangular waveguides

Fundamentals of Aperture Antennas and Arrays: From Theory to Design, Fabrication and Testing,
First Edition. Trevor S. Bird.
© 2016 John Wiley & Sons, Ltd. Published 2016 by John Wiley & Sons, Ltd.
Companion website: www.wiley.com/go/bird448

terminated in a circular conducting cylinder. Of particular interest are uniform asymptotic representations of the basic problem due to a magnetic point source on a cylinder that has a medium to large radius. Although there is an exact modal solution to this problem, it is not useful for practical array analysis except for cylinders of small radius. A periodic array approach is described for a cylinder that is similar to the infinite array approach for the plane.

A uniform asymptotic representation of diffraction is presented for analysing the array. These are uniform in that they provide a smooth transition across the shadow boundary from a region of direct illumination by the source to the shadow region where there is only a diffracted field. As with the geometrical theory of diffraction (GTD), this representation admits a ray-based interpretation of the field on the cylinder in terms of creeping waves. Also, following the general principles of GTD, the cylinder problem is treated as a canonical problem and, knowing its solution, a generalized solution can be devised for any smooth conducting convex surface.

The next section gives a brief review of the literature on the application of asymptotic techniques to the analysis of mutual coupling and radiation of apertures in convex surfaces. The reference list is a representative of the material published in the open literature. An analysis of mutual coupling is described in Section 8.3 along with some computed results that include a comparison of the coupling coefficients calculated from a present finite array analysis with measured results. Some computed radiation patterns are presented for two different structures. These results are compared with patterns obtained from an exact harmonic series solution (Golden et al., 1974). Also, in Section 8.3, methods are discussed for reducing a large 'characteristic' grating lobe in patterns of phased conformal arrays. Coupling in a periodic array is described in Section 8.4 for a concave structure that is suitable for a variety of microwave lenses that are used for array beamforming.

8.2 Radiation from a Conformal Aperture Array

Suppose an array of identical rectangular waveguides of width a and height b terminates in a conducting cylinder of radius R_o as illustrated in Figure 8.1. As a first approximation, each waveguide is assumed to support only the dominant TE_{10} mode. Without loss of generality two cases will be considered. The one shown in Figure 8.1 is for waveguides oriented with the electric field polarized in the circumferential direction and the case where the waveguides are rotated through 90° and the electric field is parallel to the axis.

8.2.1 Waveguide with E-Field Polarized in Circumferential Direction

Consider a waveguide that is oriented so that its broadwall is parallel to the axis of the cylinder as shown in Figure 8.1. On the aperture, the field can be replaced by an axial magnetic current element dM_z. It has been shown (Wait, 1959) the electric field components radiated by this axial current is of the form

$$dE_\phi(r,\theta,\phi) = \frac{dM_z}{2\pi^2 R_o} \frac{e^{-jkr}}{r} \exp(-jkz'\cos\theta) \sum_{n=-\infty}^{\infty} j^n \frac{e^{jn|\phi-\phi'|}}{H_n^{(2)'}(\gamma)} \tag{8.1a}$$

$$dE_\theta(r,\theta,\phi) = 0 = dE_r(r,\theta,\phi), \tag{8.1b}$$

where $\gamma = kR_o \sin\theta \gg 1$. The total field is obtained by integrating across the aperture of the waveguide as well as any other waveguides on the cylinder and summing the result. Assume

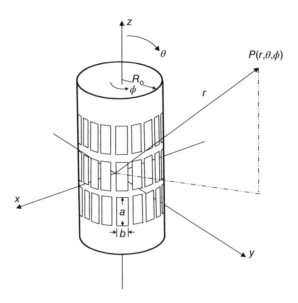

Figure 8.1 Rectangular waveguide array on circular cylinder

a ring of M equispaced apertures on the cylinder centred at locations $(R_o\phi_{oi}, z_{oi})$ $(i=1,2,\ldots M)$ on the surface, and angular extent of the unit cell of $2\pi/M$. For one such aperture, the aperture field is

$$\mathbf{E_a} = \hat{\phi} A_i \sqrt{\frac{2}{ab}} \cos\left(\frac{\pi}{a}z\right); z_{oi} - \frac{a}{2} < z < z_{oi} + \frac{a}{2}; \phi_{oi} - \Delta\phi < \phi < \phi_{oi} + \Delta\phi.$$

A_i is the amplitude of the electric field excited in the aperture and $\Delta\phi = b/2R_o$. Substituting for dM_z due to the TE$_{10}$ mode and integrating over the aperture results in

$$E_\phi(r,\theta,\phi) = A_i \frac{\sqrt{2ab}\,e^{-jkr}}{\pi^3 R_o}\frac{}{r} \exp(jkz_{oi}\cos\theta) C\left(\frac{ka}{2}\cos\theta\right) I(\theta,\phi), \qquad (8.2)$$

where

$$I(\theta,\phi) = \sum_{\nu=-\infty}^{\infty} j^n S(\nu\Delta\phi)\frac{e^{j\nu\bar{\phi}}}{H_\nu^{(2)'}(\gamma)} \qquad (8.3)$$

the function S is defined in Appendix A.4 and $\bar{\phi} = \phi - \phi_{oi}$. Eq. 8.3 converges slowly for large cylinders and for numerical evaluation the number of terms required to achieve convergence is typically $N \geq \pm(|\gamma|+5)$. As shown by Wait (1959, Chapter 9), the series in Eq. 8.3 can be converted via Watson's transformation to a more convenient form for calculation, by initially converting it to an integral in the complex ν-plane. Thus,

$$I(\theta,\phi) = -j\int_{C_\nu} \frac{\cos[\nu(\phi-\pi)]}{\sin\nu\pi} S(\nu\Delta\phi)\frac{e^{j\nu\bar{\phi}}}{H_\nu^{(2)'}(\gamma)}d\nu.$$

The contour C_ν initially runs along the horizontal axis in the complex ν-plane. To simplify the integral, use is made of a creeping wave representation. The integral is the equivalent of a series of creeping waves, which travel forward and backward around the cylinder. Taking only two terms of the series, corresponding to the two dominant creeping waves, the integral becomes

$$I(\theta,\phi) \approx Q(\theta,\phi) + Q(\theta, 2\pi - \phi),$$

where

$$Q(\theta,\phi) = \int_{C_\nu} S(\nu\Delta\phi) \frac{e^{-j\nu\left(\bar{\phi}-\pi/2\right)}}{H_\nu^{(2)'}(\gamma)} d\nu.$$

The Watson transformation is then completed by deforming C_ν around the zeros in the ν-plane. This results in

$$Q(\theta,\phi) \approx j\frac{\pi\gamma}{2} \exp\left[-j\left(\bar{\phi}-\frac{\pi}{2}\right)\right] S(\gamma\Delta\phi) \sum_{p=1}^{\infty} \frac{e^{-jM\left(\bar{\phi}-\pi/2\right)\alpha'_p}}{\alpha'_p Ai\left(\alpha'_p\right)}, \tag{8.4}$$

where $M = (\gamma/2)^{1/3}$, α'_p is the p-th zero of the derivative of the Airy integral and $Ai\left(-\alpha'_p\right)$ is the Airy function evaluated at this zero. The Airy function zeros and associated functions values are tabulated in the references (Abramowitz & Stegun, 1965, Chapter 10). The terms of the series are called creeping waves (Chan et al., 1977). Because α'_p is complex with a negative imaginary part, the creeping waves decay with angular distance $x = (\bar{\phi}-\pi/2)$. Close to the source, the series converges slowly and it is replaced instead by an equivalent function, $g(x)$, which is known as the hard acoustic Fock function, after V.A. Fock, an early investigator of diffraction by curved surfaces (Fock, 1965). Notice that $x = 0$ ($\bar{\phi} = \pi/2$) corresponds to the transition from the lit to the shadow region. When $x > 0$, which corresponds to the diffraction region, the exponentials of the acoustic Fock function decay and for large M Eq. 8.4 requires only a few terms for accurate evaluation. The region $x < 0$ corresponds to the lit region and for $x < -1$ a good approximation is

$$g(x) \approx 2 \exp\left(j\frac{x^3}{3}\right)\left(1 - \frac{j}{4x^3}\right).$$

Because of the transition at $x = 0$, it is convenient to define the modified hard acoustic Fock function as

$$G(x) = g(x) \begin{cases} \exp(-jx^3/3); & x < 0 \\ 1; & x > 0 \end{cases}. \tag{8.5}$$

Details of this function are given in Appendix G, and it is plotted in Figure 8.2. In the illumination region as $x \to -\infty$, it is seen from Figure 8.2 that $|G| \to 2$ thereby recovering the geometric optics result.

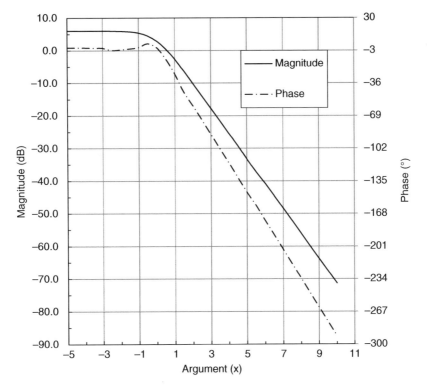

Figure 8.2 Modified hard acoustic Fock function $G(x)$, magnitude (dB) and phase

As a result of these definitions,

$$Q(\theta,\phi) \approx j\frac{\pi\gamma}{2}S(\gamma\Delta\phi)G\left(M\left(\bar{\phi}-\frac{\pi}{2}\right)\right)\exp\left(-j\left(\bar{\phi}-\frac{\pi}{2}\right)\right)$$

and

$$I(\theta,\phi) \approx j\frac{\pi\gamma}{2}S(\gamma\Delta\phi)\left[e^{-j\gamma\Phi_1}G(M\Phi_1)+e^{-j\gamma\Phi_2}\bar{G}(M\Phi_2)\right], \tag{8.6}$$

where $\Phi_1 = \bar{\phi}-\pi/2$ and $\Phi_2 = 2\pi-\bar{\phi} = 3\pi/2-\bar{\phi}$ are the angles subtended by the creeping wave paths shown in Figure 8.3. This figure shows a source and an observer of two rays indicated as 1 and 2 corresponding to three possible regions (a) illumination; (b) shadow boundary; and (c) shadow. The paths of the two creeping waves on the cylinder are also shown in Figure 8.3. The final expression for the radiated field is

$$E_\phi(r,\theta,\phi) \approx A_i\frac{\sqrt{ab/2}\gamma}{\pi^2R_0}\frac{e^{-jkr}}{r}\exp(jkz_{oi}\cos\theta)C\left(\frac{ka}{2}\cos\theta\right) \tag{8.7}$$

$$\times S(\gamma\Delta\phi)\left[e^{-j\gamma\Phi_1}G(M\Phi_1)+e^{-j\gamma\Phi_2}G(M\Phi_2)\right].$$

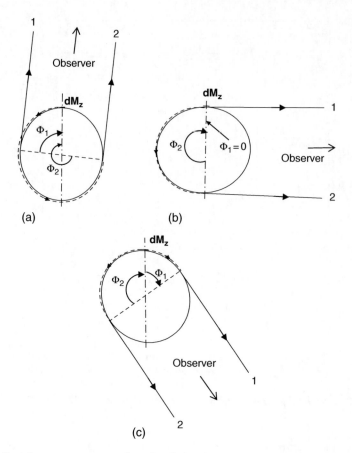

Figure 8.3 Creeping wave representation of radiation from a magnetic source on a circular cylinder. (a) Direct illumination $\Phi_1 < 0$ and $\Phi_2 < 0$. (b) Shadow boundary $\Phi_1 = 0$ and $\Phi_2 = 0$. (c) Shadow region $\Phi_1 > 0$ and $\Phi_2 > 0$

A slightly improved form of Eq. 8.6 for axially oriented waveguides is adopted to ensure a smooth transition of the hard acoustic Fock function from the illuminated to the shadow region. These expressions are listed below.

Illuminated region ($\Phi_1 < 0$ or $\Phi_2 < 0$):

$$
E_\phi(r,\theta,\phi) \approx A_i \frac{e^{-jkr}}{r} \sin\theta \left(P_z(\theta,\bar{\phi}) e^{-j\gamma M \sin \Phi_{1,2}} G(M \sin \Phi_{1,2}) \right.
$$

$$
\left. + P_z\left(\theta,\frac{\pi}{2}\right) e^{-j\gamma M \sin \Phi_{2,1}} G(M \sin \Phi_{2,1}) \right); \quad \Phi_{1,2} < 0, \quad \Phi_{2,1} > 0.
$$

(8.8a)

Shadow region ($\Phi_1 \geq 0$ and $\Phi_2 > 0$):

$$E_\phi(r,\theta,\phi) \approx A_i \frac{e^{-jkr}}{r} \sin\theta P_z\left(\theta,\frac{\pi}{2}\right)\left(e^{-j\gamma M\Phi_1}G(M\Phi_1) + e^{-j\gamma M\Phi_2}G(M\Phi_2)\right), \qquad (8.8b)$$

where

$$P_z(\theta,x) = \left(jk/\pi^2\right)\sqrt{ab/2}\,C((ka/2)\cos\theta)S((kb/2)\sin\theta\sin x)\exp(jkz_{oi}\cos\theta)$$

is the pattern function in the azimuth direction for z-directed magnetic source excitation.

As an example, Figure 8.4 shows the elevation radiation pattern as a function of the azimuthal angle for a 45-element array of rectangular waveguides ($a = 2.286$, $b = 1.016$ cm) in a cylinder of radius $R_o = 12.624$ cm at a frequency of 9.5 GHz when only the central element is excited. The waveguides are oriented with their broadwall parallel to the z-axis as shown in Figure 8.1 (i.e. circumferential polarization). Also shown for comparison are the patterns of a single isolated waveguide in the same cylinder. The impact of the array is clearly shown.

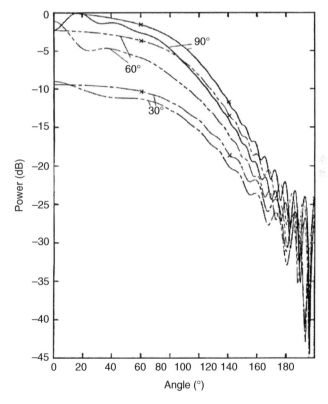

Figure 8.4 Elevation patterns of 45-element array ($a = 2.286$ cm and $b = 1.016$ cm) in a cylinder with a radius $R_o = 12.624$ cm versus azimuthal angle. Circumferential E-polarization at 9.5 GHz. × single waveguide

To steer the beam in a cylindrical conformal array, the path length from the cylinder to a fictitious aperture plane needs to be included in the phase of the excitation coefficients. Thus, a beam is formed in the direction (r, θ_b, ϕ_b) in the far-field, if element i at (R_o, ϕ_{oi}, z_{oi}) on the cylinder, has a complex amplitude given by

$$a_i = A(\phi_{oi}, z_{oi}) \exp[jk(R_o(1 - \sin\theta_b \cos(\phi_{oi} - \phi_b)) - z_{oi} \cos\theta_b],$$

where A is the illumination function (e.g. cosine). The aperture plane is at angle θ_b and the first term in the exponential is the additional distance a ray travels from the centre of element i to the aperture.

In a second example, the E-plane pattern produced by a 44-element azimuthal array that has been excited and phased to produce a beam on boresight is shown in Figure 8.5. The waveguide dimensions are $a = 2.4$ cm and $b = 0.4$ cm. In this example, $R_o = 45.635$ cm and the azimuthal spacing is $2.2432°$ (1.7866 cm or 0.56 λ). The pattern in Figure 8.5 includes a large grating lobe in the vicinity of $128°$. The position of the n-th grating lobe of a cylindrical array is approximately given by

$$\sin\left(\frac{\phi_{nGL}}{2}\right) = -\left(\frac{n\lambda}{2d}\right) \sin\theta_o,$$

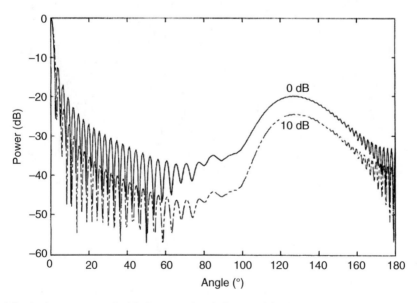

Figure 8.5 E-plane patterns of a 44-element azimuthal array with edge tapers of 0 and 10 dB at 9 GHz. Dimensions $a = 2.4$ cm and $b = 0.4$ cm, $R_o = 45.635$ cm, and element spacing $2.2432°$

where d is the element spacing and θ_o is the elevation angle. As has been described earlier in Chapter 7, grating lobes can be reduced by tapering the illumination appropriately. This is demonstrated in Figure 8.5 where a 10 dB edge taper has been applied to the excitation.

8.2.2   Waveguide with E-Polarized in Axial Direction

For completeness, the equivalent expression is given here for a TE_{10} mode excited waveguide that has its electric field polarized in the z-direction (magnetic current is ϕ-directed), that is, the broadwall is parallel to the circumferential direction. The incremental electric field components radiated by the magnetic current dM_ϕ in this case are (Wait, 1959):

$$dE_\theta(r,\theta,\phi) = -j\frac{dM_\phi}{2\pi^2 R_o}\frac{e^{-jkr}}{r}\exp(jkz'\cos\theta)\mathrm{cosec}\theta\sum_{n=-\infty}^{\infty}j^n\frac{e^{jn|\phi-\phi'|}}{H_n^{(2)}(\gamma)} \tag{8.9a}$$

$$dE_\phi(r,\theta,\phi) = -j\frac{dM_\phi}{2\pi^2 R_o\gamma}\frac{e^{-jkr}}{r}\exp(jkz'\cos\theta)\cot\theta\sum_{n=-\infty}^{\infty}j^n n\frac{e^{jn|\phi-\phi'|}}{H_n^{(2)'}(\gamma)} \tag{8.9b}$$

and

$$dE_r(r,\theta,\phi) \approx 0. \tag{8.9c}$$

For a TE_{10} mode excited aperture with width a and height b centred at (R_o,ϕ_{oi},z_{oi}), the aperture field is of the form

$$\mathbf{E}_A = \hat{z}A_i\sqrt{\frac{2}{ab}}\cos\left[\frac{\pi R_o}{a}(\phi-\phi_{oi})\right].$$

Making the substitution for the magnetic current and integrating over the aperture, the components of the electric field in the far-zone are:

$$E_\theta(r,\theta,\phi) = -jA_i\frac{\sqrt{2ab}}{\pi^3 R_o}\frac{e^{-jkr}}{r}\mathrm{cosec}\theta\, S\left(\frac{kb}{2}\cos\theta\right)\exp(jkz_{oi}\cos\theta))$$

$$\times\sum_{n=-\infty}^{\infty}j^n C\left(n\frac{a}{2R_o}\right)\frac{e^{j\nu\bar{\phi}}}{H_\nu^{(2)}(\gamma)} \tag{8.10a}$$

$$E_\phi(r,\theta,\phi) = -jA_i\frac{\sqrt{2ab}}{\pi^3 kR_o^2}\frac{e^{-jkr}}{r}\mathrm{cosec}\theta\cot\theta\, S\left(\frac{kb}{2}\cos\theta\right)\exp(jkz_{oi}\cos\theta))$$

$$\times\sum_{n=-\infty}^{\infty}j^n nC\left(n\frac{a}{2R_o}\right)\frac{e^{j\nu\bar{\phi}}}{H_\nu^{(2)'}(\gamma)}. \tag{8.10b}$$

The series in Eq. 8.10 can be evaluated asymptotically using the Watson transformation as described earlier for the circumferentially polarized aperture. The reader is referred to the

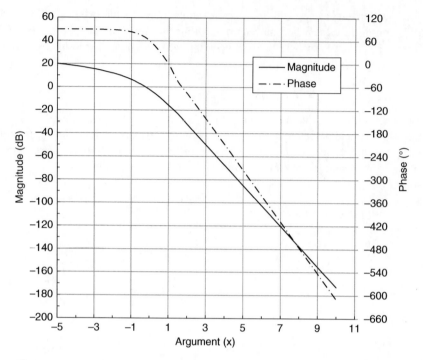

Figure 8.6 Modified soft acoustic Fock function $F(x)$, magnitude (dB) and phase

references for details of the next steps (e.g. Wait, 1959). For the present case, an appropriate acoustic Fock function is required. This is the modified soft acoustic function, which is defined as

$$F(x) = f(x) \begin{cases} \exp(-jx^3/3); & x < 0 \\ 1; & x > 0 \end{cases},$$

where $f(x)$ is the soft acoustic Fock function that is defined in Appendix G. The modified soft function is plotted in Figure 8.6 over the range of arguments $-5 < x < 10$.

The solution for axially polarized waveguides is a described by the following equations.

Illuminated region ($\Phi_1 < 0$ or $\Phi_2 < 0$):

$$E_\theta(r,\theta,\phi) = -jA_i \frac{e^{-jkr}}{r} \frac{1}{M} \left(P_\phi(\theta,\bar\phi) e^{-j\gamma M\Phi_{1,2}} F(M \sin \Phi_{1,2}) \right.$$
$$\left. + P_\phi(\theta,\pi/2) e^{-j\gamma M\Phi_{2,1}} F(M \sin \Phi_{2,1}) \right) \tag{8.11a}$$

$$E_\phi(r,\theta,\phi) = -A_i \frac{e^{-jkr}}{r} \cos\theta \left(\cos \Phi_{1,2} P_\phi(\theta,\bar\phi) e^{-j\gamma M\Phi_{1,2}} G(M \sin \Phi_{1,2}) \right.$$
$$\left. + P_\phi(\theta,\pi/2) e^{-j\gamma M\Phi_{2,1}} G(M \sin \Phi_{2,1}) \right). \tag{8.11b}$$

Shadow region ($\Phi_1 \geq 0$ and $\Phi_2 > 0$):

$$E_\theta(r,\theta,\phi) = -jA_i \frac{e^{-jkr}}{r} \frac{P_\phi(\theta,\pi/2)}{M} \left(e^{-j\gamma M\Phi_1} F(M\Phi_1) + e^{-j\gamma M\Phi_2} F(M\Phi_2)\right) \qquad (8.12a)$$

$$E_\phi(r,\theta,\phi) = -A_i \frac{e^{-jkr}}{r} \cos\theta P_\phi(\theta,\pi/2)\left(e^{-j\gamma M\Phi_1} G(M\Phi_1) + e^{-j\gamma M\Phi_2} G(M\Phi_2)\right), \qquad (8.12b)$$

where

$$P_\phi(\theta,x) = \frac{jk}{\pi^2}\sqrt{\frac{ab}{2}} C\left(\frac{ka}{2}\sin\theta \sin x\right) S\left(\frac{kb}{2}\cos\theta\right) \exp(jkz_{oi}\cos\theta)$$

is the pattern function in the azimuthal direction for ϕ-directed magnetic source excitation. Once again $\gamma = kR_o \sin\theta$, G is defined by Eq. 8.5 and $M = (\gamma/2)^{1/3}$. The plot in Figure 8.6 shows that $F(x)$ rapidly decays with increasing positive arguments, which indicates that the E_θ-component level falls significantly with increasing angle into the shadow region.

8.2.3 Historical Overview of Asymptotic Solutions for Conformal Surfaces

In the previous section, some basic techniques were introduced for evaluating the radiation from apertures in conformal surfaces. There are several basic problems inherent to this work, which has significant historical interest. This early work is summarized in this section.

The problem of diffraction by a smooth convex surface on which a source is located has been studied since the early days of radio. A review of the literature concerned with this problem is contained in the references (e.g. Pistolkors, 1947; Logan, 1959; Bowman et al., 1963; Fock, 1965; Mittra & Safavi-Naini, 1979; James, 1986). It is apparent from these references that most of the early work involved asymptotic analysis of the radiation field and, with a few notable exceptions, asymptotic investigation of the surface field held limited practical interest.

The formal solution of the fields due to a source on a circular cylinder was first obtained by Silver and Saunders (1950), although earlier publications by other authors described special solutions. As the series for the surface fields converges slowly, even for cylinders of moderate radius, it was not until later that the surface field solution was studied in any depth. Hasserjian and Ishimaru (1962) obtained asymptotic solutions for the surface field due to axial and circumferential slots. This work was motivated by the need to determine mutual coupling between slots on a cylinder. Asymptotic solutions for the currents on a sphere had been obtained earlier by Fock (1965), Franz (1954) and Wait (1956). A significant feature of the paper by Hasserjian and Ishimaru (1962) is that they derived a small argument approximation that is valid near the shadow boundary by expressing the contour integral for the field as an inverse Laplace transform (Bromwich integral) the power series of which is easily evaluated. This is now a standard tool for determining the small argument approximation to functions which arise in the asymptotic analysis of the current on large conducting bodies, the so-called surface Fock functions which are defined in Appendix G.

Asymptotic solutions for the radiation from apertures in cylinders have been covered by numerous authors including Sensiper (1957), Bailin and Spellmire (1957), Wait (1959) and Lee and Safavi-Naini (1976). One of the earliest attempts to extend the cylindrical solution to arbitrary structures is that of Goodrich et al. (1959) who generalized the argument used by Fock in the vicinity of the shadow boundary. This technique is essentially the GTD approach where a solution for an arbitrary convex surface is obtained from 'canonical' problems such as the cylinder or the sphere. A GTD solution for radiation from apertures in arbitrary convex surfaces was presented by Pathak and Kouyoumjian (1974). A short-coming of this solution is that ray torsion, due to the surface having two different radii of curvature, is not included. Neglect of ray torsion is usually not important in the principal plane radiation patterns of structures with at least one plane of symmetry; however, it can be important in other planes. Subsequently, improved solutions were developed, notably the work of Lee and Safavi-Naini (1976), Mittra and Safavi-Naini (1979) and Pathak et al. (1981). These solutions are uniform asymptotic

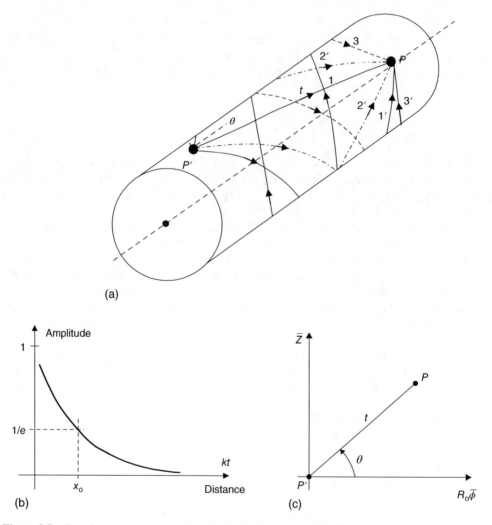

(a)

(b)

(c)

Figure 8.7 Creeping waves on a circular cylinder. (a) Ray paths of the first six creeping waves between P and P'. (b) Amplitude of creeping wave on a ray path at angle θ, where $x_o = 4M^2/3$. (c) A projected view of a ray path on the cylinder

descriptions which represent the field everywhere in terms of continuous functions, the acoustic Fock functions. The original solution given by Mittra and Safavi-Naini (1979) does not reduce to the geometric optics result although this is easily rectified. The solutions presented by Pathak and Wang (1978, 1981) and later improvements by the author (Bird, 1985a) are in general use at present, because excellent agreement with practice has been obtained for a wide range of surfaces.

The asymptotic solution obtained by Hasserjian and Ishimaru (1962) for the surface field on a cylinder retains only the lowest order terms in $q = (j/kt)$ and $(1/M)$. As shown in Fig. 7.35, t is the path length between the source and field points on the cylinder and $M = (kR_t/2)^{1/3}$ is the magnification factor. R_t is the radius of curvature of the surface in the ray direction. This first order solution predicts, for example, zero field along the axis for an axial magnetic point source and hence zero H-plane coupling for arrays of TE_{10} mode excited circumferentially polarized waveguides. Only by including q^2 and higher-order exact solution terms one can accurately represent the field parallel to the magnetic point source. Other workers (e.g. Chang et al., 1976; Chan et al., 1977) have provided solutions for the cone, for example, which include all terms up to q^2. Both solutions are based on an asymptotic evaluation of the rigorous Green's function solution for each geometry. In later work, Boersma and Lee (1978) obtained a solution containing terms up to q^3 and $1/M^2$, using a GTD-style approach. They modified the asymptotic solution for a source on a conducting sphere such that it provided accurate results for the cylinder as well. In another approach, Pathak and Wang (1981) described an asymptotic solution for surface field on an electrically large conducting convex surface. This solution was devised from asymptotic solutions for the cylinder and the sphere and, therefore, recovers these results as limiting cases. It contains all terms up to q^3 and $1/M$. All solutions mentioned above recover the solution for a magnetic source on a plane in the limit that the radii of curvature become infinitely large. However, each solution predicts a different limiting value for the field in the vicinity of the axis (paraxial region) of a cylinder excited by a circumferential magnetic point source. This discrepancy occurs because the exact solution is dependent on some $1/M^2$ terms in the region of the axis. It can be shown that an asymptotic surface field solution for the circular cylinder that retains terms up to q^3 and $1/M^2$ gives the correct limiting value. Other significant references on apertures in conformal surfaces include Shapira et al. (1974a, b), Lee and Mittra (1977, 1979), Golden et al. (1974), Steyskal (1977) and Hessel et al. (1979).

8.3 Mutual Coupling in Conformal Arrays

Consider an array of identical rectangular waveguide of width a and height b that terminate in an arbitrary surface. It is assumed that each waveguide supports only the dominant TE_{10} mode. Starting from Eq. 7.31 and introducing Eq. 7.33 for a cylinder, the normalized mutual admittance of these modes (i.e. mode 1) in the surface is

$$y_{ij}(1,1) = -\frac{1}{A_i A_j} \iint_{D_i} \frac{1}{Y_j} (\mathbf{E_j} \times \hat{\rho}) \cdot \mathbf{H}_{ij}(R) dS_j, \tag{8.13}$$

where

$$\mathbf{E}_i = A_i \hat{p} \sqrt{\frac{2}{ab}} \cos\left(\frac{\pi x_i}{a}\right)$$

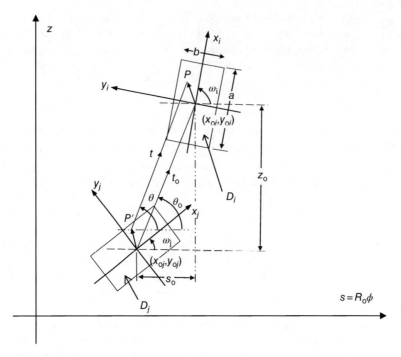

Figure 8.8 Geometry of mutual coupling analysis on a cylinder of radius R_o

with $\hat{p} = \left(\hat{\phi}\cos\omega_i + \hat{z}\sin\omega_i\right)$ the polarization of the electric field, $Y_i = Y_{10} = Y_o\beta_{10}/k$ is the TE_{10} mode wave admittance and β_{10} is its propagation constant. In the above, ω_i is the angle of the aperture relative to the axis of the cylinder as shown in Figure 8.8. The arrangement where all elements are aligned in the circumferential direction ($\omega_i = 90°$) is shown in Figure 8.1. In addition,

$$H_{ij}(\mathbf{R}) = \iint_{D_j} \begin{bmatrix} 0 & 0 & 0 \\ 0 & G_{\phi\phi}^{(h)} & G_{z\phi}^{(h)} \\ 0 & G_{\phi z}^{(h)} & G_{zz}^{(h)} \end{bmatrix} \cdot \mathbf{M_j} dS_j, \tag{8.14}$$

where the matrix in Eq. 8.14 is a representation of the dyadic Green's function $\overline{\overline{\mathbf{G}}}^{(h)}(\mathbf{R}|\mathbf{R}')$ in Eq. 7.33. The zero elements in the dyadic correspond to the radial components which vanish because of the boundary condition on the cylinder surface. Also, because of symmetry $G_{z\phi}^{(h)} = G_{\phi z}^{(h)}$. Continuing, the magnetic current on aperture i is $\mathbf{M_i} = \mathbf{E_i} \times \hat{p}$ and, therefore, Eq. 8.13 becomes

$$y_{ij}(1,1) = -\frac{2k}{abY_o\beta_{10}} \iint_{D_i} dS_i \cos\left(\frac{\pi x_i}{a}\right) \cdot \iint_{D_j} dS_j \cos\left(\frac{\pi x_j}{a}\right)$$

$$\times \left[G_{\phi\phi}^{(h)} \cos\omega_i\cos\omega_j + G_{zz}^{(h)} \sin\omega_i\sin\omega_j + G_{\phi z}^{(h)} \sin\left(\omega_i + \omega_j\right) \right]. \tag{8.15}$$

The three independent components of the dyadic function for the cylinder, namely, $G_{\phi\phi}^{(h)}$, $G_{zz}^{(h)}$ and $G_{\phi z}^{(h)}$, correspond to the magnetic field on the surface of a metallic surface, due to a magnetic point source of unit strength also on the surface. These components can be expressed in closed form as follows (Felsen & Marcuvitz, 1973; Bird, 1985a):

$$G_{\phi\phi}^{(h)}(\phi_i, z_i | \phi_j, z_j) = -\frac{jY_0}{(2\pi)^2 R_0} \int_{-\infty}^{\infty} d\xi \sum_{n=-\infty}^{\infty} \exp(j(n\bar{\phi} - \xi\bar{z})) \left(\frac{k}{h}\right)$$

$$\times \left(\frac{H_n^{(2)'}(hR_0)}{H_n^{(2)}(hR_0)} - \left(\frac{n\xi}{kz}\right)^2 \frac{H_n^{(2)}(hR_0)}{H_n^{(2)'}(hR_0)}\right)$$

$$\tag{8.16a}$$

$$G_{\phi z}^{(h)}(\phi_i, z_i | \phi_j, z_j) = \frac{jY_0}{(2\pi)^2 R_0} \int_{-\infty}^{\infty} d\xi \sum_{n=-\infty}^{\infty} \exp(j(n\bar{\phi} - \xi\bar{z}))$$

$$\times \left(\frac{n\xi}{khR_0}\right) \frac{H_n^{(2)}(hR_0)}{H_n^{(2)'}(hR_0)}$$

$$\tag{8.16b}$$

$$G_{z\phi}^{(h)}(\phi_i, z_i | \phi_j, z_j) = G_{\phi z}^{(h)}(\phi_i, z_i | \phi_j, z_j) \tag{8.16c}$$

$$G_{zz}^{(h)}(\phi_i, z_i | \phi_j, z_j) = \frac{jY_0}{(2\pi)^2 R_0} \int_{-\infty}^{\infty} d\xi \sum_{n=-\infty}^{\infty} \exp(j(n\bar{\phi} - \xi\bar{z})) \frac{h}{k} \frac{H_n^{(2)}(hR_0)}{H_n^{(2)'}(hR_0)}, \tag{8.16d}$$

where

$h = \sqrt{k^2 - \xi^2}$, $\bar{\phi} = \bar{\phi}_i - \phi_j$ and $\bar{z} = z_i - z_j$. Note that when $|\xi| > k$, $h = -j\sqrt{|k^2 - \xi^2|}$.

The dyadic components given in Eq. 8.16 can be substituted into Eq. 8.15, and a closed form result can be obtained for the mutual admittance on a cylindrical surface. The result can be time consuming to compute numerically for cylinders of moderate to large radius (i.e. $kR_0 > 10$). However, in that latter case, approximate solutions for the dyadic components can be derived by asymptotic methods. From these it is then relatively straight-forward to compute elements of the admittance matrix from Eq. 8.15 by numerical integration. A similar approach is possible for apertures on a conducting sphere or a more general surface (Hessel et al., 1979; Wills, 1986).

The coupling coefficients are the elements of the scattering matrix for an array on the cylinder or sphere, and this matrix is obtained in much the same way as for a planar array. The element-by-element approach is used to calculate admittances between two apertures at a time. This is possible because the apertures are in a conducting cylinder. The self-admittances can be calculated directly from Eq. 8.15, or if the surface has a moderate to large radius of curvature, the admittance of a single aperture in a ground plane is a good approximation (Bird, 1988). Once the admittance matrix $\mathbf{y}$ has been found, the scattering matrix can be computed as for the planar case from $\mathbf{S} = 2(\mathbf{I} + \mathbf{y})^{-1} - \mathbf{I}$, where $\mathbf{I}$ is the unit matrix. The complex amplitudes of the electric and magnetic fields are $\mathbf{A} = (\mathbf{a} + \mathbf{b})$ and $\mathbf{B} = (\mathbf{a} - \mathbf{b})$, respectively, and the amplitudes are related through $\mathbf{b} = \mathbf{S}\mathbf{a}$ as described in Section 7.3.3.

8.3.1 Asymptotic Solution for Surface Dyadic*

An asymptotic solution for the surface field at (R_o, ϕ, z) on the surface of the cylinder due to a source situated at (R_o, ϕ', z') also on the surface is expressed in the form

$$\underline{\underline{G}}^{(h)}(\phi, z|\phi', z') \sim \sum_{p=0}^{\infty} \left[\underline{\underline{G}}\left(\bar{\phi} + 2p\pi, \bar{z}\right) + \underline{\underline{G}}\left(2\pi - \bar{\phi} + 2p\pi, \bar{z}\right) \right], \qquad (8.17)$$

where $\bar{\phi} = \phi - \phi'$ and $\bar{z} = |z - z'|$. The dyadic function $\mathbf{G}$ is the field due to a single surface ray in one direction while the summation corresponds to creeping waves travelling along helical paths between the field and source points. Figure 8.7 illustrates creeping waves on a cylinder. In this section an approximate asymptotic solution is obtained for each component of the surface field dyadic of a cylinder. Each term of Eq. 8.17 can be interpreted as a surface wave and the field on the cylinder as a standing wave produced by contra-rotating surface waves, which propagate from the source to the field point along helical ray paths, as depicted in Figure 8.7a. These waves follow the curvature of the cylindrical surface and appear to 'creep' from the source to the field point. These creeping waves attenuate as they propagate requiring, in any practical calculation (see Fig. 8.7b), the consideration of only two contra-rotating creeping waves encircling the cylinder just once. That is,

$$\underline{\underline{G}}^{(h)}(\phi, z|\phi', z') \sim \underline{\underline{G}}\left(\bar{\phi}, \bar{z}\right) + \underline{\underline{G}}\left(2\pi - \bar{\phi}, \bar{z}\right).$$

Several approximate asymptotic solutions have been obtained, and five of them were compared with the exact solution for a range of parameters (Bird, 1984). The asymptotic solution becomes identical to the exact solution for a point source on a plane conductor when the radius of the cylinder approaches infinity.

Let t and θ be co-ordinates relative to the source on the projected surface of the cylinder, Figure 8.7c, where $t = \sqrt{\bar{z}^2 + \left(R_o\bar{\phi}\right)^2}$ and $\theta = \tan^{-1}\left(\bar{z}/R_o\bar{\phi}\right)$. An accurate solution for the three non-zero components of the dyadic are as follows (Bird, 1985a):

$$G_{zz}(t, \theta) \sim v(\beta)G_o(t) \left[\cos^2\theta + q\left(2 - 3\cos^2\theta\right) \right] + C_{zz}^{(1)}(t, \theta), \qquad (8.18a)$$

where

$$C_{zz}^{(1)}(t, \theta) = G_o(t)q \left[v(\beta)\left(\frac{31}{72}\sin^2\theta - \frac{5}{24}\right) + v_1(\beta)\left(\frac{11}{60} - \frac{17}{36}\sin^2\theta\right) \right.$$
$$\left. + v_2(\beta)\left(\frac{1}{24}\sin^2\theta + \frac{1}{40}\right) + \frac{j\beta}{5}v_0^{(1)}(\beta) \right],$$

where $\beta = kt/2M^2 = kt\left(\cos^2\theta/\sqrt{2kR_o}\right)^{2/3}$ is the distance parameter, $G_o(t) = -\left(k^2/2\pi\eta_o\right)q$ $\exp(-jkt)$ and $q = j/kt$. The functions $v(\beta)$, $v_1(\beta)$, $v_2(\beta)$ and $v_0^{(1)}(\beta)$ are the surface Fock functions, which are detailed in Appendix G. The remaining asymptotic dyadic components are

$$G_{\phi\phi}(t,\theta) \sim v(\beta)G_o(t)\left(\sin^2\theta + q\left(2 - 3\sin^2\theta\right)\right) + C_{\phi\phi}^{(1)}(t,\theta) \tag{8.18b}$$

and

$$G_{\phi z}(t,\theta) \sim v(\beta)G_o(t)\sin\theta\cos\theta(1 - 3q) + C_{\phi z}^{(1)}(t,\theta), \tag{8.18c}$$

where

$$C_{\phi\phi}^{(1)}(t,\theta) = G_o(t)q\left[\sec^2\theta[u(\beta) - v(\beta)] + v(\beta)\left(\frac{8}{9}\tan^2\theta - \frac{31}{72}\sin^2\theta\right)\right.$$
$$\left. + v_1(\beta)\left(\frac{17}{36}\sin^2\theta - \frac{43}{45}\tan^2\theta\right) + v_2(\beta)\left(\frac{1}{15}\tan^2\theta - \frac{1}{24}\sin^2\theta\right) + \frac{j\beta}{5}v_0^{(1)}(\beta)\tan^2\theta\right]$$

and

$$C_{\phi z}^{(1)}(t,\theta) = -G_o(t)q\sin\theta\cos\theta\left[v(\beta)\left(\frac{5}{9}\sec^2\theta - \frac{31}{72}\right)\right.$$
$$\left. + v_1(\beta)\left(\frac{17}{36} - \frac{28}{45}\sec^2\theta\right) + v_2(\beta)\left(\frac{1}{15}\sec^2\theta - \frac{1}{24}\right) + \frac{j\beta}{5}v_0^{(1)}(\beta)\sec^2\theta\right].$$

A more accurate result is possible by the addition of higher-order terms (Bird, 1984), where terms up to third order in q are required. This full solution was used to study the effect of the aperture orientation in the cylinder on the self-admittance (Bird, 1988). This work showed that for moderately large cylinders the self-admittance is relatively weakly dependent on orientation and the self-admittance of the waveguide in a ground plane is an excellent approximation compared with experiment.

As a verification of the above solution, some results are described for mutual coupling between waveguides in a cylinder. As a first example, Figure 8.9 shows the computed magnitude and phase of the mutual admittances of apertures coupling in the H-plane at 9 GHz as a function of angle between of two WG-16 rectangular waveguides ($a = 2.286$ cm, $b = 1.016$ cm) on a cylinder of radius 5.057 cm. A single TE_{10} mode is assumed in each aperture and a 9×9 regular grid of integration points is used with Simpson's rule in the computation. The modal solution results are taken from an extensive data compilation on the topic (Lee & Mittra, 1977).

Another result is given this time for a 29-element array of WR-90 X-band waveguides that terminate in a cylinder (pictured in Figure 1.1i). The waveguides are arranged with their principal electric field oriented in the circumferential direction in the manner illustrated in Figure 8.1. The cylinder has a radius of 126.24 mm and the waveguide azimuthal and axial spacing's are $\phi_o = 11.61°$ and $x_o = 28.40$ mm (Wills, 1983). The scattering parameters are computed from the mutual admittances that were obtained from a single mode approximation for the aperture field, and these are compared with experimental results. Each waveguide is fed with a coaxial line-to-waveguide stepped end-launcher. Figures 8.10a and b show a quadrant of the array for experimental (a) and computed coupling (b) coefficients, respectively, that were obtained at 9.5 GHz when the centre waveguide was driven and the other waveguides were terminated in matched loads. The theoretical values were computed using the asymptotic

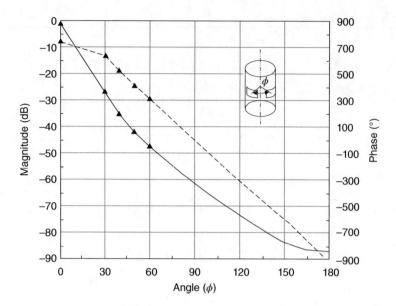

Figure 8.9 *H*-plane mutual admittance of two circumferential waveguides in a cylinder as a function of angle between their centres. Cylinder radius $R_o = 5.057$ cm, waveguide dimensions $a = 2.286$ cm, $b = 1.016$ cm. Asymptotic solution for admittance solid curve: magnitude in dB; dashed: phase; and ▲ modal (exact) solution

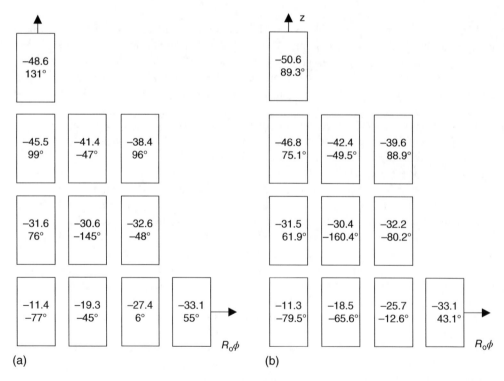

Figure 8.10 Coupling coefficients (magnitude (dB), phase (deg.)) in a quadrant of a 29-element conformal array at a frequency of 9.5 GHz. Identical WR90 waveguides terminate in a cylinder of radius 126.24 mm. Element spacing: axial 28.4 mm and azimuth 25.6 mm. (a) Measured (Wills, 1983). (b) Finite array analysis with asymptotic solution

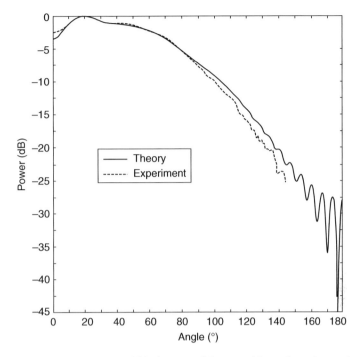

Figure 8.11   *E*-plane radiation pattern of 29-element axial waveguide conformal array (see Figure 8.10) at 10 GHz when central element is excited and other waveguides terminated in matched loads

formulae Eq. 8.18 and a single mode approximation for the aperture field. There is good agreement between the single-mode theory and experiment except for the distant elements in the *H*-plane where the coupling level is low. However, the agreement is expected to improve if more modes are assumed for the aperture field. Results computed from an infinite array analysis implemented by Wills (1986) are also in good agreement with the finite array analysis except on the array periphery and at several waveguides in the *H*-plane.

The radiation pattern in the *E*-plane of the 29-element conformal array of axially oriented waveguides described earlier is shown in Figure 8.11. The central waveguide is excited, and the remaining elements are terminated in matched loads. It is seen that reasonable agreement is obtained between theoretical and experimental results except mainly at wide angles.

8.4   Coupling in a Concave Array: Periodic Solution[*]

Two-dimensional conformal periodic arrays can be handled through the natural periodicity of the circular geometry and two examples are illustrated in Figure 8.12. Two-dimensional microwave lenses are an important class of beamforming networks for array antennas. Practical examples include the Ruze and Rotman lenses for linear arrays and the R-2R lens for cylindrical arrays, a part of which is a cylindrical lens (Ruze, 1950; Rotman & Turner, 1963). A basic realization consists of a parallel plate region fed by coaxial probes mounted

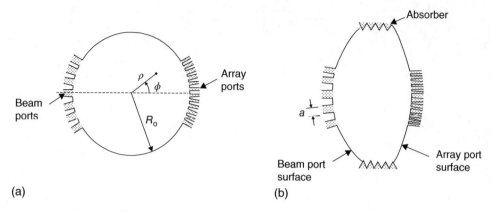

Figure 8.12 Waveguide-fed parallel plate concave arrays. (a) Circular lens. (b) Rotman lens.
Source: Reproduced with permission of the Institution of Engineers, Australia

close to the lens contour or by waveguides terminated in the lens contour. In a typical micro-wave lens design, the input (beam) and output (array) contours are chosen so that a beam is formed when one or more inputs are excited. According to geometric optics, this is possible if all ray path lengths are identical from the feed port to the radiated wave-front by means of the array port contour. This cannot be achieved in general for each ray and there is an associated path length error. Usually the two lens contours are designed to minimize these errors for all beams. In this section a simplified model for a concave array is developed based on a periodic array on a cylinder.

Consider an array of N apertures of width a that terminates in a concave metallic cylinder of radius R_0. A unit cell is shown in Figure 8.13. Suppose ψ is the phase shift per unit cell around the cylinder and $\phi_0 = 2\pi/N$ is its angular extent. When the axis of the cylinder is along the z-direction, the field in the cylinder is periodic and of the form

$$E_z(\rho,\phi) = k^2 \sum_{n=-\infty}^{\infty} D_n J_{\nu_n}(k\rho) \exp\left(j\nu_n\phi\right) \tag{8.19a}$$

$$H_\phi(\rho,\phi) = \frac{-jk^2}{\eta_0} \sum_{n=-\infty}^{\infty} D_n J'_{\nu_n}(k\rho) \exp\left(j\nu_n\phi\right), \tag{8.19b}$$

where $\nu_n = (\psi + 2\pi n)/\phi_0$. For electrical periodicity, it is required that $\psi = (2\pi/N)\ell$; $\ell = -(N/2)+1,\ldots,0,\ldots,(N/2)$. Therefore, $\nu_n = \ell + nN$.

Boundary conditions on the surface of the cylinder $\rho = R_0$ require $\hat{z}E_z = E_A\hat{z}$ over the aperture in the unit cell $-\phi_0/2 < \phi < \phi_0/2$ (see Figure 8.13). This requires that

$$E_A = k^2 \sum_{n=-\infty}^{\infty} D_n J_{\nu_n}(kR_0) \exp\left(j\nu_n\phi\right) \tag{8.20}$$

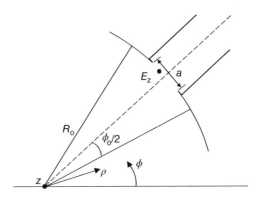

Figure 8.13 Unit cell of concave periodic array

for $-\phi_o/2 < \phi < \phi_o/2$. Taking the inverse Fourier transform results in

$$D_n = \frac{N}{2\pi k^2} \frac{1}{J_{\nu_n}(kR_o)} \int_{-\phi_o/2}^{\phi_o/2} E_A(R_o,\phi') \exp\left(-j\nu_n\phi'\right)d\phi'. \tag{8.21}$$

Introducing Eq. 8.21 into Eqs. 8.19 gives the field inside the cylinder as

$$E_z(\rho,\phi) = \frac{N}{2\pi} \sum_{n=-\infty}^{\infty} \frac{J_{\nu_n}(k\rho)}{J_{\nu_n}(kR_o)} \int_{-\phi_o/2}^{\phi_o/2} E_A(R_o,\phi') \exp(j\nu_n(\phi-\phi'))d\phi'$$

$$H_\phi = -\frac{jN}{2\pi\eta_o} \sum_{n=-\infty}^{\infty} \frac{J'_{\nu_n}(k\rho)}{J_{\nu_n}(kR_o)} \int_{-\phi_o/2}^{\phi_o/2} E_A(R_o,\phi') \exp\left(j\nu_n(\phi-\phi')\right)d\phi'.$$

Let the field in the aperture be represented by the following modal expansions:

$$E_A\hat{z} = \hat{z}\sum_{p=1}^{\infty} A_p e_p(\phi)$$

and

$$H_A\hat{\phi} = \hat{\phi}\sum_{p=1}^{\infty} B_p Y_p e_p(\phi),$$

where $e_p = \sqrt{2/\phi_o}\cos\left(p\pi\phi/\phi_o\right)$, Y_p is the admittance of mode p, while A_p and B_p are the mode coefficients of the electric and magnetic fields. Continuity of the magnetic field component H_ϕ at $\rho = R_o$ requires that

$$\sum_{p=1}^{\infty} B_p Y_p e_p = -\frac{jN}{2\pi\eta_o} \sum_{n=-\infty}^{\infty} \frac{J'_{\nu_n}(K)}{J_{\nu_n}(K)} \exp(j\nu_n\phi) \sum_{p=1}^{\infty} A_p \int_{-\phi_o/2}^{\phi_o/2} e_p(\phi') \exp(-j\nu_n\phi')) d\phi'$$

$$(8.22)$$

$$= -\frac{jN}{2\pi\eta_o} \sum_{n=-\infty}^{\infty} \frac{J'_{\nu_n}(K)}{J_{\nu_n}(K)} \exp(j\nu_n\phi) \sum_{p=1}^{\infty} A_p I_p(-\nu_n),$$

where

$$I_p(\nu) = \int_{-\phi_o/2}^{\phi_o/2} e_p(\phi') \exp(j\nu\phi') d\phi'$$

$$= \frac{2\sqrt{2\phi_o}}{p\pi} (-1)^{(p-1)/2} C\left(\frac{\phi_o\nu}{2}\right).$$

$K = kR_o$, and the function $C(x)$ is defined in Appendix A.4. Multiplying both sides of Eq. 8.22 by e_q and integrating across the aperture results in

$$B_p Y_p = -\frac{jN}{2\pi\eta_o} \sum_{q=1}^{\infty} A_q \sum_{n=-\infty}^{\infty} \frac{J'_{\nu_n}(K)}{J_{\nu_n}(K)} I_p(-\nu_n) I_q(\nu_n).$$

That is,

$$B_p = \sum_{p=1}^{\infty} Y_{pq} A_q,$$

where Y_{pq} is the mutual admittance of elements p and q in the apertures for the ℓ-th phase sequence. This is expressed as follows:

$$Y_{pq}(\ell) = -\frac{jN}{2\pi\eta_o Y_p} \sum_{n=-\infty}^{\infty} \frac{J'_{\nu_n}(K)}{J_{\nu_n}(K)} I_p(-\nu_n) I_q(\nu_n).$$

Suppose the aperture field consists of only the fundamental TE_{10} mode, that is, $p = 1 = q$ then

$$Y_{11}(\ell) = -\frac{j4N\phi_o}{\pi^3}\left(\frac{k}{\beta_{10}}\right) \sum_{n=-\infty}^{\infty} \frac{J'_{\nu_n}(K)}{J_{\nu_n}(K)} C\left(\frac{\nu_n\phi_o}{2}\right)^2,$$

$$(8.23)$$

where $Y_p = \beta_{10}/(k\eta_o)$ is the mode admittance, and β_{10} is the mode propagation constant. Recall that $\nu_n = \ell + nN$. When $\nu_n > 2kR_o$, the following asymptotic formula can be used for the ratio of the Bessel function derivative with itself (see Appendix B.1). Thus,

$$\frac{J'_{\nu_n}(K)}{J_{\nu_n}(K)} \approx \sqrt{\left(\frac{\nu_n}{K}\right)^2 - 1}.$$

The series in Eq. 8.23 converges rapidly for $2/\phi_0 \ll \nu_n$ as ν_n becomes large resulting in

$$\frac{J'_{\nu_n}(K)}{J_{\nu_n}(K)} C \left(\frac{\nu_n \phi_0}{2}\right)^2 \approx \frac{1}{K} \left(\frac{\pi}{\phi_0}\right)^4 \frac{\cos^2 (\nu_n \phi_0/2)}{\nu_n^3}.$$

The active reflection coefficient can be used to determine the coupling to the other waveguides. The active reflection coefficient in the present case is defined by

$$\Gamma(\ell) = \frac{1 - Y_{11}(\ell)}{1 + Y_{11}(\ell)}$$

$$= \exp\left(2j \arg\left(1 - Y_{11}(\ell)\right)\right),$$

since the admittance Y_{11} is purely imaginary. The coupling coefficient between waveguide n and the central (0-th) element is given by

$$S_{n0} \approx \frac{1}{N} \sum_{\ell = -L_{\min}}^{L_{\max}} \Gamma(\ell) \exp\left(j\frac{2\pi n\ell}{N}\right)$$

$$= \frac{1}{N} \sum_{\ell = -L_{\min}}^{L_{\max}} \exp\left[2j\left(\frac{\pi n\ell}{N} + \arg(1 - Y_{11}(\ell))\right)\right], \tag{8.24}$$

where $L_{\max} = \text{Int}(N/2)$, $L_{\min} = N - (1 + \text{Int}(N/2))$ and the function $\text{Int}(x)$ is the integer value of $x \leq x$. The limits of ℓ are chosen to make the middle element to be located at $\ell = 0$.

As an example of the results given by this formulation and Eq. 8.24, the coupling coefficients in the H-plane at 9.5 GHz are shown in Figure 8.14 for 60 element array of X-band waveguides (width $a = 0.724\lambda$ corresponding to WG-16 waveguide at 9.5 GHz) located in a cylinder of radius $R_0 = 8\lambda$. The elements are spaced a distance 0.84λ or $\phi_0 = 6°$ apart. A geometric optics estimate for the coupling in the H-plane based on a ray analysis can be obtained and is

$$|S_{21}(\phi_a)|^2 \approx \frac{8ka}{\pi^3} \left(\frac{k}{\beta_{10}}\right)\left(1 + \frac{\beta_{10}}{k}\right)^2 \left[\sin\left(\frac{\phi_a}{2}\right) C\left(\frac{ka}{2} \sin\frac{\phi_a}{2}\right)\right]^4 \left(\frac{1}{ks}\right), \tag{8.25}$$

where ϕ_a is the angle at the centre of the circle between the source and receiving elements, and $s = 2R_0 \sin(\phi_a/2)$ is the distance between the waveguide centres. The results predicted by Eq. 8.25 are shown in Figure 8.14 as the GO solution. Eq. 8.25 gives a reasonable estimate providing the receiving waveguide is not too near the source waveguide.

The circular lens, shown in Figure 8.12a, can be analysed using the above formulation or the one based on the mutual admittance that is obtained from an asymptotic formation of the surface field as described in the references (Ishihara et al., 1978; Bird, 1985b; Parini & Lee-Yow, 1986). An indication of the level of coupling that can be expected in circular lens, the power coupled across the lens, is shown in Figure 8.15 for a lens of diameter 25 cm. The lens is excited by ten WG-16 waveguides (width 2.286 cm). Five waveguides are located on the

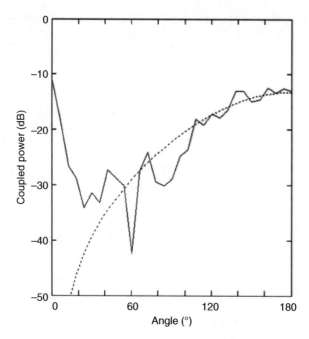

Figure 8.14 Coupled power versus angle in a circular lens of radius $R_o = 8\lambda$ at 9.5 GHz due to an array of 60 WG-16 rectangular waveguides terminated in a cylinder and oriented in their *H*-plane. Solid curve: periodic solution and dashed curve: GO solution

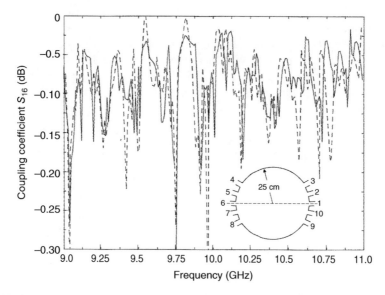

Figure 8.15 Magnitude of coupling coefficient between ports 1 and 6 versus frequency for a closed circular lens with radius $R_o = 25$ cm. Solid line: theory; dashed line: experiment (Parini & Lee-Yow, 1986). *Source*: Reproduced with permission from the Institution of Engineers, Australia

beam (input) side as shown in the inset in Figure 8.15 and another five on the array (output) side. Figure 8.15 shows the computed and measured amplitude of the coupling coefficient S_{16}. Despite the strong interactions around the lens, quite good agreement has obtained for this and the other coupling coefficients including the input reflection coefficient.

8.5 Problems

P8.1 Calculate and plot the radiation patterns in the E- and H-planes at 28 GHz for a waveguide with inside dimensions 7.112×3.556 mm that is terminated in a cylinder of diameter of 100 mm. The narrow wall of the waveguide is parallel to the axis of the cylinder (i.e. axial polarization).

P8.2 Verify that the excitation coefficient of an element of an array on a cylinder that produces a beam in the direction (r, θ_b, ϕ_b) is given by $a = A(\phi_o, z_o) \exp [jk(R_o(1 - \sin\theta_b \cos (\phi_o - \phi_b)) - z_o \cos \theta_b]$, where the centre of the element is at (R_o, ϕ_o, z_o) and A is the illumination function.

P8.3 Commencing with the far-field radiated by an axially oriented waveguide in Eq. 8.8, plot the E-plane radiation pattern of a 33-element equispaced conformal array of these elements on a cylinder of radius $R_o = 25\lambda$. The waveguide dimensions are width $a = 0.7\lambda$ and height $b = 0.4\lambda$, and the element spacing is 0.5λ. Apply an excitation to steer the array in the azimuth direction to an angle of $30°$ from boresight.

P8.4 For the same array configuration and cylinder radius as in P8.3, increase the element spacing to 1λ and demonstrate the formation of two grating lobes.

P8.5 From Eq. 8.18, show that as the radius of a cylinder becomes increasingly large the asymptotic solution for the surface field approaches that of an infinite metallic plane.

P8.6 Simplify the self-admittance of a rectangular waveguide aligned in the axial direction $(\omega_i = \pi/2)$ in a cylinder given by Eq. 8.15. In particular, consider the integral in the vicinity of $h = 0$ and the evaluation of the Hankel functions. Evaluate the ratio of the Hankel function and its derivative in this region. Describe what happens when the radius of the cylinder becomes very large (i.e. $R_o \rightarrow \infty$).

P8.7 Using the definition of the hard surface Fock function, $v_n^{(m)}(z)$, given by Eq. G.5 in Appendix G, where the argument z is complex, show that its derivative can be obtained from itself and the next order function as follows: $v_n^{(m)'}(z) = (n + 1/2)\left(v_n^{(m)}(z) - v_{n+1}^{(m)}(z)/z\right)$.

References

Abramowitz, M. and Stegun, I.A. (1965): 'Handbook of mathematical functions', Dover Inc., New York.

Bailin, L.L. and Spellmire, R.J. (1957): 'Convergent representations for the radiation fields from slots in large circular cylinders', IRE Trans. Anntenas Propag., Vol. **AP-5**, pp. 374–382.

Bird, T.S. (1984): 'Comparison of asymptotic solutions for the surface field excited by a magnetic dipole on a cylinder', IEEE Trans. Antennas Propag., Vol. **AP-32**, pp. 1237–1244.

Bird, T.S. (1985a): 'Accurate asymptotic solution for the surface field due to apertures in a conducting cylinder', IEEE Trans. Antennas Propag., Vol. **AP-33**, pp. 1108–1117.

Bird, T.S. (1985b): 'Mutual coupling in waveguide-fed parallel plate lenses', Convention Digest, IREECON, Melbourne, pp. 936–939.

Bird, T.S. (1988): 'Admittance of rectangular waveguide radiating from a conducting cylinder', IEEE Trans. Antennas Propag., Vol. **AP-36**, pp. 1217–1220.

Boersma, J. and Lee, S.W. (1978): 'Surface field due to a magnetic dipole on a cylinder: asymptotic expansion of exact solution', Department of Electrical Engineering, Electromagnetics Laboratory, University of Illinois at Urbana-Champaign, Champaign, IL. Report No. 78-17.

Bowman, J.J., Senior, T.B.A. and Uslenghi, P.L.E. (1963): 'Electromagnetic and acoustic scattering by simple shapes', North-Holland Publishing Company, Amsterdam, the Netherlands.

Chan, K.K., Felsen, L.B., Hessel, A. and Shmoys, J. (1977): 'Creeping waves on a perfectly conducting cone', IEEE Trans. Antennas Propag., Vol. AP-25, pp. 661–670.

Chang, Z.W., Felsen, L.B. and Hessel, A. (1976): 'Surface ray methods for mutual coupling in conformal arrays on cylinder and conical surfaces', Polytechnic Institute of New York, Brooklyn, NY. Final Report (prepared under contract N00123-76-C-0236).

Felsen, L.B. and Marcuvitz, N. (1973): 'Radiation and scattering of waves', Prentice-Hall, Englewood Cliffs, NJ.

Fock, V.A. (1965): 'Electromagnetic diffraction and propagation problems', Pergamon Press, London, UK.

Franz, W. (1954): 'Uber die Green'sche function des zylinders und der kugel', Z. Naturforsch., Vol. 9A, pp. 705–716.

Golden, K.E., Stewart, G.E. and Pridmore-Brown, D.C. (1974): 'Approximation techniques for the mutual admittance of slot antennas on metallic cones', IEEE Trans. Antennas Propag., Vol. AP-22, pp. 43–48.

Goodrich, R.F., Kleinman, R.E., Maffett, A.L., Schonsted, A.L., Siegel, K.M., Chernin, M.G., Shrank, H.E. and Plummers, R.E. (1959): 'Radiation from slot array and cones', IRE Trans. Anntenas Propag., AP-7, pp. 213–222.

Hasserjian, G. and Ishimaru, A. (1962): 'Excitation of a conducting cylindrical surface of large radius of curvature', IRE Trans. Anntenas Propag., Vol. AP-10, pp. 264–273.

Hessel, A., Lin, Y.L. and Shmoys, J. (1979): 'Mutual admittance between circular apertures on a large conducting sphere', Radio Sci., Vol. 14, pp. 35–41.

Ishihara, T., Felsen, L.B. and Green, A. (1978): 'High frequency fields excited by a line source located on a perfectly conducting concave cylindrical surface', IEEE Trans. Antennas Propag., Vol. AP-26, pp. 757–767.

James, G.L. (1986): 'Geometrical theory of diffraction for electromagnetic waves', 3rd ed., Peter Peregrinus Ltd., London, UK.

Lee, S.W. and Safavi-Naini, S. (1976): 'Asymptotic solution of surface field due to a magnetic dipole on a cylinder', Department of Electrical Engineering, Electromagnetics Laboratory, University of Illinois at Urbana-Champaign, Champaign, IL. Report No. 76-11.

Lee, S.W. and Mittra, R. (1977): 'Mutual admittance between slots on a cylinder or cone', Department of Electrical Engineering, Electromagnetics Laboratory, University of Illinois at Urbana-Champaign, Champaign, IL. Report No. 77-24.

Lee, S.W. and Mittra, R. (1979): 'GTD solution of slot admittance on a cone or cylinder', Proc. IEE, Vol. 126, pp. 487–492.

Logan, N.A. (1959): 'General research in diffraction theory', Missiles and Space Division, Lockheed Aircraft Corporation, Sunnyvale, CA. Vol. l: Rep. LMSD-288087 and Vol. 2: Rep. I. MSC-288088.

Mittra, R. and Safavi-Naini, S. (1979): 'Source radiation in the presence of smooth convex bodies', Radio Sci., Vol. 14, pp. 217–237.

Parini, C.G. and Lee-Yow, L.M.C.E. (1986): 'Mutual coupling effects in waveguide Rotman lens beam forming networks', Military microwaves '86; Proceedings of the Conference, Brighton, UK, June 24–26, pp. 273–278.

Pathak, P.H. and Kouyoumjian, R.G. (1974): 'An analysis of the radiation from apertures in curved surfaces by the geometrical theory of diffraction', Proc. IEEE., Vol. 62, pp. 1438–1447.

Pathak, P.H. and Wang, N. (1978): 'An analysis of the mutual coupling between antennas on a smooth convex surface', Department of Electrical Engineering, The Ohio State University Electro-Science Laboratory, Columbus, OH. Rep. 784583-7, October (prepared under contract N62269-76-C-0554 for Naval Air Development Centre (AD/A 065 591) (NADC-79046-30)).

Pathak, P.H., Wang, N., Burnside, W.D. and Kouyoumjian, R.G. (1981): 'A uniform GTD solution for the radiation from sources on a convex surface', IEEE Trans. Antennas Propag., Vol. AP-29, pp. 609–621.

Pathak, P.H. and Wang, N. (1981): 'Ray analysis of mutual coupling between antennas on a convex surface', IEEE Trans. Antennas Propag., AP-29, pp. 911–922.

Pistolkors, A.A. (1947): 'Radiation from longitudinal and transverse slots in a circular cylinder', J. Tech. Phys. (USSR), Vol. 17, pp. 365–385.

Rotman, W. and Turner, R.F. (1963): 'Wide-angle lens for line source applications', IEEE Trans. Antennas Propag., Vol. AP-11, pp. 623–632.

Ruze, J. (1950): 'Wide-angle metal plate optics', Proc. IRE, Vol. 38, pp. 53–59.

Sensiper, S. (1957): 'Cylindrical radio waves', IRE Trans. Antennas Propagat., Vol. AP-5, pp. 56–70.

Shapira, J., Felsen, J. and Hessel, A. (1974a): 'Ray analysis of conformal antenna arrays', IEEE Trans. Antennas Propag., Vol. **AP-22**, pp. 49–63.

Shapira, J., Felsen, L.B. and Hessel, A. (1974b): 'Surface ray analysis of mutually coupled arrays on variable curvature cylindrical surfaces', Proc. IEEE, **62**, pp. 1482–1492.

Silver, S. and Saunders, W.K. (1950): 'The external field produced by a slot on an infinite circular cylinder', J. Appl. Phys., Vol. **21**, pp. 153–158.

Steyskal, H. (1977): 'Analysis of circular waveguide arrays on cylinders', IEEE Trans. Antennas Propag., Vol. **AP-25**, pp. 610–616.

Wait, J.R. (1956): 'Currents excited on a conducting surface of large radius of curvature', IRE Trans. Microw. Theory Tech., Vol. **MTT-4**, pp. 143–145.

Wait, J.R. (1959): 'Electromagnetic radiation from cylindrical structures', Pergamon Press, London, UK.

Wills, R.W. (1983): 'Analysis of radiation from waveguide arrays on metallic cylinders including mutual coupling effects', 13th European Microwave Conference, Nuremberg, West Germany, 3–8 September, pp. 383–388.

Wills, R.W. (1986): 'Mutual coupling between waveguide antenna elements on conformal surfaces', Report. Plessey Electronic Systems Ltd., Addlestone, Surrey, UK.

9

Reflectarrays and Other Aperture Antennas

9.1 Introduction

There are a variety of other aperture antennas which find use in specialized applications that have not yet been considered. Some of these antennas that will be described in this chapter are illustrated in Figure 9.1. The first type shown in Figure 9.1a is the reflectarray. In design and operation, reflectarrays are inherently a combination of arrays and reflectors. They were first developed in the 1960s to achieve a high gain in combination with a low profile and with a potential ability to modify the beam direction without the complexity of a complex RF beamformer, which is usually associated with scanning arrays (Berry et al., 1963). Since the 1980s, reflectarrays have increasingly been made with microstrip patch elements on a substrate with a metal ground plane. These microstrip reflectarrays are illuminated by a feed and consist of a number of discrete resonant elements that are conventionally passive and arranged on a plane or conformal to another surface such as a cylinder. By appropriate phasing of the elements, a reflectarray can be a source of directive radiation (Huang & Encinar, 2008).

Another type of aperture antenna that gives directive radiations is a lens. Some lens antennas are illustrated in Figure 9.1b. A lens was first used as an antenna for radio by Oliver Lodge in 1889. In the same way as in optics, a lens can create directive radiation from a point source or conversely receive energy and focus it to a point. A lens is often used in combination with other aperture antennas such as a horn or reflector.

The third type of aperture antenna described here, which is illustrated in Figure 9.1c, is the Fabry–Pérot resonator antenna. This can be highly efficient and flexible as a radiator, but in

Fundamentals of Aperture Antennas and Arrays: From Theory to Design, Fabrication and Testing,
First Edition. Trevor S. Bird.
© 2016 John Wiley & Sons, Ltd. Published 2016 by John Wiley & Sons, Ltd.
Companion website: www.wiley.com/go/bird448

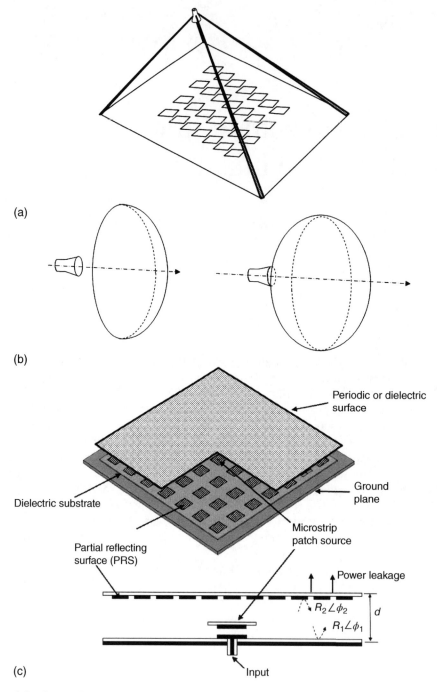

(a)

(b)

(c)

Figure 9.1   Some other aperture antennas. (a) Reflectarray; (b) plano-convex and spherical lenses and (c) Fabry–Pérot resonator antenna

its basic form, it is a narrowband device. The Fabry–Pérot cavity was first used in optics to create an intense beam of light. This cavity was invented by the physicists Charles Fabry and Alfred Pérot in the 1890s (Born & Wolf, 1959). In common with practice in optics where light is injected between two mirrors one of which is slightly transparent, radio frequency versions operate in a similar fashion as shown in Figure 9.1c. The operation of a Fabry–Pérot resonator antenna is described in this chapter and as well as some techniques that are used to broaden its bandwidth.

9.2   Basic Theory of Reflectarrays

A reflectarray consists of discrete resonant elements that are typically arranged on a planar surface. An important attribute of an element in this application is phase compensation. It is vital for reflectarrays to have this attribute so that in a similar manner to a paraboloidal reflector it can focus and redirect rays from a source towards an intended direction or in receive mode into a feed horn. When the elements are excited at their resonant frequency, the scattered energy can be highly directive as for a solid reflector antenna. In this section a simplified model of the array elements is developed. In this model, mutual coupling has been neglected although the principles described in the previous sections could be adopted for a more accurate representation of the antenna radiation.

Typically, a reflectarray is a planar surface, but in principle, it could be conformal with other surfaces to suit the application. Whatever the surface profile, a fundamental property of a reflectarray element located at (x_i, y_i) is to have a phase shift Ψ_{ij} such that

$$k\left(\rho_{ij} - \mathbf{r}_{ij} \cdot \hat{r}_b\right) - \Psi_{ij} = 2N\pi; \quad i = 1, \ldots, N_x; \quad i = 1, \ldots, N_x, \tag{9.1}$$

where ρ_{ij} is the radial distance from the feed at F to the element located at (x_i, y_i), as shown in Figure 9.2a, and the total number of elements is $N_e = N_x + N_y$. The vector $\mathbf{r}_{ij}$ extends from O to the centre of element i, $\hat{r}_b$ is a unit vector in the direction of the main beam, and N is an integer selected such that the phase is in the range 0–360°. Individual patches are usually designed to resonate at or near half the guided wavelength in the substrate (i.e. $\sim \lambda/2\sqrt{\varepsilon_r}$, where ε_r is the dielectric constant of the substrate above the ground plane), and the spacing between elements is chosen to be half the free-space wavelength, that is, $\lambda/2$. The most common ways to introduce the phase shift Ψ_{ij} into element ij is to either change its size (see Figure 9.2b), to add a short or open circuited stub or to shunt load the element with an active element such as a varactor diode. The advantage of the last-mentioned technique is that the reflectarray can be made adaptable to enable the beam to be steered electronically or to make the surface reconfigurable. Whatever phase compensation technique is employed, it is helpful for design to know the dependence of the phase response on variable parameters such as dimensions, length of stub or diode bias. Linear polarization is often handled with rectangular elements, while in circular polarization square, two orthogonal stubs or circular patches can be used. An early difficulty with reflectarrays was the very narrow bandwidth available mainly due to phase errors. This has now been largely overcome by the use of thicker substrates, stacked patches and improved modelling techniques.

An element can be modelled in a simple way as a transmission line or more accurately with a commercial full-wave software package. A simple narrowband radiation model

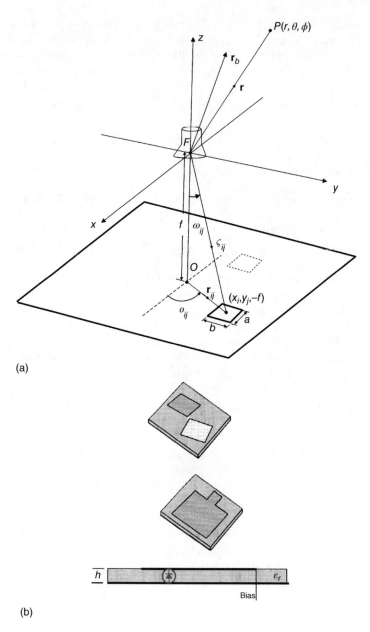

(a)

(b)

Figure 9.2 Geometry of the reflectarray. (a) Feed illumination and geometry and (b) typical phase compensating patch elements

of a single rectangular patch is shown in Figure 9.2a. It consists of an array of patches with a uniform current that has been excited by a feed horn, the pattern of which can be calculated from basic array theory. The uniform current is assumed to be set up on element i located at (x_i, y_i) by the feed located at F. In spherical polar co-ordinates relative to F, this is

$$\mathbf{J}_{sij} = 2\hat{z} \times \mathbf{H}_{i}^{inc} \exp(-jkh\sqrt{\varepsilon_r}) \cos(jkh\sqrt{\varepsilon_r}) \exp(j\Psi_{ij}), \tag{9.2}$$

where $\mathbf{H}_i^{inc}$ is the incident magnetic field from the feed. It is assumed that the element is located on a substrate of thickness h with a dielectric constant ε_r that is backed by a ground plane. Also each patch has associated with it an additional phase factor Ψ_{ij} due to techniques such as those shown in Figure 9.2b. The element is assumed in the far-field of the feed, and therefore, its radiated electric field is given by

$$\mathbf{E}_{ij}^{inc} = \frac{E_o \exp(-jk\rho_{ij})}{\rho_{ij}} \left[\hat{\psi} A(\psi_{ij}) \cos \xi_{ij} - \hat{\xi} B(\psi_{ij}) \sin \xi_{ii} \right],$$

where E_o is a constant. Relative to the feed point F, the centre of the patch i has spherical polar co-ordinates given by $(\rho_{ij}, \psi_{ij}, \xi_{ij})$, where $\rho_{ij} = \sqrt{x_{ij}^2 + y_{ij}^2 + f^2}$, $\psi_{ij} = \tan^{-1}\left(\sqrt{x_{ij}^2 + y_{ij}^2}/\rho_i\right)$ and $\xi = \tan^{-1}\left(y_{ij}/x_{ij}\right)$. The magnetic field at the centre of the patch is

$$\mathbf{H}_{ij}^{inc} = \frac{1}{\eta_o} \hat{\rho} \times \mathbf{E}_{ij}^{inc} = \frac{E_o \exp(-jk\rho_{ij})}{\eta_o \rho_{ij}} \left[\hat{\psi} B(\psi_{ij}) \sin \xi_{ij} + \hat{\xi} A(\psi_{ij}) \cos \xi_{ij} \right].$$

The power radiated by the feed can be calculated from

$$P_f = \frac{1}{2} \int_0^{\pi/2} d\psi \sin \psi \int_0^{2\pi} d\xi \rho^2 \left(\mathbf{E}^{inc} \times \mathbf{H}^{inc} \right) \cdot \hat{\rho}$$

$$= \frac{E_o \pi}{4\eta_o} \int_0^{\pi/2} d\psi \sin \psi \left(|A(\psi)|^2 + |B(\psi)|^2 \right). \tag{9.3}$$

In the special case of a symmetric feed pattern, $A(\psi) = B(\psi) = \cos^n \psi$, Eq. 9.3 simplifies to $P_f = E_o \pi / \eta (2n + 1)$. The surface current is now expressed as

$$\mathbf{J}_{sij} = -\frac{2E_o}{\eta_o \rho_{ij}} \exp\left[-jk(h\sqrt{\varepsilon_r} + \rho_i) + j\Psi_{ij}\right] \cos(kh\sqrt{\varepsilon_r}) \mathbf{G}_{ij}, \tag{9.4}$$

where the vector $\mathbf{G}_i$ has the following components when expressed in the global co-ordinate system (x, y, z)

$$G_{ijx} = A(\psi_{ij}) \cos^2 \xi_{ij} + B(\psi_{ij}) \cos \psi_{ij} \sin^2 \xi_{ij}$$

and

$$G_{ijy} = \left[A(\psi_{ij}) - B(\psi'_{ij}) \cos \psi_{ij} \right] \frac{\sin 2\xi_{ij}}{2}.$$

The field radiated by this current on element i to a far-field point can be found from Eq. 6.31. The total field consists of this radiation plus the field scattered by the dielectric substrate and the ground plane. For the moment, the scattered field from the substrate is ignored assuming the patches are densely packed and they are the only scatterer. Further, for simplicity, it is assumed here that the current is constant over the surface of the patches. Thus, for patch i,

$$\mathbf{E}_{ij} \approx -\frac{jk\eta e^{-jkr}}{4\pi} \frac{1}{r} \left(\mathbf{J}_{sij} - (\mathbf{J}_{sij}\cdot\hat{r})\hat{r}\right) \iint_{S_{ij}} \exp\left(jk\hat{r}\cdot\mathbf{r}_{ij}\right) dS', \tag{9.5}$$

where $\mathbf{r}_{ij} = \hat{x}x_{ij} + \hat{y}y_{ij}$. As noted previously, the second term in Eq. 9.5 cancels the radial components produced by the first term. The remaining components are

$$E_{ij\theta} = -\frac{e^{-jkr}}{r}\cos\theta\left(F_{ijx}\cos\phi + F_{ijy}\sin\phi\right) \tag{9.6a}$$

$$E_{\phi ij} = -\frac{e^{-jkr}}{r}\left(-F_{ijx}\sin\phi + F_{ijy}\cos\phi\right), \tag{9.6b}$$

where

$$F_{ij}(u,v) = \frac{jk\eta e^{-jkf\cos\theta}}{4\pi} \int_{x_{ij}-a_i/2}^{x_{ij}+a_i/2} dx' \int_{y_{ii}-b_i/2}^{y_{ij}+b_i/2} dy' \mathbf{J}_{sij} \exp(jk(ux' + vy'))$$

$$= -\frac{jkE_o}{2\pi\rho_{ji}} \exp\left[-jk(h\sqrt{\varepsilon_r} + R_i) - f\cos\theta + j\Psi_{ij}\right]\cos\left(kh\sqrt{\varepsilon_r}\right)\mathbf{G}_{ij}I_{ij}(u,v)$$

with $u = \sin\theta\cos\phi$ and $v = \sin\theta\sin\phi$. The integral in u, v simplifies as follows:

$$I_{ij}(u,v) = \int_{x_{ij}-a/2}^{x_{ij}+a/2} dx' \int_{y_{ij}-b/2}^{y_{ij}+b/2} dy' \exp(jk(ux' + vy'))$$

$$= \exp\left[jk\left(ux_{ij} + vy_{ij}\right)\right](-ab)S\left(\frac{kua}{2}\right)S\left(\frac{kvb}{2}\right).$$

S is the sinc function, which is defined in Appendix A.4. Now assume Ψ_{ij} is chosen so that Eq. 9.1 is satisfied. That is, $k\rho_{ij} - \Psi_{ij} - k\mathbf{r}_{ij}\cdot\hat{r}_b = 2N\pi$. Thus,

$$F_{ij} = \frac{jkE_o a_i b_i}{2} \exp\left[jk\left(ux_{ij} + vy_{ij} - f\cos\theta - \mathbf{r}_{ij}\cdot\hat{r}_b - h\sqrt{\varepsilon_r}\right)\right]\mathbf{G}_{ij}S\left(\frac{kua}{2}\right)S\left(\frac{kvb}{2}\right)\cos\left(kh\sqrt{\varepsilon_r}\right). \tag{9.7}$$

Suppose the centre of the beam has co-ordinates (θ_b, ϕ_b), the beam direction is

$$\hat{r}_b = \hat{x}\sin\theta_b\cos\phi_b + \hat{y}\sin\theta_b\sin\phi_b + \hat{z}\cos\theta_b$$

$$= \hat{x}u_b + \hat{y}v_b + \hat{z}\cos\theta_b.$$

Since $r_i = \hat{x}x_i + \hat{y}y_i$,

$$\mathbf{r}_i\cdot\hat{r}_b = x_i u_b + y_i v_b.$$

This relationship, together with Eq. 9.6, results in the following electric field components radiated by element i:

$$E_{ij\theta} \approx -\alpha \frac{e^{-jkr}}{r} \frac{1}{\rho_{ij}} \exp\{jk[x_{ij}(u-u_b)+y_{ij}(v-v_b)-f\cos\theta)]\}$$

$$\times S\left(\frac{uka}{2}\right)S\left(\frac{vkb}{2}\right)\cos\theta(G_{ijx}\cos\phi+G_{ijy}\sin\phi) \tag{9.8a}$$

$$E_{ij\phi} \approx -\alpha \frac{e^{-jkr}}{r} \frac{1}{\rho_{ij}} \exp\{jk[x_{ij}(u-u_b)+y_{ij}(v-v_b)-f\cos\theta)]\}$$

$$\times S\left(\frac{uka}{2}\right)S\left(\frac{vkb}{2}\right)(-G_{ijx}\sin\phi+G_{ijy}\cos\phi), \tag{9.8b}$$

where $\alpha = (jkE_o ab)/2\pi \exp(-jkh\sqrt{\varepsilon_r})\cos(kh\sqrt{\varepsilon_r})$. The total field is obtained by summing over the fields radiated by all such elements, namely, $i,j=1,\ldots N_e$. The gain can be computed from Eqs. 3.48, 9.8, and 9.3.

As an example, consider a reflectarray with focal length 5λ consisting of nine square patches, on a square 3×3 grid with a spacing of 0.5λ and a central element located at the origin. A beam is required in the vertical (z-) direction. The reflectarray is illuminated by a horn that is linearly polarized in the x-direction and has an axisymmetric pattern given by $A(\theta')=B(\theta')=\cos^{20}\psi$, which has a half-power beamwidth of $21.3°$. In this case, the vector $\mathbf{G}_{ij}$ in Eq. 9.4 has the following components:

$$G_{ijx}=A(\psi_{ij})[(\cos\psi_{ij}+1)\sin^2\xi_{ij}-1] \quad\text{and}\quad G_{ijy}=A(\psi_{ij})(1-\cos\psi_{ij})\frac{\sin 2\xi_{ij}}{2}. \tag{9.9}$$

The patches are chosen to have a sidelength of 0.35λ and are on a substrate of thickness 0.02λ and dielectric constant $\varepsilon_r=2$. The resulting principal radiation patterns are shown in Figure 9.3. For comparison also shown in Figure 9.3 is the radiation pattern of a uniformly illuminated $2\lambda \times 2\lambda$ square aperture corresponding to the maximum extent of the reflectarray. The 10 dB-beamwidth and first sidelobe level are seen to be comparable. The gain of this square aperture is 13.6 dBi, while for the nine-element reflectarray, it is calculated to be 5.7 dBi. The large gain difference is due to the very high spillover from the feed beyond the reflectarray. For a practical reflectarray, there will be many more elements on the substrate and they will better extend across the field of view of the feed. In the example above, if the feed taper is increased by increasing the power n of the cosine amplitude function, the aperture efficiency increases.

9.3 Extensions to the Basic Theory

The radiation pattern analysis given above for the reflectarray and resulting in Figure 9.3 is a very simplified one. It could be improved by including the reflections from the grounded substrate, using a better representation for the currents on the patches, and including coupling between the patches (Pozar et al., 1997). As well, the phase compensation through Eq. 9.1 will not be met exactly, and there are losses in the substrate and in the metallic patches. These are in

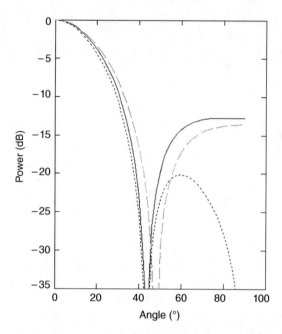

Figure 9.3 Radiation patterns of a reflectarray of 3×3 patches with spacing of 0.5λ. Feed pattern: $\cos^{20}\theta$. Patch dimensions: $0.35\lambda \times 0.35\lambda$, $h = 0.005\lambda$ and $\varepsilon_r = 2$. Solid curve: E_θ in $\phi = 0°$ plane, short dash: E_ϕ in $\phi = 90°$ plane and long dash: pattern of $2\lambda \times 2\lambda$ uniformly illuminated aperture

addition to the usual losses experienced by reflectors such as feed spillover and mismatch. Mutual coupling between the patches could be included as described in Section 7.3.5.7. The field scattered back from the dielectric interface and the ground plane could be included by means of the approximation

$$\mathbf{E}^{scatt} \approx (\mathbf{\Gamma}_{12} + \mathbf{T}_{21}\mathbf{T}_{12}\exp(jk2d\sqrt{\varepsilon_{r2}}))\mathbf{E}^{inc}\exp\frac{-jkr}{r}, \tag{9.10}$$

where $\mathbf{E}^{inc}$ is the incident field from the feed, $\mathbf{\Gamma}_{12}$ is the reflection coefficient matrix for a wave reflected from the substrate (region 2 with dielectric constant ε_{r2}), $\mathbf{T}_{12}$ is the transmission matrix for the forward-wave from free-space (region 1) into the substrate and $\mathbf{T}_{21}$ is the transmission matrix for the reverse travelling wave from the substrate into free-space. The elements of the reflection and transmission matrices are obtained from the Fresnel coefficients for wave interaction parallel ($\parallel$) and perpendicular ($\perp$) to the plane of incidence, for example,

$$\mathbf{\Gamma}_{12} = \begin{bmatrix} \Gamma_{12}^{\parallel} & 0 \\ 0 & \Gamma_{12}^{\perp} \end{bmatrix}$$

where $\Gamma_{12}^{\parallel}$ and $\Gamma_{12}^{\perp}$ are given by Eqs. 3.92. In the implementation of Eq. 9.10, the field $\mathbf{E}^{inc}$ must be resolved into components that are parallel ($\parallel$) and perpendicular ($\perp$) to the plane of incidence. Thus,

$$
\mathbf{E}^{\text{scatt}} \approx
\begin{bmatrix} c_x & d_x \\ c_y & d_x \\ c_z & d_z \end{bmatrix}
\begin{bmatrix} \Gamma_{12}^{\parallel} + T_{21}^{\parallel} T_{12}^{\parallel} \exp\!\left(jk2d\sqrt{\varepsilon_{r2}}\right) & 0 \\ 0 & \Gamma_{12}^{\perp} T_{21}^{\perp} T_{12}^{\perp} \exp\!\left(jk2d\sqrt{\varepsilon_{r2}}\right) \end{bmatrix} \cdot
$$

$$
\times
\begin{bmatrix} a_x & a_y & a_z \\ b_x & b_y & b_z \end{bmatrix}
\begin{bmatrix} E_x^{\text{inc}} \\ E_y^{\text{inc}} \\ E_z^{\text{inc}} \end{bmatrix}
\exp\frac{-jkr}{r},
$$

(9.11)

where the unit vectors $\hat{a}$ and $\hat{b}$ are parallel and perpendicular to the plane of incidence given by $\hat{n} \times (\mathbf{s}_l \times \mathbf{r}) = 0$. Thus, from elementary geometry, it is found that $\hat{a} \cdot \mathbf{s}_l = 0$, $\hat{n} \times \hat{a} = 0$ and $\hat{b} = (\hat{n} \times \mathbf{s}_l)/|\hat{n} \times \mathbf{s}_l|$, where $\hat{a} \cdot \hat{a} = 1$, $\hat{b} \cdot \hat{b} = 1$ and $\hat{a} \cdot \hat{b} = 0$. Also, $\hat{c} = \left(\hat{x}/a_x + \hat{y}/a_y + \hat{z}/a_z\right)\left(a_x a_y a_z\right)$ and $\hat{d} = \left(\hat{x}/b_x + \hat{y}/b_y + \hat{z}/b_z\right)\left(b_x b_y b_z\right)$.

To obtain the total field, the scattered field, given by Eq. 9.11, is added to the field components given by Eqs. 9.8. Note that if the reflection and transmission coefficients are small, Eq. 9.11 makes a small contribution to the total. This is generally not the case as there is usually significant reflection from the substrate, typically at the -10 dB level or even greater.

Nevertheless, as an illustration that the formulation given by Eq. 9.8 can yield useful results, Figure 9.4 shows a comparison of the results of the simplified model and an analysis that includes both scattering and mutual coupling. The reflectarray is 15.24 cm square and operates at 77 GHz

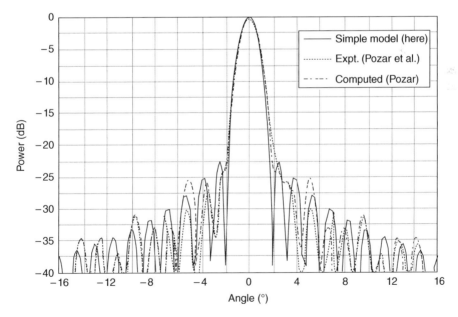

Figure 9.4 Principal plane radiation pattern at 77 GHz of a reflectarray with 76×76 rectangular patches of dimensions $a = b = 0.276\lambda$, on a rectangular grid $0.506\lambda \times 0.506\lambda$ and focal length $f = 39\lambda$. Solid curve: model described, Eq. 9.8, with an axisymmetric cosine feed and radix $n = 10$; dotted curve: experiment (Pozar et al., 1997) and long dash: computed with accurate model (Pozar et al., 1997)

(Pozar et al., 1997). It was fabricated on a 0.127 mm thick duroid substrate ($\varepsilon_r = 2$). It consists of 5776 linearly polarized identical square patches of sidelength 1.076 mm that are spaced 1.978 mm apart. The feed radiates an axisymmetric field and has a $\cos^{10}\psi$ pattern function, which gives a −9 dB edge illumination when fed at a distance of 154.2 mm from the array. Both theoretical models are seen to be in reasonable agreement with the experimental results. The mutual coupling between the elements is expected to be modest due to the element spacing and dielectric loss. The measured gain was 36 dBi, which is slightly less than the gain of 36.7 dBi that is computed from the simplified model with the axisymmetric feed. A uniformly illuminated aperture of the same maximum dimensions has a gain of 42.4 dBi. As well as some spillover, a reduction in gain is expected due to ohmic loss and random fabrication phase errors.

9.4   Other Aperture Antennas

9.4.1   Lenses

A lens is another useful technique for achieving directive radiation (Brown, 1953; Cornbleet, 1976; Bodnar, 2007). At this point, an axisymmetric dielectric medium is considered about the z-direction. Such a lens can have two surfaces – one on the side of the incident field and the second surface for the emerging wave. As with all media transitions, reflection and refraction occurs at both surfaces, which obey Snell's laws. For simplicity, consider a lens with one curved surface on the side of a feed antenna and a planar surface (a plano-convex lens) on the exit face as shown in Figure 9.5. Due to the symmetry, the wave propagation can be considered a two-dimensional problem. For a spherical (or cylindrical) source of waves incident on a lens with refractive index n situated in a vacuum, the optical paths through the lens to the aperture plane located a distance Δ from the exit face are given by

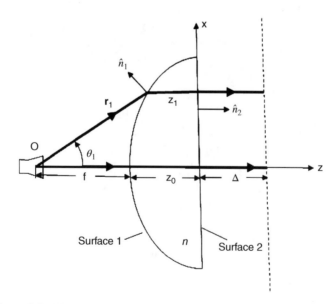

Figure 9.5   Dielectric lens with hyperbolic input face and planar exit surface

$$r_1 + nz_1 + \Delta = f + nz_0 + \Delta$$

or

$$r_1 = f + n(r_1 \cos \theta_1 - f).$$

That is, a general location on surface 1 from O is given by

$$r = \frac{(n-1)f}{n \cos \theta - 1}. \tag{9.12}$$

This is the equation of a hyperbola with eccentricity n with O as the focus. To obtain the field on the aperture plane, conservation of power from the source to the aperture plane is used. Suppose the source radiates the incremental power $P(\theta)\sin \theta \, d\theta \, d\phi$, where $P(\theta)$ is the feed power pattern, and the corresponding power collected on a segment of the aperture at a radius ρ is $P_a(\rho)\rho \, d\rho \, d\phi$. Thus, for conservation of power, $P(\theta) \sin \theta d\theta = P_a(\rho)\rho d\rho$, which can be re-written as

$$\frac{P_a(\rho)}{P(\theta)} = \frac{\sin \theta}{\rho(d\rho/d\theta)}. \tag{9.13}$$

For the hyperbolic lens, this relationship results in

$$\frac{P_a(\rho)}{P(\theta)} = \frac{1}{r\left(\dfrac{dr}{d\theta}\sin \theta + r \cos \theta\right)} \tag{9.14}$$

$$= \frac{(n\cos\theta - 1)^3}{[(n-1)f]^2(n - \cos \theta)}.$$

The field radiated by the lens is obtained from Eq. 9.14 by means of Eq. 3.20. Assuming the magnetic field is related to the electric field through $\mathbf{H}_a = (1/\eta_0)\hat{z} \times \mathbf{E}_a$, the electric field in the aperture is given by

$$\mathbf{E}_a = \hat{x}A(\theta)\sqrt{\left|\frac{P_a(\rho)}{P(\theta)}\right|} \tag{9.15}$$

for a linearly polarized axisymmetric source with an amplitude function given by $A(\theta)$. Such a pattern is produced by a corrugated horn operating at the balanced hybrid condition or a circular waveguide feed with a corrugated flange. One of the problems with this lens is that amplitude of the aperture field given by Eq. 9.15 falls quickly due to the decay of the power function ratio under the square root sign. For example, a lens with $n = 1.5$ and $f = 10\lambda$ at an angle of $\theta = 45°$, which is close to the edge of the lens, the edge of the illumination has dropped 25 dB below the illumination of the feed at that angle. One way around this is to design both lens surfaces or to reverse the lens.

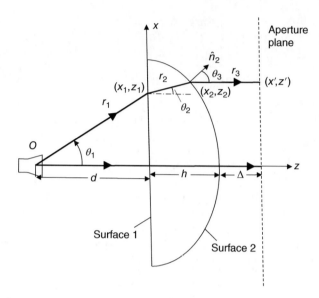

Figure 9.6   Design of lens profile on surface 2 for uniform aperture illumination

In the design of lens profiles, the input surface can be chosen depending on the wavefront of the source such as a spherical surface. A limiting case of this is a plane surface. If the refractive index of the lens is close to 1, the reflection back to feed is small and the reflection coefficient is approximately $\Gamma \approx (n-1)/(n+1)$. Thus, for a polythene lens with $n = 1.5$, $\Gamma \approx -14\,\text{dB}$. Nevertheless, there are several techniques available to reduce this reflection further. One approach is to include a matching pad in the second surface, which is designed to give a uniform exit amplitude. This is found by requiring, as best as possible, a uniform amplitude and phase across the aperture of the lens. In this approach, shown in Figure 9.6, ray tracing from the feed-phase centre, through the lens and thence to the aperture plane, results in a path length given by

$$L = d\sec(\theta_1) + n\sqrt{(x_2 - x_1)^2 + (z_2 - z_1)^2} + (h + \Delta - z_2)\sec\theta_3, \tag{9.16}$$

where $\theta_2 = \sin^{-1}((1/n)\sin\theta_1)$ and θ_3 is found by an application of Snell's law at surface 2 (see Figure 9.6). The total phase shift from the feed to the aperture plane is kL. Eq. 9.13 is valid for the present uniform phase lens configuration. Making the transformation $\partial\rho/\partial\theta = (\partial\rho/\partial t)(1/(\partial\theta/\partial t))$, where t is the radial distance from the axis at the curved surface, for a lens with a planar surface 1 and a shaped surface 2, the power relationship becomes

$$\frac{P_a(t)}{P(\theta)} = \frac{x_1(t)}{\rho(t)^2\sqrt{x_1^2(t) + d^2}} \frac{[(t + (z(t) + d)(\partial z/\partial t)]}{[((z(t) + d) - t(\partial z/\partial t)]} \tag{9.17}$$

where $z(t)$ is the lens profile and $\rho(t) = \sqrt{(z(t) + d)^2 + t^2}$. The co-ordinate $x_1(t)$ on surface 1 is found by tracing the ray back from the aperture point through the lens to the surface 1 and solving for the intersection points on the lens surfaces. If the feed radiation is axisymmetric, the

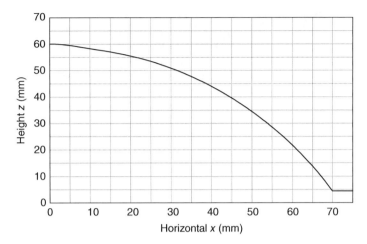

Figure 9.7 Profile of polyethylene lens that was synthesized to provide uniform aperture illumination. Dimensions are in millimetres (mm)

radiation pattern of the lens is obtained from Eqs. 9.15 and 9.17 by means of Eq. 6.10 and the transformation integral Eq. 6.12 for axisymmetric illumination.

As an example of some results that can be achieved, Figure 9.7 shows the synthesized profile of a polyethylene lens with a refractive index $n = 1.5$. A best-fit equation to these data is given by

$$z(t) = -0.0001t^3 - 0.0019t^2 - 0.1688t + 60.066 \qquad (9.18)$$

Typically, the error in this approximation is less than 0.5% except at the outer extremity of the lens. The diameter of this lens is 140 mm, the distance to surface 1 is $d = 76.67$ mm and the focal length is 136.67 mm. The lens is fed by a circular waveguide that has a flange containing two parasitic rings to produce an axisymmetric radiation pattern. A function that approximates the feed pattern is $A(\theta) = \cos^3\theta$, which results in a −7.7 dB edge illumination at 42° on the lens's rim. The prototype lens and the feed are pictured in Figure 9.8a. The results computed by the method outlined above are shown in Figure 9.8b along with experimental results in the 45°-plane at 40 GHz (Chandran & Hayman, 1994). Both patterns have been normalized to the measured gain of 35.0 dBi and are seen to be in quite good agreement given the approximation of the feed illumination.

Spherical lenses have found several applications due to their wide-angle beam scanning capability without pattern deterioration and their capacity to produce multiple beams. They can also operate over a wide frequency range, and there is no feed blockage. There are two main types of inhomogeneous spherical lens where each source has a single focus. The first is the Luneburg lens, which is distinguished by being radially inhomogeneous. In addition, any two points on a great circle through the lens, which are on opposite sides of a radius vector, pass through the centre of symmetry. This lens has a refractive index that varies with radial distance r from the centre given by

$$n(r) = \sqrt{2 - \left(\frac{r}{R}\right)^2}. \qquad (9.19)$$

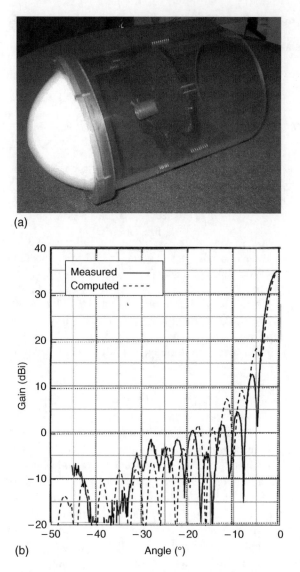

Figure 9.8 A 140 mm planar-convex lens antenna with a gain of 35 dBi at 40 GHz. (a) Prototype and (b) measured and computed patterns in the 45° plane. Solid curve: measured (Chandran & Hayman, 1994); dash curve: this theory

Geometric optics indicates that a point source which is located on the surface radiates a plane wave diametrically opposite the source. A hemispherical lens on a ground plane can be fed from a point source, which creates a virtual source from the absent hemisphere in the image below the ground plane. It can be shown that in general for a spherical lens that the path of shortest distance follows the ray path given by the differential equation (Collin & Zucker, 1969).

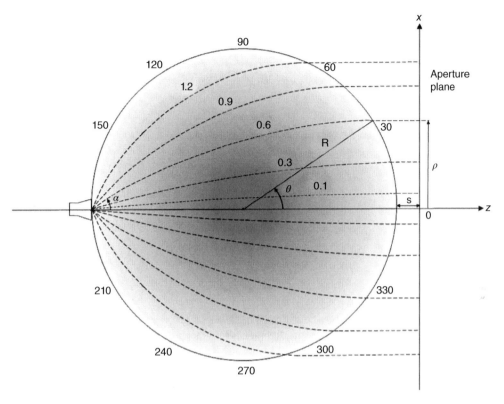

Figure 9.9 Ray paths (dashed) in a spherical Luneburg lens from the source to the aperture plane. The labels on the rays refer to the angle (in radians) of incidence θ_i from the source

$$\left(\frac{\kappa\,dr}{r\,d\theta}\right)^2 = n^2(r)r^2 - \kappa^2, \tag{9.20}$$

where κ is a parameter specific to a ray path. The ray paths in a Luneburg lens follow elliptical paths given by

$$r(\theta,\alpha) = \frac{R\sin\alpha}{\sqrt{\sin^2\theta + \sin^2(\theta-\alpha)}}, \tag{9.21}$$

where (r, θ) are polar co-ordinates R is the radius, and α is the launch angle between the initial ray at the source and the diameter of the sphere. The ray parameter in Eq. 9.20 for the Luneburg lens can be shown to be $\kappa = R\sin\alpha$. A plot of some typical ray paths in the lens is shown in Figure 9.9. From Eq. 9.21 it can be shown that a ray launched at an angle α intersects the surface on opposite of the lens at an angle $\theta = \alpha$. The amplitude distribution can be readily estimated by the methods used earlier in this section. The ray that leaves the feed at an angle α exits the lens and intersects with the aperture at a distance ρ from the centre where $\rho = R\sin\alpha$. The electrical path length of a ray from the feed, through the lens and to the aperture plane shown in Figure 9.9 is a constant as can be found by evaluating the phase shift given by

$$\Phi(\theta) = k \int_{\tan^{-1}(\sin(\alpha))}^{\pi} n(r(\theta,\alpha)) \sqrt{r^2(\theta,\alpha) + \left(\frac{dr(\theta,\alpha)}{d\theta}\right)^2}\, d\theta$$
$$= kR\left(1 + \frac{\pi}{2}\right), \tag{9.22}$$

where Eqs. 9.20 and 9.21 are used in Eq. 9.22. The power radiated by the feed between angles α and $\alpha + d\alpha$ leaves the aperture between ρ and $\rho + d\rho$, where $d\rho = R \cos \alpha d\alpha$. Making use of Eq. 9.13 for the Luneburg lens, it is found that the power in the aperture is given by

$$\frac{P_a(\rho)}{P(\alpha)} = \frac{\sin \alpha}{\rho(d\rho/d\alpha)}$$
$$= \frac{\sec \alpha}{R^2}, \tag{9.23}$$

where $P(\alpha)$ is the power pattern of the source. The factor $\sec \alpha$ appears because the rays are not uniformly spread across the aperture but become closely bunched as $\alpha \to \pi/2$. For a maximum feed incidence of $\alpha = \pi/2$, the ray travels an angular distance of $\pi/2$ along the surface of the lens. As the electrical path length travelled by each ray is constant, as verified by Eq. 9.22, no additional phase factor is required in the aperture field distribution. Taking all these factors into account, and making use of Eq. 9.15, the aperture field produced by an x-directed axisymmetric feed with an amplitude pattern $\sqrt{P(\alpha)} = \cos^n \alpha$ is

$$\mathbf{E}_{ax} = \hat{x} \frac{E_o}{R\sqrt{\cos \alpha}} \cos^n \alpha.$$

This equation is substituted into Eq. 6.10 to obtain the radiation pattern of the lens. Once again the feed is axisymmetric and the transformation integral, Eq. 6.12, can be used. The latter gives

$$N_x(\theta,\phi) = 2\pi E_o \int_0^R \cos^{n-1/2}\left[\sin^{-1}(\rho/R)\right] J_0(k\rho \sin \theta) \rho\, d\rho.$$

As an example, consider a Luneburg lens with a radius $R = 14.423$ cm that is fed by an axisymmetric feed with an $n = 1$ radix at a frequency of 16.65 GHz. The co-polar component of the electric field in the 45°-plane is plotted in Figure 9.10, the form of which is characteristic of the geodesic Luneburg lens. The computed half-power beamwidth is 3.94°, which corresponds to HPBW $= 1.12\lambda/D$, where D is the diameter, and the gain is 33.8 dBi. These results are comparable to experimental results of a lens of a similar diameter (Ap Rhys, 1970).

A second type of spherical lens is the Maxwell fish-eye lens is one for which any two points on the same radius that pass through the centre are inverse curves relative to a circle of radius R.

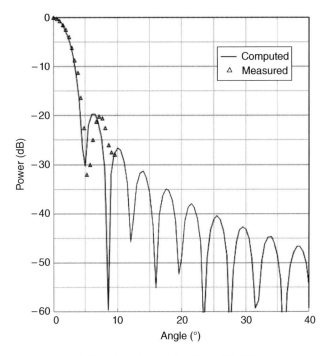

Figure 9.10 Radiation pattern of a Luneburg lens with radius of 14.438 cm at a frequency of 16.65 GHz. The lens is fed by an axisymmetric cosine source with a power of $n = 1$. Solid line: computed; Δ experiment (Ap Rhys, 1970)

The Maxwell fish-eye is also radially inhomogeneous and has a refractive index that varies from its centre as follows:

$$n(r) = \frac{2}{\left[1 + (r/R)^2\right]}. \tag{9.24}$$

This lens has the property that it creates an image at a point diametrically opposite of the point source on the sphere. The ray paths are segments of circular arcs which create figures shaped like the eye of a fish. The ray paths are great circles and an image of the source on the surface is formed on the opposite side of the lens. The ray path on a great circle is given by

$$r^2(\theta, \alpha) + 2R\cot(\alpha r(\theta, \alpha)) \sin\theta = R^2, \tag{9.25}$$

where (r, θ) are polar co-ordinates in the plane and α is the launch angle between the ray and the axis that passes through the source phase centre and the origin at the centre of the lens. The roots of Eq. 9.25 are

$$r(\theta, \alpha) = R\left[-\sin\theta\cot\alpha \pm \sqrt{(\sin\theta\cot\alpha)^2 + 1}\right].$$

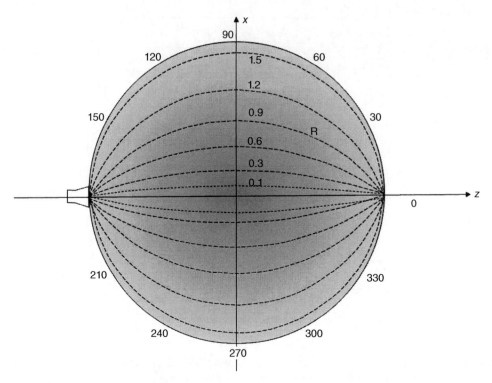

Figure 9.11   Maxwell fish-eye lens and ray paths from a point source for launch angles α from 0.1 to 1.5 rad

A valid root for the Maxwell lens is chosen so that $|r(\theta,\alpha)| \leq R$. Ray paths are illustrated in Figure 9.11 for a source located at $\theta = \pi$. At the plane midway through the lens, this source creates a plane phase front. If the lens is halved, the resulting mid-point surface will be a constant phase aperture which equals $\Phi(\pm\pi/2) = 3kR\pi/2$. As a result, a directive beam is produced from this semi-circular lens. Only rays in the sector $\pi/2 < \alpha < 3\pi/2$ are valid. Some mismatch can occur at the aperture plane due to the refraction index difference between the lens dielectric and free-space. To reduce this, a matching layer coating on the surface can be used to improve the match. Compared to the Luneburg lens, the fish-eye lens cannot steer the beam by moving the feed and hence cannot be used to produce multiple beams. Finally, it is noted that the techniques used above for analysing the Luneburg lens, for example, the radiated fields, can also be applied to the Maxwell fish-eye lens.

9.4.2   Fabry–Pérot Resonator Antennas

The Fabry–Pérot cavity was conceived for optics as an arrangement of two highly reflecting mirrors (with some small transitivity) to create a standing wave resonator (Born & Wolf, 1959). A typical Fabry–Pérot resonator antenna is shown in Figure 9.1c. It consists of a bottom surface that is usually a good reflector and top surface that is of a partial reflecting surface (PRS) material both of which are excited by a small radiator such as a microstrip patch. The gain of this antenna is determined, up to a limit, by the lateral dimensions of the two surfaces, the reflection coefficient of the PRS material, and the gain of the small radiator (or feed antenna). This small antenna couples energy to the cavity formed by the PRS material and the ground plane. The operating frequency of the PRS antenna can be derived by referring to Figure 9.1c. This figure shows the PRS material with a reflection coefficient of $R_2 \angle \phi_2$ placed a distance d above a bottom surface that may be phase agile, which has a reflection coefficient $R_1 \angle \phi_1$. These reflection coefficients are functions of angle of incidence to the surface, but for simplicity, it will be assumed that the rays in the cavity are normal to both surfaces. If a wave originates from the lower patch antenna that is normally incident on the PRS, then the field will be maximized at the output face of the antenna when (Wang et al., 2006)

$$(\phi_1 + \phi_2) - 2kd = -2m\pi; \quad m = 1, 2, \dots. \tag{9.26}$$

Considering the lowest resonance frequency (i.e. $m = 1$), this operating frequency is related to the cavity height and reflection phase of the PRS and phase agile surface as shown

$$f = \frac{c}{2d} \left[\left(\frac{\phi_1 + \phi_2}{2\pi} \right) + 1 \right]. \tag{9.27}$$

The standard microwave resonator does not have a phase agile lower surface and usually consists of a metallic plate for which $\phi_1 = \pi$. More advanced antennas allow the reflection phase of either surface to be varied with tuneable elements, the operating frequency of which may be reconfigured over a considerable range. This allows the frequency band of the antenna to be increased, but it may not provide a large enough instantaneous bandwidth. Various types of PRS materials have been used including dielectric sheets, a periodic surface such as an electromagnetic bandgap (EBG) or a reflectarray. As well, an inherent disadvantage of the Fabry–Pérot resonator antenna is its typically narrowband operation (3%). To overcome this, the lower surface can be tuned to make the antenna reconfigurable and to operate over a wide frequency range (Weily et al., 2008). One of the problems is that the reflection coefficient of the top surface varies significantly over the frequency band and especially at frequencies where significant but partial reflection occurs. A frequency selective surface (FSS), for example, exhibits this behaviour near the FSS resonance frequency. Although the reflection phase of a conventional FSS decreases with frequency at most frequencies, it can be made to increase with frequency over a frequency band that is close to the FSS's resonance frequency. Advantage can be taken of this phenomenon to design a PRS with increasing phase over a selected frequency band. Such a surface has been developed on a single dielectric slab, with arrays of periodic elements such as dipoles, slots, patches, rings etc., printed on both surfaces (Ge et al., 2012). The directivity of the resonator antenna changes with area up to a dimension that will depend on the bottom and top surfaces. With a simple probe or microstrip source, an EBG top surface and a metallic bottom plate, the directivity has been shown to vary in a linear manner with aperture width/length from about 2λ to $\sim 6.5\lambda$, at the centre frequency. This means that the maximum directivity possible is about 27 dBi although this can be increased by employing a more complex source of wider extent such as an array. Above the upper limit the directivity reduces slightly, and converges to a value just below the maximum. The minimum width is limited by the size of the source and the sidelobe level as these are higher for smaller aperture dimensions.

9.5   Problems

P9.1 A reflectarray with a focal length of 10λ consists of 81 square patches, on a square 9×9 grid with a spacing of 0.5λ. The central element of the grid is located at the centre of the beam from the feed. Assume that the basic requirement for a reflectarray array that is given by Eq. 9.1 has been included in the patch design. In addition, the patch has dimensions $0.4\lambda \times 0.4\lambda$ and the array is on a substrate with $\varepsilon_r = 2$ and $h = 0.005\lambda$. The reflectarray is illuminated by a horn that is linearly polarized in the x-direction and with an axisymmetric pattern given by $A(\Theta) = B(\Theta) = \cos^{20}\Theta$.

   a. What is the illumination level at the edge of the patch array?

   b. Obtain the far-field patterns radiated by this reflectarray in the boresight direction.

   c. Calculate the scattered field from Eq. 9.11 where

$$\Gamma_{12} = \begin{bmatrix} 0.17 & 0 \\ 0 & 0.16 \end{bmatrix} \quad T_{12} = \begin{bmatrix} 0.83 & 0 \\ 0 & 0.84 \end{bmatrix}, \quad \text{and} \quad T_{21} = \begin{bmatrix} -1.17 & 0 \\ 0 & -1.16 \end{bmatrix}$$

The dielectric material in region 2 has a relative permittivity $\varepsilon_{r2} = 1$. Add this field to the result in (b) to find the total field.

P9.2 Use Eq. 9.8 to obtain the ideal gain of a reflectarray antenna.

P9.3 A reflectarray has been designed by the method of Berry et al. (1963), which has been formed from short sections of square waveguides of width $a = 0.75\lambda$. The open-ended waveguides are arranged in a hexagonal array of 37 elements with spacing between elements $s = 0.8\lambda$. The focal length is $f = 6\lambda$. Neglecting mutual coupling, obtain the radiation pattern in the far-field when the waveguides are fed with an axisymmetric source that gives a 6 dB taper at the centre of the elements most distant from the centre. Use the array factor of a hexagonal array given by Eq. 7.15. Compare the radiation pattern obtained with that of a parabolic reflector of the same diameter and focal length.

P9.4 Suggest ways of increasing the bandwidth of a reflectarray of patches. In particular, devise an approach that could achieve operation over a 20% bandwidth at X-band.

P9.5 Verify that the path length from the source to the aperture plane via the plano-convex lens in Figure 9.6 and Eq. 9.18 is almost constant for rays at angles up to 45° from the source.

P9.6 Use Eq. 9.18 to obtain the aperture distribution for the lens shown in Figure 9.8a. Use an axisymmetric feed with a $\cos^3\theta$ amplitude.

P9.7 By means of the field correlation theorem, Eq. 3.62, show that the reflection coefficient at surface 2 of the hyperbolic lens shown in Figure 9.6 is given by

$$\Gamma = \frac{\displaystyle\int_0^{\psi_c} A^2(\psi)\exp(-2jkr(\psi))\sin\psi\,d\psi}{\displaystyle\int_0^{\pi} |A(\psi)|^2 \sin\psi\,d\psi},$$

where $r(\psi) = (n-1)f/(n\cos\theta - 1)$ and $A(\theta)$ is the amplitude pattern function for an axisymmetric feed.

P9.8 Design a Luneburg lens for an application requiring a directive radiation pattern and a half-power beamwidth of 5°. Assume the lens is fed with a horn with an axisymmetric radiation pattern and the highest refractive index material available has a value of 2. What diameter lens is required? Describe a way of achieving the desired variation in refractive index.

P9.9   Show that on a plane at the mid-way point of a Maxwell fish-eye lens, that:
   a. The distance from the centre is given by $\rho(\alpha) = \pm R \tan \alpha/2$, where $\theta = \pm \pi/2$.
   b. Use your favourite software to verify that the path length of a ray from a point source located at $\theta = \pi$ to this plane is independent of launch angle α.
   c. Use the distance given in (a) to verify that the amplitude of the aperture field of a hemispherical Maxwell fish-eye lens that is fed with a axisymmetric spherical source is

$$A_a(\rho) = 2\sqrt{\frac{P(\alpha)\left(\cos^2\alpha/2\right)}{R}\frac{n(\rho)}{1+n(\rho)}},$$

where $P(\alpha)$ is the power distribution of the source and $n(\rho)$ is the refractive index.

P9.10 A quarter-wave matching layer is added to the aperture face of the hemispherical Maxwell fish-eye lens described in P9.9.
   a. What dielectric constant of material is required for this layer if the design is based on the average power mismatch between the lens aperture and free-space?
   b. Describe better ways of achieving a good match over a band of frequencies.

P9.11 Verify that the field is maximized at the output face of a Fabry–Pérot cavity resonator antenna when condition Eq. 9.26 is satisfied.

P9.12 A one-dimensional Fabry–Pérot cavity resonator is formed by a dielectric sheet of thickness τ and dielectric constant ε_{r2} in free-space. Assume that a source in region 1 (see Figure P9.1) creates planes waves between the dielectric and a ground plane, which propagate in the z-direction. The total electric field in the three regions shown in Figure P9.1 consists of plane waves in the form $E_{xi} = A_i \exp(-jk_iz) + B_i \exp(jk_iz)$; $i = 1,2,3$, where i refers to the region number, A_i and B_i are the amplitudes of the forward and reverse travelling waves. Region 1 is characterized by a propagation constant $k_i = k\sqrt{\varepsilon_{ri}}$, dielectric constant ε_{ri} and wave impedance. In the free-space regions, $\varepsilon_{r1} = \varepsilon_{r3} = 1$ and $\eta_1 = \eta_3 = \eta_o$. By satisfying the boundary conditions, obtain:
   a. The resonance frequency
   b. The aperture field in region 3 at $z = \tau$

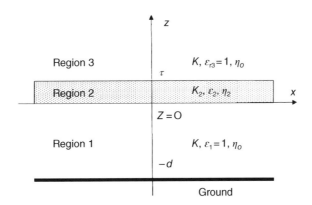

Figure P9.1   Dielectric sheet above a ground plane forming a one-dimensional Fabry–Pérot cavity

References

Ap Rhys, T.L. (1970): 'The design of radially symmetric lenses', IEEE Trans. Antennas Propag., Vol. **AP-18**, No. 4, pp. 497–506.

Berry, D.G., Malech, R.G. and Kennedy, W.A. (1963): 'The reflectarray antenna', IEEE Trans. Antennas Propag., Vol. **AP-11**, pp. 645–651.

Bodnar, D.G. (2007): 'Lens antennas', (Chapter 18). In Volakis, J.L. (ed) 'Antenna engineering handbook', 4th ed., McGraw-Hill, New York.

Born, M. and WOLF, E. (1959): 'Principles of optics', Pergamon Press, London, UK.

Brown, J. (1953): 'Microwave lenses', Methuen & Co. Ltd., London, UK.

Chandran, T. and Hayman, D.B. (1994): 'Ka-band lens antenna measurements'. CSIRO Division of Radiophysics, Sydney, Australia. Research Report RPP3738.

Collin, R.E. and Zucker, F.J. (1969): 'Antenna theory', Parts 1 & 2, McGraw-Hill, New York.

Cornbleet, S. (1976): 'Microwave optics', Academic Press Inc., London, UK.

Ge, Y., Esselle, K.P. and Bird, T.S. (2012): 'The use of simple thin partially reflective surfaces with positive reflection phase gradients to design wideband, low-profile EBG resonator antennas', IEEE Trans. Antennas Propag., Vol. **AP-60**, No. 2, pp. 743–750.

Huang, J. and Encinar, J.A.(2008): 'Reflectarray antennas', IEEE Press, Wiley-Interscience, Hoboken, NJ.

Pozar, D.M., Targonski, S.D. and Syrigos, H.D. (1997): 'Design of millimeter wave microstrip reflectarrays', IEEE Trans. Antennas Propag., Vol. **AP-45**, pp. 287–296.

Wang, S., Feresidis, A.P., Goussetis, G. and Vardaxoglou, J.C. (2006), 'High-gain subwavelength resonant cavity antenna based on metamaterial ground planes', IEE Proc. Microwaves Antennas Propag., Vol. **153**, No. 1, pp. 1–6.

Weily, A.R., Bird, T.S. and Guo, Y.J. (2008): 'A reconfigurable high gain partially reflecting surface antenna', IEEE Trans. Antennas Propag., Vol. **AP-56**, No. 11, pp. 3382–3390.

10

Aperture Antennas in Application

The already theory of aperture antennas as described in the earlier chapters is only part of the story in creating aperture antennas. In this chapter, the reader is introduced to another side of antenna engineering, that of fabrication, measurement and testing. The approach adopted here is one of introduction to the basic principles as there are texts that are dedicated to all or parts of these topics (Hollis et al., 1970; Levy, 1996; Schofield & Breach, 2007). The principles of fabrication, measurement and test are summarized in the next two sections. In the final section, the fabrication and testing of several practical aperture antennas are described. These antennas are a lightweight horn for space flight, a dual reflector producing multiple beams and a radio telescope that operates with multiple beams to reduce observation time.

10.1   Fabrication

Vital to the application of aperture antennas is their construction and testing, which often involves techniques that are suitable for the purpose within budget constraints. The fabrication of aperture antennas can be slightly different from other antennas, due to their geometry and often large dimensions. A variety of techniques are used, and some will be described in this section.

10.1.1   Machining

Prototype aperture antennas are often made from an aluminium billet or plate by machining with a machining centre or a hand lathe. A variety of aluminum types are available and the one selected should suit the application. For example, a slab of 6061-T6 aluminum can be suitable for a test reflector because of its machinability and stability including temperature. Another example, is an aluminium alloy such as 5086 can be useful for producing profiled circular horns

Fundamentals of Aperture Antennas and Arrays: From Theory to Design, Fabrication and Testing,
First Edition. Trevor S. Bird.
© 2016 John Wiley & Sons, Ltd. Published 2016 by John Wiley & Sons, Ltd.
Companion website: www.wiley.com/go/bird448

by boring from a solid block of the metal. With boring the horn does not need to halved, which can lead to cross-polarization, and also other horns can be bored in the same block to form an array. As well as aluminium another possible material is brass. However, it tends to result in components that are heavy and tarnish with time due to repeated handling and oxidation. The accuracy required varies and depends mainly on the operating wavelength of the prototype. For accurate comparison with theory, for example, a surface accuracy of horn components is normally better than $\lambda/50$. Under normal operating conditions, accuracy needs normally to be better than $\lambda/20$. The production of a quantity of aperture antennas by machining will probably be uneconomical except in specialized applications. Casting components such as horns from an alloy of aluminum is possible when quantity is required although surface quality and loss can be issues depending on the technique. Some success has been achieved by making these antennas from plastic by means of a mould. After the plastic base structure has been produced, a metallization layer is applied either by spraying or dipping in a molten metal bath.

10.1.2   Printing

Aperture antennas, and especially those based on patches, can be made by etching or depositing metal on a dielectric substrate or mould. The dielectric substrate should be chosen so that the dielectric loss is low. Some dielectrics that have low loss may not be conducive to depositing metal, and often this can be overcome by coating the dielectric with another material that may provide better adhesion than the base substrate. The patches can also contribute significant loss by the high current densities on the edges. The accuracy required for most patches is not high and a resolution of $\lambda/10$ is usually sufficient.

10.1.3   Mould Formation

Some aperture antennas such as reflectors and horns can be made from low melting point metal or a plastic material by means of a mould on which a high-conductivity metal can be deposited. The mould should allow for some shrinkage of the deposit, and its surface should be slightly more accurate than the required accuracy of the antenna to allow for unevenness of the metal deposits.

10.1.4   Electroforming

This technique is a way of making metal items by forming a thin skin through electrodepositing on a base object known as a mandrel. This compares with the previously mentioned technique of electroplating a mould. After electroforming the mandrel is usually removed. Compared with electroplating, the deposits obtained in electroforming are usually much thicker. An electrolytic bath is used to deposit nickel or another similar metal onto a conductive surface such as stainless steel. Once the deposited material has been built up to a desired thickness, the elector is separated from the mandrel. The pyramidal horns shown in the array pictured in Figure 1.1c were produced by electroforming.

10.1.5   Lightweight Construction

Even with careful design, the weight of the aperture antenna may be too heavy for the intended application or possibly too costly to mount. In these instances, a lightweight construction approach may be necessary. Some reduction in weight could be achieved by thinning the walls or the profile, but this may still result in an unacceptably high weight antenna. In some applications, such as space flight, a very low weight construction is necessary to achieve the lowest possible payload cost and still achieve a high performance. To achieve a much lower weight, a new design may be required with lightweight manufacture in composite materials in mind.

Before undertaking this new design, the greatest contributors to the weight should be identified. Structural methods should be adopted where possible, such as thin walls and fins, although this usually leads to a compromise between weight and mechanical strength. The effect of the changes should be examined to see if they impact the overall performance.

Before the design is completed and fabrication commenced, a tolerance study should be made of the antenna geometry. The purpose is to indicate the most sensitive part of the antenna to change and to error. Tolerances should be assigned to various subassemblies although these should be achievable with practical fabrication methods.

A lightweight antenna can be constructed from various composite materials such as carbon fibre reinforced plastic (CFRP) materials. It is recommended that this work is undertaken by experienced personnel that are equipped to handle such materials. If a metal coated surface is required, such as the interior surfaces of a CFRP horn, all active surfaces can be plated with copper, for example, to increase the electrical conductivity. The outer surfaces can be produced in a one-piece mould. Assembly of any intricate parts with the skin should be accomplished with specially designed fixtures that ensure concentricity or correct spacing according to the design. After all parts are bonded together, the interior surface can be plated with a continuous layer of copper. The thickness of the metal layer is usually >10 skin-depths (δ) at the centre frequency to ensure losses are negligible. The skin depth is the distance a normally incident plane wave travels in a metal before its amplitude decays to $1/e$ of its value at the surface. This depth is given by $\delta = 1/\sqrt{\omega \mu_0 \sigma}$, where σ is the conductivity of the metal. Copper has a conductivity $\sigma = 5.8 \times 10^7$ S, and, therefore, the skin depth in copper is $2.06/\sqrt{f}$ μm where f is the frequency in GHz. In the present example, the copper thickness was >10 μm.

An example of lightweight construction, is the corrugated horn shown in Figure 4.43 which was fabricated using the technique described in the previous paragraph. A prototype antenna was machined initially from aluminium and this weighed 110 kg. The operational antenna was fabricated from CFRP and other lightweight materials which resulted in a horn weighing less than 10 kg. This horn will be described in a later section.

10.1.6 Pressing and Stretch Forming of Reflector Surfaces

Large reflector surfaces are usually composed of many small panels, which are manufactured separately (Levy, 1996). A technique often used to produce reflector panels or complete reflectors involves the use of a number of thin metal sheets. Usually aluminium is used and is forced down onto a mould under pressure to form the desired shape. The pressure must be applied evenly to achieve the desired reflector shape which is curved in two dimensions. To save time and cost, the mould may consist of studs that are adjusted to the desired curvature under computer control (e.g. Parsons & Yabsley, 1985). In this approach, the metal sheet is often drawn down under a vacuum to prevent denting or marking as illustrated in Figure 10.1. A backing structure is then attached to the panel usually with a suitable adhesive to help bond the sheet in the desired shape and also to take up any gap between the panel and the backing. The backing structure may be metallic beams or made from a honey-comb structure. The use of any rivets or

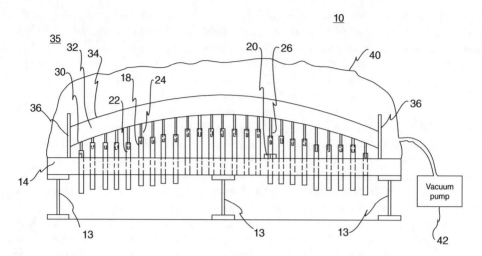

Figure 10.1 An adjustable mould and reflector surface with a backing structure formed under vacuum. *Source*: US patent (Kommineni et al., 1988)

similar for further attachment should be avoided to prevent deformation of the shape and raised areas on the panel surface. After the metal sheets have been held in the desired contour for a predetermined time, the vacuum is released. The sheet and its backing structure should be now permanently deformed into the specified shape.

In another approach, a reflector can be formed from several successive layers of cloth and fibreglass placed over a mould to a depth of suitable strength. Hardening of the fibreglass produces a rigid structure, which can be accurately shaped by polishing or filled to overcome any surface imperfections. The reflector is then flame sprayed with a metallic spray to a suitable thickness, typically to 4–6 skin depths to produce a low loss reflective surface.

An example of reflector panels produced by the first approach is described below and shown in Figure 10.12. The reflector panel surface accuracy is 0.3 mm rms or about $\lambda/80$ at 12 GHz.

10.1.7 Assembly and Alignment

An important part in the final stages of fabrication is the assembly and alignment of various parts of the antenna. Within limits, this can be done with micrometer, tape and surveying tools although sometimes special alignment rods and jigs are manufactured along with the antenna components. However, there are many parts of the assembly that require greater accuracy and techniques such as laser ranging and photogrammetry can greatly assist the antenna engineer (Schofield & Breach, 2007). Some of these methods (e.g. lasers) extend to operational correction in real time for deformations due to ambient temperature, differential solar radiation and the gravitational deformation as the antenna scans to different positions.

Distance measurement methods by optics are typically time-of-flight, interferometry or triangulation (Hodges & Greenwood, 1971). A time-of-flight system measures the round trip time between a light pulse emission and the return of the pulse echo as a result of reflection from the target. Therefore, the distance is found by multiplying the velocity of light by the trip time divided by 2 to give the one-way distance to the object. Initial alignment and test are important for accurate results. An initial test could be undertaken with a calibrated target before the actual measurement commences. Commercial laser range finders are usually equipped with angle encoders to enable the definition of the co-ordinates of a measurement point relative to the ranger. Scanning of several points at a time is possible and is carried out manually or automatically. Reflector surface verification measurements can be undertaken by theodolite (Levy, 1996) or laser triangulation.

Photogrammetry is a three-dimensional co-ordinate measurement technique that uses photographs as the fundamental medium for metrology of objects that have been fitted together. In operation, a single camera system can be used to record digital images of the object from several different locations. The system measures the location of high contrast retro-reflective target points that are placed on the object. By imaging the object from several different locations, the points on the object are seen from enough geometrically diverse locations to support determination of their spatial locations by triangulation.

There are several photogrammetry systems commercially available. One such system called V-STARS is capable of producing *XYZ* co-ordinate point data to an accuracy of $1:120\,000$ (0.025 mm on a 3 m object).

10.2   Measurement and Testing

After the antenna has been fabricated, it is important to test a prototype or a sample of the antennas manufactured. Some aspects of the antenna can be measured and assessed using standard microwave network measurement methods such as reflection coefficient, insertion loss and transmission coefficient. These can be obtained with a vector network analyser and calibrated standard. As well, a variety of radiation pattern measurement techniques are possible, and these are summarized in Figure 10.2.

10.2.1   Far-Field Measurement

The aim of far-field measurement is to measure the distant radiation pattern on a sphere of constant radius. This measurement can be made in transmission or reception. The latter is considered here. The equipment needed is a rotator in azimuth and/or elevation (Hollis et al., 1970). A source antenna is located at a distance at least equal to the Rayleigh distance $R = 2D^2/\lambda$ as shown as (3) in Figure 10.2, where D is the maximum dimension of the antenna under test (AUT) and λ is the shortest wavelength (highest frequency) in the band of interest. The equipment set-up is shown in Figure 10.3.

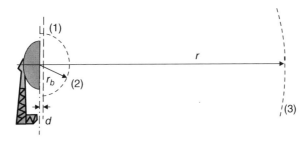

Figure 10.2   Techniques employed for aperture antenna radiation pattern measurement. (1) Planar near-field; (2) holographic distance or spherical near-field and (3) far-field

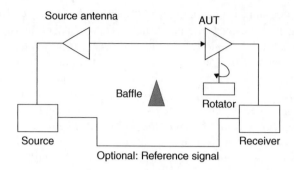

Figure 10.3   Far-field measurement equipment

The AUT is placed on a rotator with its phase centre as close as possible to the axis or axes of rotation in the vertical and horizontal directions. This can be done with a measuring tape and string or more accurately with photogrammetry. The phase centre can be estimated using the technique described in Section 3.5.5 or obtained through measurement as outlined below. The AUT should be located on the rotator according to this estimate of phase centre. It is then rotated and the received signal recorded with a sensitive receiver. This will give a first estimate of the radiation pattern. If the phase is recorded as the antenna is rotated, this can be used to measure the phase centre. The AUT is moved backwards and forwards along its axial direction on the rotator as depicted in Figure 10.3. Its position should be recorded and a note made of this on the antenna. Further adjustment both forwards and backwards will indicate a position for which the phase is most uniform. This is the location of the phase centre. When the AUT is positioned at its phase centre, the radiation patterns should be recorded.

The H-plane co-polar pattern is obtained when the AUT is rotated in a plane perpendicular to the electric field direction as shown in Figure 3.7. While still rotating in the same plane, but with the source polarization rotated by 90°, the pattern that is then obtained is the cross-polar pattern in the H-plane. The E-plane pattern is obtained by physically rotating the AUT and the source polarization through 90° and making a measurement in the plane shown in Figure 3.7. The radiation patterns in another plane, such as in the 45° plane, may be obtained by rotating the AUT through 45° and aligning the polarization in the same plane as the pattern to be measured. The cross-polar pattern in the 45° plane is then found by maintaining the source in the same position as for the co-polar measurement but with the AUT now rotated through ±90° relative to the previous location. Thus, the cross-polar pattern is obtained in the plane perpendicular to the plane of parallel to the polarization.

Antenna gain can be measured by several ways. The most common technique used is gain comparison. The maximum signal is recorded with a reference antenna (or standard) for which gain is accurately known either by other measurement or with a well-characterized antenna, such as a horn, from which the geometry allows an accurate estimate of gain. The maximum signal of the AUT is recorded under the same conditions. The gain of the AUT is calculated as follows:

$$\text{Gain(AUT, dB)} = \text{gain(reference, dBi)} - \text{signal(reference, dBi)} + \text{signal(AUT, dB)}. \quad (10.1)$$

For example, suppose a reference horn has a gain of 20 dBi. If the maximum signal received from the AUT is 17.5 dB while from the reference the measured signal is 11.2 dB, the gain of the AUT given by Eq. 10.1 is 26.3 dBi.

Another technique that is used is the 'three antenna measurement method'. The basis is the Friis transmission formula (Kraus & Carver, 1973). When a transmitting antenna with gain G_{tx} radiates a power P_{tx} in watts to a second antenna at a distance r, the power received is

$$P_{rx} = P_{tx} G_{tx} G_{rx} \left(\frac{\lambda}{4\pi r}\right)^2, \qquad (10.2)$$

where G_{rx} is the receiving antenna gain and λ is the wavelength.

The signal of the AUT and two reference antennas is then measured as described above for the gain comparison method. However, on this occasion each of the three antennas is used in turn as the source antenna. All three combinations are used and this results in three equations as follows:

$$\text{Gain(AUT, dBi)} + \text{Gain(Ref1, dBi)} = 20\log_{10}\left(\frac{4\pi r}{\lambda}\right) + 10\log_{10}\left(\frac{\text{Power}\,Rx\,\text{AUT}}{\text{Power}\,Tx1}\right) \quad (10.3a)$$

$$\text{Gain(Ref2, dBi)} + \text{Gain(Ref1, dBi)} = 20\log_{10}\left(\frac{4\pi r}{\lambda}\right) + 10\log_{10}\left(\frac{\text{Power}\,Rx\,\text{Ref2}}{\text{Power}\,Tx1}\right) \quad (10.3b)$$

$$\text{Gain(AUT, dBi)} + \text{Gain(Ref2, dBi)} = 20\log_{10}\left(\frac{4\pi r}{\lambda}\right) + 10\log_{10}\left(\frac{\text{Power}\,Rx\,\text{AUT}}{\text{Power}\,Tx2}\right). \quad (10.3c)$$

where Gain (AUT, dBi) refers to the gain in dBi of the AUT, Power Rx Ref$k$ is the received power with antenna k as a reference and PowerTx k refers to when reference antenna k is used as the source antenna. The gains of the three antennas are found from the simultaneous solution of Eqs. 10.3.

Circular polarization can be measured by means of several techniques. One of the simplest is to rotate the receiving antenna at a controlled speed of rotation typically 20–30 revolutions per minute. Assuming $E_r \approx 0$, this enables the two field components to be represented as is now shown for the incident circularly polarized field given as

$$\mathbf{E}(\theta, \phi) = \hat{R}\left(\frac{E_\theta + jE_\phi}{\sqrt{2}}\right) + \hat{L}\left(\frac{E_\theta - jE_\phi}{\sqrt{2}}\right) \qquad (10.4)$$

where E_θ and E_ϕ are the non-zero components of the far-field and also the time and distance dependence given by $\exp[j(\omega t - kr)]/r$ has been suppressed. The unit vectors in Eq. 10.4 are given by $\hat{R} = (\hat{\theta} - j\hat{\phi}/\sqrt{2})$ and $\hat{L} = (\hat{\theta} + j\hat{\phi}/\sqrt{2})$. Note that Eq. 10.4 is an alternative representation of $\mathbf{E}(\theta, \phi) = \hat{\theta}E_\theta(\theta, \phi) + \hat{\phi}E_\phi(\theta, \phi)$, which can be verified by substituting the unit vectors $\hat{R}$ and $\hat{L}$ into Eq. 10.4 and simplifying the expression. The phase angle $\exp(\pm j\pi/2)$ relative to the phasor reference ωt refers to right (+) or left (−) hand circular polarization corresponding, respectively, to clockwise or counter-clockwise rotation as viewed for a wave propagating away from the transmitter (this is reversed for a receiver). A measurement is performed of either component by rotating the receiving antenna in either directions $\hat{R}$ or $\hat{L}$. The former gives a measurement of $(E_\theta + jE_\phi)/\sqrt{2}$ while the latter gives $(E_\theta - jE_\phi)/\sqrt{2}$.

10.2.2 Near-Field Measurement

The far-field region of some microwave antennas such as reflectors and horns can occur several kilometres or more away from the source depending on the antenna dimensions and operating frequency. A possible option is to use an interstellar source. Such measurements require a sensitive receiver for them to be successful. The moon can also be used, but care needs to be taken due to the angle it subtends with respect to a point on the earth's surface. Constraints on measurements made outdoors and a desire for an all-weather facility in which measurements can be performed close to the workplace of most personnel and under controlled conditions have led to the development of other methods of measuring far-field patterns.

One particularly powerful technique, which was developed in the early 1960s as the necessary computational power became available, is probe compensated near-field measurement (NFM) (Burnside et al., 2007). Today, this technique is very widely used in research as well as in industry. The technique requires the near-field of the AUT to be sampled on a specified surface at about 5–10 wavelengths away (as in Figure 10.2a), and these data are numerically transformed to obtain the far-field. Some effects of the sampling probe on the measurement need to be included in this transformation, as will be described.

Probe-compensated NFM has been shown to be time and cost effective, and results can be obtained that are at least comparable to those obtained with a far-field range. Stray reflections can be controlled by the use of microwave absorber and this, plus the error involved in positioning the probe, usually limits the useful frequency range of NFM from low UHF to around 1 THz. NFM is more complicated than other techniques, it must be automated and staff must be highly skilled. Also the probe antenna must be extensively and accurately calibrated at all operating frequencies. The measurement is made under computer control, and the same computer can also be used to transform the near-field data to the desired region in real time, which can be an advantage. This may not be possible with electrically large antennas as these measurements can take several hours to complete.

The three principal surfaces surrounding the test antenna that are used for near-field probing are the plane, cylinder and sphere, as illustrated in Figure 10.4. The planar approach is the most common, despite the need to accurately position the probe on a plane, because the far-field calculation is easily and efficiently implemented (Paris et al., 1978). However, the cylindrical and spherical schemes can provide more of the far-fields in the far-zone, the latter most completely. The planar technique on the other hand becomes more inaccurate as the angle between the test and probe antennas increases, although this can be overcome if it forms part of a cylindrical scan. The planar scan can provide far-field information out to the angle subtended by the plane at the AUT, which is

$$\theta_{max} = \tan^{-1}\left(\frac{x_{max}}{d}\right),$$

where d the probe is distance and x_{max} is the maximum extent of the planar scan in direction x.

If a sample probe does not influence the near-field, the far-field of the test antenna is obtained directly from the solution to the field equations in the appropriate separable co-ordinate system. For the planar configuration shown in Figure 10.4a, the far-field intensity is the Fourier transform of the near-field on a plane at a distance d in front of the test antenna. That is, the far-field is (Paris et al., 1978)

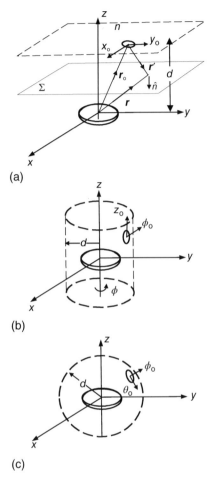

Figure 10.4 Near-field measurement surfaces. (a) Planar scan at distance d and an intermediate surface Σ; (b) cylindrical scan of radius d and (c) spherical scan of radius d

$$E(\theta,\phi) = \frac{jk_z}{2\pi}\exp(jk_z d)\int\limits_{-\infty}^{\infty}\int e(x_o,y_o,d)\exp\left[j\left(k_x x_o + k_y y_o\right)\right]dx_o dy_o, \qquad (10.5)$$

where $e(x_o,y_o,d)$ is the aperture field, k_x, k_y and k_z are the rectangular components of the wave vector in the far-field direction, namely, $\mathbf{k} = \left(\hat{x}k_x + \hat{y}k_y + \hat{z}k_z\right) = k(\hat{x}\sin\theta\cos\phi + \hat{y}\sin\theta\sin\phi + \hat{z}\cos\theta)$ where $k = 2\pi/\lambda$.

The probe affects the measurement in two ways. Its output is a weighted average of the field over its aperture, and there is some interaction between the test and probe antennas. Compensation for these effects is possible although some of the latter is ignored in current near-field techniques. The theory of probe compensation is briefly described here.

Let Eq. 10.5 be the near-field of the test antenna in the absence of the probe on a surface Σ between the aperture and the plane at $z = d$ as shown in Figure 10.4. Also let $(\mathbf{e}', \mathbf{h}')$ be the near-fields of the probe antenna. Assuming the scattered fields are small compared to the primary fields, the reciprocity theorem (see Section 2.1.4) applied to the situation depicted in Figure 10.4a can be shown to give

$$\iint_{\Sigma} (\mathbf{e} \times \mathbf{h}' - \mathbf{e}' \times \mathbf{h}) \cdot \hat{n} dS = v_{\alpha}(\mathbf{r}_{o}). \tag{10.6}$$

$v_{\alpha}(\mathbf{r}_{o})$ is proportional to the open-circuit received voltage at the probe, which is rotated an angle α about its axis. The derivation of the far-fields from Eq. 10.6 depends upon the chosen measurement surface, but the aim is the same – to express the near-fields of the test and probe antennas in terms of quantities from which the far-fields can be calculated directly. The planar case is by far the simplest for, as Eq. 3.18 shows, the near-field is the inverse Fourier transform of the far-field. In the cylindrical and spherical cases, the near-field is expressed as a continuous spectrum of, respectively, cylindrical and spherical vector wave functions. For example, the near-field in the cylindrical case is

$$\mathbf{e}(d,\theta,\phi) = \sum_{n=-\infty}^{\infty} \int_{-\infty}^{\infty} dk' [a_{n}(k')\mathbf{M}_{n}(\phi,z) + b_{n}(k')\mathbf{N}_{n}(\phi,z)], \tag{10.7}$$

where $\mathbf{M}_{n}(\phi, z)$ and $\mathbf{N}_{n}(\phi, z)$ are the n-th cylindrical vector wave functions evaluated on a cylinder of constant radius d. The aim is to determine the coefficients $a_{n}(k')$ and $b_{n}(k')$ to obtain the far-field. The cylindrical wave functions involve Bessel functions, which rapidly increase with $|n|$. However, the coefficients decay at a faster rate. The series in Eq. 10.7 can be truncated when $|n| > kd$ as the terms become negligibly small. An explicit solution for the unknown coefficients can be obtained after considerable additional manipulation (Leach & Paris, 1973).

In the planar case, Eq. 10.6 results in

$$-E_{\theta}(\theta,\phi) E_{\theta}'(\theta, -(\phi+\alpha)) + E_{\phi}(\theta,\phi) E_{\phi}'(\theta, -(\phi+\alpha)) = F(\theta) V_{\alpha}(\theta,\phi), \tag{10.8}$$

where $F(\theta) = ((k\omega\mu)/8\pi^{2})\cos\theta \exp(jkd\cos\theta)$, while $E_{\theta}'(\theta,\phi)$ and $E_{\phi}'(\theta,\phi)$ are the electric field intensity in the far-field of the probe antenna. For a general probe, these data are obtained from measurement and need to be stored. The transform of the measured data for a probe at a rotation angle α is

$$V_{\alpha}(\theta,\phi) = \iint_{-\infty}^{\infty} v_{\alpha}(x_{o}, y_{o}) \exp\left[j(k_{x}x_{o} + k_{y}y_{o})\right] dx_{o} dy_{o}. \tag{10.9}$$

Eq. 10.8 contains two unknown functions and thus measurements are required at two probe orientation angles. The two resulting equations can be solved simultaneously to obtain the far-field functions.

In the cylindrical case, equations for the coefficients a_{n} and b_{n} in Eq. 10.7 are obtained similarly, and these also involve Fourier transforms of the measured data for each function. In the spherical case, Figure 10.4c, probe correction is more complex, and only one of the integrals involved in the transformation is a Fourier integral and, hence, the other must be evaluated by numerical integration. As the planar method is simpler, without loss of generality, the remainder of the discussion of NFM is limited to this technique.

Through selection of a suitable probe antenna, the transformation and the equations for the unknown functions can be simplified, and this can substantially reduce the amount of probe data stored in memory. Consider a probe antenna with physical circular symmetry, such as a thin wall circular waveguide, which is linearly polarized in the x-direction. It has been shown in Section 4.4.1 that the field components of this antenna take the form

$$E'_\theta(\theta,\phi) = A(\theta)\cos\phi \quad \text{and} \quad E'_\phi(\theta,\phi) = -B(\theta)\sin\phi, \tag{10.10}$$

where the functions $A(\theta)$ and $B(\theta)$ are the complex patterns in the E- and H-planes, respectively. These functions characterize the antenna and only data for these planes need be measured and stored in memory. Precisely manufactured circular aperture antennas, such as a thin wall TE_{11} mode waveguide, are represented very accurately by Eq. 10.10. In addition, this probe has the advantage that exact expressions for functions $A(\theta)$ and $B(\theta)$ are available (Weinstein, 1969).

When Eq. 10.10 are introduced into Eq. 10.8 and measurements are taken at probe orientations $\alpha = 0°$ and $90°$, the far-field functions of the test antenna can be calculated from

$$E_\theta(\theta,\phi) = \frac{F(\theta)}{A(\theta)}[V_0(\theta,\phi)\cos\phi - V_{90}(\theta,\phi)\sin\phi] \tag{10.11a}$$

$$E_\phi(\theta,\phi) = \frac{F(\theta)}{B(\theta)}[V_0(\theta,\phi)\sin\phi + V_{90}(\theta,\phi)\cos\phi], \tag{10.11b}$$

where $V_0(\theta,\phi)$ and $V_{90}(\theta,\phi)$ are the two sets of measured two-dimensional data that have been transformed by means of Eq. 10.9.

Equations 10.11 are in a convenient form for computing great circle cuts in the far-field. In the special case that measured data distributions are separable in the probe co-ordinates, it can be shown from Eqs. 10.11 that the principal plane cuts are one-dimensional Fourier transforms of the measured data in those principal planes.

A planar NFM system based on the principles described above is shown in Figure 10.5. The probe is assumed to have been characterized as described above. In one implementation, the probe was positioned by means of belt-driven gears which are driven individually by stepper motors under computer control. After a short settling time, the output of the probe was sampled. A sensitive vector receiver or network analyser was used to measure the signal from the probe. The output of the receiver was converted to binary-format for more efficient storage.

The main computational function to be performed by the computer is the computation of the discrete Fourier transform of the rectangular array of data. The fast Fourier transform (FFT) (Brigham, 1974) can be used to reduce the number of multiplications required for a one-dimensional transform of N points from the order of N^2 (the number required for a direct

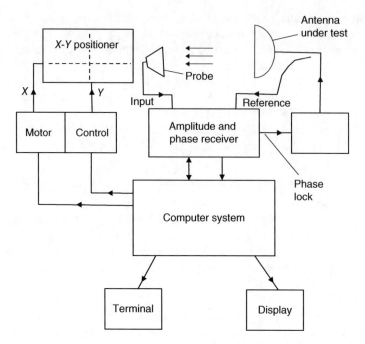

Figure 10.5 Planar near-field measurement instrumentation

application of the transform formula) to the order of $N \log_2 N$. For a two-dimensional transform over MxN points, the repeated application of the one-dimensional FFT to produce a two-dimensional FFT requires of the order of $NM \log_2 M + MN \log_2 N$ multiplications. An efficient implementation of multi-dimensional Fourier transforms can reduce the required number of multiplications below that for repeated application of one-dimensional transforms. However, in this application, the critical process governing the time to produce results is the acquisition of raw data and display of the computed fields.

In operation, observations can be made at predetermined points across a linear scan in the near-field plane, and the received data placed in random access memory. While the next scan is being performed, the FFT of the observations in the previous scan could be computed. When all scans have been made, the transform of the final scan can be completed. The remainder of the two-dimensional transform can be performed by taking transforms over the stored data in the direction normal to the scan direction in the plane. A final task for the system is the display of results in two-dimensional format and plotting of selected pattern cuts.

The antenna gain can be obtained from the computed far-fields by means of Eqs. 10.11 from its basic definition (Eq. 3.48). The total radiated power is found by making use of reciprocity, through connecting the power source to the input of the AUT and measuring transmitted power.

The field in the aperture plane can be computed from the calculated far-field pattern by computing the inverse Fourier transform. If the probe affects have been removed, the computed aperture field will be quite accurate. The aperture field is very useful to have available for aligning or adjusting the performance of the antenna.

10.2.3　Intermediate-Field Measurement

Antenna radiation patterns can be obtained using the Fresnel-zone holographic technique if the antenna is too large for a near-field range or the far-field range is too short compared to the Rayleigh distance $R = 2D^2/\lambda$ (Keen, 1978; Poulton, 1983). A schematic of a typical measurement set-up for a Fresnel zone measurement is shown in Figure 10.6. The signal received from the antenna and a reference signal, which is taken from a splitter placed just before the source horn, is fed into a vector microwave receiver. A computer logs the data and drives the antenna in azimuth and elevation to produce a two-dimensional map on a hemispherical surface. The number of samples required is determined by the Nyquist criterion which ensures there is no aliasing of the predicted aperture field. The spacing, in radians, must be $< \lambda/D$, where D is the diameter of the AUT. To compensate for drift in the signal, calibration points should be made at regular intervals. This is equally true of NFMs and, therefore, should be programmed into the measurement procedure.

With a planar aperture A in the x–y plane and with the assumption of an electric field backed by a magnetic conductor, the far-fields may be computed from the following Fresnel approximation (Poulton, 1983) obtained from Eq. 3.10:

$$\mathbf{E}(u,v)\exp[jq(u,v)] = -\frac{jk}{4\pi}\frac{e^{-jkr}}{r}\int_A \mathbf{E_a}(x',y')\,\exp\left[-jk\frac{\left(x'^2+y'^2\right)}{r_b}\right]\exp[-j2\pi(ux'+vy')]dS',$$

$$(10.12)$$

where $\mathbf{E_a}$ is the aperture field in the (x', y') plane, $u = \sin\theta\cos\phi$, $v = \sin\theta\sin\phi$, r_b is the distance from the aperture to the source when $u = 0 = v$, and $q(u, v)$ is a phase term compensating for offsets d_1, d_2, h and x_0 all shown in Figure 10.7.

Co- and cross-polarization data can be taken with the set-up in Figure 10.7 and then processed using the following steps (Poulton, 1983):

1. Correct for phase and amplitude drifts by interpolating between the calibration points.
2. Compensate for the offset in the centre of rotation if the antenna is set above and forward of the centre of rotation.
3. The aperture field is calculated by applying a Fourier transform with a quadratic phase correction as given in Eq. 10.12.
4. A circle with a radius one wavelength larger than the reflector can be used as a mask and the field set to zero outside it to remove measurement artefacts.
5. A second Fourier transform is taken to create a two-dimensional map of the far-field, covering the desired angular range in both azimuth and elevation.
6. Cardinal and inter-cardinal plane pattern cuts are then interpolated from the two-dimensional maps.

The typical equipment required for radiation pattern measurement at an intermediate range is illustrated in Figure 10.6. In much the same way as for full NFM, the heart of the measurement is a vector receiver and a computer controller. As an example of the radiation patterns obtained

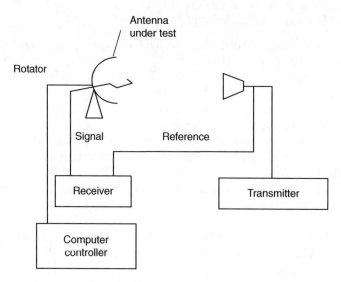

Figure 10.6 Intermediate radiation pattern measurement set-up

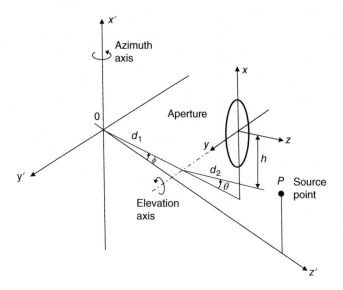

Figure 10.7 Fresnel zone measurement geometry

through an intermediate-field measurement, Figure 10.8 shows the co-polar pattern of a radar dish antenna that was refurbished for dual-polarization operation (Keenan et al., 1998). As both amplitude and phase information was available, the reflector surface errors were calculated by transforming the data back to the aperture and expressing the phase variations into physical deviations from the expected surface profile of the paraboloid.

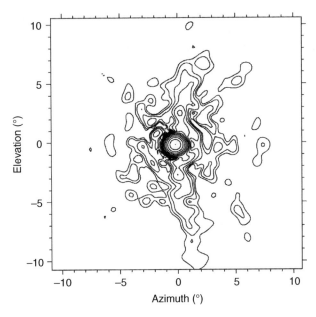

Figure 10.8 Co-polar far-field of a meteorological radar antenna measured using the intermediate distance technique. Plot contours are at 3 dB intervals. *Source*: Reproduced from Hayman et al. (1998)

10.3 Modern Aperture Antennas

Some examples of recently developed aperture antennas will now be described. They are included to support some of the topics discussed previously and are used also to illustrate where improvements have been made in recent years to the state-of-the-art. They should also indicate possible future enhancements. The first example is the design and fabrication of a compact, low-sidelobe horn (Granet et al., 2000). This type of antenna has several applications from a feed in a reflector antenna to lightweight reference antenna for a point-to-point communication system. The second example is of a shaped reflector antenna which extends the concepts described in Chapter 6 (Hay et al., 2001). The aim was to achieve multiple beams with several feed horns located in the focal region of the antenna. The final example is an array of horns for feeding a radio telescope (Staveley-Smith et al., 1996).

10.3.1 Compact Low-Sidelobe Horns

Horn antennas are often required for applications requiring a moderate gain, low sidelobes and a compact geometry for lightweight applications or to enable the source to be located close to a secondary radiating aperture. In geostationary satellite applications, a horn antenna is frequently used to provide full earth coverage for telemetry and command signals transmission or reception as well as conventional communication traffic. In addition with an increasing number of satellites orbiting the earth, minimizing interference with other satellites has become more important than in the past. To achieve this, the amount of sidelobe energy

Figure 10.9 A profiled rectangular horn designed for maximum efficiency and for operation in the frequency range 11.7–12.2 GHz. *Source*: Reproduced with permission from CSIRO

should be as low as possible, both for co- and cross-polarized signals while at the same time illuminating the earth efficiently at the required power level over the 17.4° angle subtended by the earth. Another important factor for any satellite application is keeping the weight of the antenna as low as possible as well as making the size manageable. Therefore, the horn should be as compact as possible. Usually there is a trade-off between performance and size/weight of the horn.

In designing a compact horn there are two main approaches. As described in Section 4.5.3, these are profiling, or tapering, the horn or introducing steps. A stepped horn has the advantage that the horn can be made up of a number of uniform sections, which may be easier to machine. Stepped horns have been used in many forms from a series of irises to corrugations. A systematic approach to designing profiled horns has been described (Bird & Granet, 2013). This approach uses numerical optimization in concert with an accurate analysis method. It is described in Section 4.5.3 and it has the advantage that most a priori requirements can be included in a performance index of the optimization method. The approach also has the advantage that it can be applied to design other antenna geometries.

A rectangular horn designed by optimization for maximum efficiency is illustrated in Figure 10.9. The performance index was optimized to maximize gain and to minimize the reflection coefficient as set out in Section 4.5.3.1. The final performance of the square-profiled horn in summarized in Table 10.1.

Table 10.1 Summary of profiled square-horn performance

Parameter	Property	
Aperture size	50.21×50.21 mm $2 \times 2\lambda_m$ (at 11.95 GHz)	
Total horn length	80 mm ($3.2\lambda_m$ at 11.95 GHz)	
Return loss	>25 dB	
Cross-polar maximum	<−25 dB	
Efficiency	11.70 GHz	96.5%
	11.95 GHz	96.8%
	12.20 GHz	96.1%

Figure 10.10 Low-sidelobe horn manufactured in carbon fibre and copper coated is its interior for high performance. *Source*: Reproduced with permission from CSIRO

In another example, a horn was designed to have low sidelobes by means of a sine raised to power p profile as described in Section 4.5.3.2 (Granet et al., 2000). Simulations with accurate computer software showed that a design with $p = 0.8$ gave best overall results for sidelobe level and gain. To verify the concept and test the design, a prototype was initially constructed from aluminium. This prototype weighed about 100 kg. Following successful tests on the prototype, a flight-suitable lightweight version was constructed from CFRP, Figure 10.10. All materials were rated <0.01% vacuum condensable material. The horn body consisted of the mode generator, flare section and an input probe exciter; all three corrugated parts were constructed separately. The ring corrugations were first manufactured using flat stock and cut to size. The outer skin of the mode generator and flare section was produced in a one-piece mould. A specially designed fixture ensured that the corrugations were concentric and were correctly spaced. When the construction was completed the interior was plated with a continuous layer of copper (thickness 4–5 skin depths). The worst case manufacturing error is estimated to be $<0.00018\lambda$.

Table 10.2   Performance of low sidelobe corrugated horn with $p = 0.8$ profile

Parameter	Performance
Centre frequency (f_c)	S-band (λ_c)
Specified bandwidth	1%
Half-power beamwidth	17.4°
Gain at (f_c)	20.9 dBi
First sidelobe level	−36 dB below peak
Peak cross-polarization	<−40 dB
Aperture diameter	4.6λ_c
Length	5.6λ_c

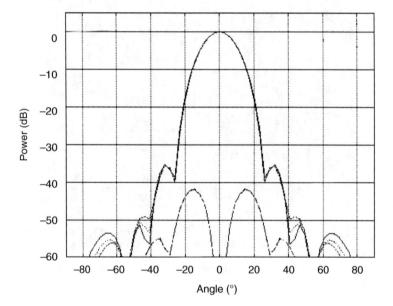

Figure 10.11   Radiation patterns of low-sidelobe corrugated horn. Solid line: theory; and dash line: experiment

The total weight of the horn and exciter was 9.12 kg, which is less than a tenth of the weight of the prototype. The completed horn was subject to a complete range of tests including electrical, mechanical vibration and thermal in the range −30 to +60°C to assess the effect of temperature on the sidelobe performance. The performance of the low-sidelobe horn is summarized in Table 10.2, and some measured radiation patterns are shown in Figure 10.11 along with the computed results. The experimental results overlap the predictions made with software that uses the mode matching method which indicates excellent agreement with the design. The first sidelobe level was at the −35 dB level relative to the peak and reflection coefficient was <−20 dB over a 5% bandwidth.

10.3.2   Multibeam Earth Station

As the number of geostationary satellites increases year-by-year, in some applications it is advantageous to be able to address several satellites at one time with an earth station antenna. To access these and other new satellites, pay-tv operators and service providers such as teleports are installing more earth station antennas. As an alternative to a standard earth station antenna, where one dish receives the signal from a single satellite, a multibeam earth station offers significant advantages. These advantages include that a multibeam earth station occupies less real estate than several dishes, maintenance can be simplified, a new satellite can be accommodated by the addition of a new feed system and services from several satellites can be multiplexed.

A reflector surface in the shape of a torus can be used to access several satellites by placing several feeds in the focal region because it has a line focus. A limitation of the torus reflector is that the sidelobes and cross-polarization produced can be unacceptable in some situations such as in transmission due to inherent aberrations created by the toroidal surface. An alternative approach is to shape the surfaces of the reflector as described in Chapter 6 to provide multiple beams over a limited angular range. One such design shown in Figures 10.12 and 10.13 is briefly described (Hay et al., 2001).

A multibeam antenna was designed using two reflectors and with up to 19 feed horns to cover a 40° field of view, each viewing a satellite with a possible separation of 2°. Unlike a conventional reflector antenna, which has a single focal point where the feed must be positioned, the multibeam antenna has a focal surface on which the feed horns are located. The reflectors are specially shaped and strategically positioned to maximize the field-of-view of the geostationary satellite arc. The antenna that resulted is shown in Figures 10.12 and 10.13. In Figure 10.12, three such antennas are combined to cover a total of 120° coverage of the geostationary arc.

The resulting multibeam antenna design has shaped reflectors arranged in a Cassegrain configuration, and each feed horn illuminates part of an extended subreflector. The entire main

Figure 10.12   Multibeam earth stations at SES-Astra in Luxembourg. Each antenna covers a 40-degree field-of-view of the geostationary satellite arc. *Source*: Reproduced with permission from SES-ASTRA

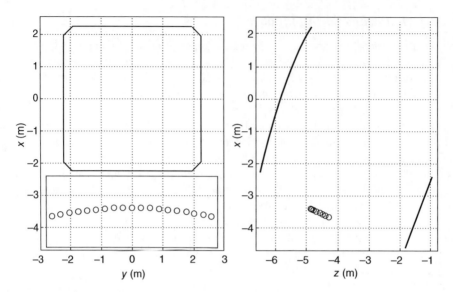

Figure 10.13 Geometry of the 4.5 m diameter multibeam antenna for Ku-band. *Source*: Reproduced with permission from CSIRO

reflector is used by each beam. Both reflectors are under illuminated to produce low sidelobes so as to meet the sidelobe specification. To help achieve low sidelobes, profiled corrugated horns were also used. A compact corrugated feed horn was designed using accurate electromagnetic modelling software as described in Section 4.5.3. The horns were designed to give low signal spillover and low cross-polarization. To create the initial shaped reflector surfaces, a geometric-optics ray-tracing programme was used.

The main reflector is about 4.5 m square, while the subreflector is about 5.6 m wide by 2.3 m high. The final reflector shapes are found by means of an accurate physical-optics-based optimizer to achieve low sidelobes over a wide field-of-view while maintaining good aperture efficiency. To ensure that the electrical performance of the overall antenna was satisfactory, the complete antenna was modelled with an accurate computer software package, which combines physical optics modelling of both reflectors and mode matching for the feeds. Antenna feed horns placed on the focal surface in transmit mode produce directional beams over the ±20° field of view. The sidelobes of each beam satisfy the ITU envelope of $29-25\log\theta$ dBi, where θ is the angle from the beam maximum. In the specification 10% of the sidelobe power is permitted to exceed this envelope. In the forward direction, the sidelobes are below the required envelope, but for angles >100° several peaks are expected to exceed—5 dBi due to the spillover from the subreflector. The subreflector is significantly larger than in conventional dual reflector antennas and the feed horn sidelobes illuminate this reflector, which in turn produce the spillover. Both reflectors were fabricated from shaped reflector panels by means of the adjustable mould method that was described in Section 10.1.6. A bed-of-bolts shown in Figure 10.1 was set under computer control for each panel on the antennas. The final accuracy of the panels was < 0.025λ.

The feed system that is used in conjunction with the wideband-corrugated horn covers the required frequency range of 10.7–12.75 GHz. In a typical arrangement, the horn is connected to an orthomode transducer that allows reception of two orthogonal polarizations, and filters are

Table 10.3 Gain and noise temperature of 4.5 m multibeam antenna at two frequencies and beam positions, measured and simulated results

Frequency (GHz)	Beam	Gain (dBi)		Antenna temperature at 30° elevation (°K)	
		Theory	Experiment	Theory	Experiment
12.75	0°	51.4	50.8 ± 0.5	14	15 ± 4
11.20	0°	51.0	50.5 ± 0.5	20	23 ± 4
12.75	18°	51.4	51.0 ± 0.5	14	17 ± 4

provided in each port to reduce interference at transmit band frequencies. The feed system was placed in a cylindrical enclosure which was then attached to a beam supporting all feed systems. The mount allows translation and rotation of the feeds for alignment purposes. As the antenna may have to operate in freezing conditions, de-icing equipment was also included. Hot air blowers were placed on the back of each reflector, and resistive tape heaters were placed around the aperture of each horn to prevent ice and water forming on the radome.

To obtain permission to transmit into commercial satellite networks using an antenna, such as a multibeam earth station, verification testing of radiation, gain and noise temperature performance is normally required. The fixed structure of a multibeam antenna does not allow for all the conventional on-site tests to be made. Therefore, in conjunction with several major satellite operators, a protocol was developed to verify that the performance was acceptable for transmit mode of operation. This protocol involved the following steps: (i) assembly of the antenna at the site of manufacture followed by measurement of the antenna structure using photogrammetry; (ii) detailed radiation pattern testing on an antenna range; (iii) disassembly of the antenna and shipment to the final site; (iv) re-assembly and re-measurement by means of photogrammetry; and (v) limited on-site tests including measurement of gain and on-axis cross-polar isolation.

Photogrammetry was used to verify the assembled accuracy compared with the design. For instance, the antenna shown in Figure 10.12 had the following accuracy: main reflector surface, 0.3 mm rms; subreflector surface, 0.5 mm rms; and inter-reflector alignment, 1.6 mm rms.

Verification testing at the site of manufacture can be performed before installation, as long as the antenna is sufficiently modular so that it can disassembled and transported to its final location. One way to test the performance of an earth station is to use the signal from a satellite in reception. This can be used to optimize gain and sidelobes as well as adjustment of the orientation of the ribs of the reflector panels to ensure spurious sidelobes were not created by periodic ribs. The radiation patterns were measured by means of a large rotator that could be moved in azimuth and elevation directions about the centre of the beam. At the same time, a beacon signal at 12.748 GHz was recorded from a locally operating Ku-band geostationary satellite. Receive tests were completed for several sample beams and transmit tests were also made for two beams. The results are summarized in Table 10.3. The radiation pattern and gain performance for all measured beam positions were excellent and exceeded the requirements except for the cross-polar isolation within the 1 dB beamwidth, which had a target specification of 35 dB/K. However, the measured cross-polarization isolation within this 1 dB beamwidth was better than 30 dB which is the operating specification of most satellite operators. Antenna noise temperature was also measured at an elevation angle of 25°. These measurements gave an antenna noise temperature, T_a, of ≤21°K for the receive-only feeds and ≤39°K for the transmit/receive feed

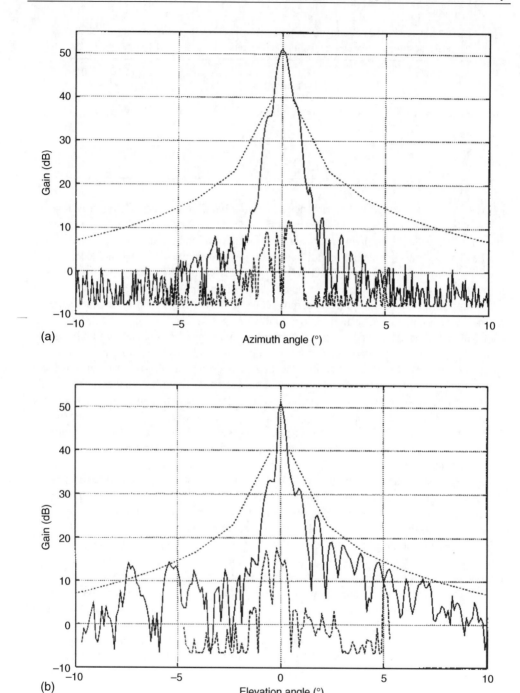

Figure 10.14 Measured radiation patterns of the 4.5 m multibeam antenna at 12.748 GHz for beam at 18.6° from boresight. (a) Azimuth pattern and (b) elevation pattern. *Source*: Reproduced with permission from CSIRO

system (with diplexer). For the frequencies of interest, the G/T_a is ≤ 31.2 dB/°K for feeds operating only in the receive-only band and ≤ 30.2 dB/°K for full bandwidth transmit/receive feeds, and this gives a significant system margin for operators above the typical specification of 29 dB/K. The measured results are summarized in Table 10.3, and a typical measured radiation pattern at 12.748 GHz for a beam +18.6° from boresight is shown in Figure 10.14.

10.3.3 Radio Telescopes

On several continents there are a number of radio observatories with large reflector antennas that are used for radio astronomy. The diameter of the telescope is important as it determines the minimum sensitivity and its ability to detect signals from distant galaxies. The objective is to observe sources sufficiently far enough away in specific frequency bands such as the OH molecule (1.4–1.6 GHz) and to monitor these signals over time. It is predicted that the signals emanate from distances far away, thought to be close in time to the creation of the universe. In the past, most single radio telescopes had a single sensitive feed horn. Over recent years, a cluster of feeds has been used, each with two orthogonal linear polarizations so that survey speeds can be reduced to a more manageable observation limits. As well, multiple feeds can be used to provide redundancy to improve the radio images. This was demonstrated early on with the Parkes multibeam feed (Bird, 1994; Staveley-Smith et al., 1996). Compared with a high-efficiency feed at the focus, an N-beam focal plane array reduces the sampling time by a factor C_t/N, where C_t is a time penalty due to the longer sample times that are required with a multibeam feed to compensate for slightly reduced antenna gain and increased noise caused by higher feed spillover. As an example, consider the design of a seven multiple beam feed cluster for the Arecibo radio telescope. This antenna uses a Gregorian reflector system. The main reflector diameter is 320 m, and the Gregorian subreflector subtends an angle of 60° with respect to the optical axis of the feed, which allows waveguide feeds to be used. The specifications of a seven-element feed cluster for this reflector system are summarized in Table 10.4.

There are two major drivers of the feed array design. The first of these is that the feed elements should efficiently illuminate the nearest reflector with a suitably low edge taper and low spillover. If the edge taper is too low, the antenna efficiency may suffer. The second driver is that the feed aperture diameter is limited in size, since the larger the diameter of each feed the further the beams are apart and the greater the scan loss. A rough estimate for the radius that is based on Eq. 6.107 is $a \approx f_{\text{eff}}\beta$, where f_{eff} is the effective focal length and β is the maximum scan angle (in radians). Figure 10.15 shows the scan loss as a feed is moved from the focus of the Gregorian system. A scan loss of about 1.5 dB was considered the upper limit and this constrained the feed centre-to-centre spacing to <270 mm. The combination of these two factors restricts a practical array in the Arecibo antenna to about seven elements (see Figure 10.16), which still gives a significant reduction in observing time. The limited size also means that the most efficient feeds for wide-angle reflectors cannot be used because their effective aperture would be made too large through the addition of external slots and chokes for higher performance. To control the aperture diameter, TE_{11}-mode stepped circular and also coaxial horns were investigated as potential candidates as array elements. Both horn types were analysed in detail using mode-matching software for arrays. A range of horn sizes were tried in assessing the various trade-offs.

Another important geometric constraint is the diameter of the input waveguide to the horn. This is usually set at the commencement of the design. The usual requirement is that only the

Table 10.4   Specifications for a multibeam reflector for radio astronomy in L-band

Frequency range	1.225–1.525 GHz
Number of elements	7-hexagonal array
Return loss	>20 dB
Polarization	Dual linear
Polarization isolation	>20 dB
Edge illumination	<−10 dB
Input waveguide	150 mm
Ground plane diameter	1000 mm

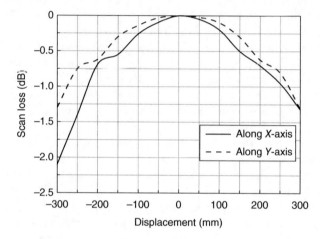

Figure 10.15   Scan loss of the Arecibo Gregorian reflector system at 1.375 GHz for feed movements in two-directions about co-ordinates C–C 260 mm and an X-offset 27 mm

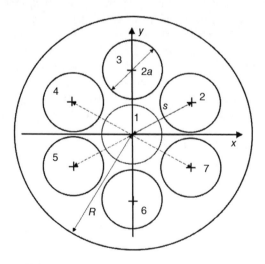

Figure 10.16   Seven-element array in a circular ground plane

fundamental TE_{11} mode can propagate in the frequency range 1.225–1.525 GHz and all other modes are cut-off. A 150 mm diameter pipe satisfies this requirement.

The horn performance was studied using mode matching software not only in isolation but also when each horn was embedded in the Gregorian system optics. A spherical wave expansion of the feed radiation patterns was used to obtain a complete antenna gain analysis of the Arecibo radio telescope optics by a combination of kinematic and electrodynamic ray tracing, and aperture field integration. Finally, a noise analysis was performed with the feed to obtain the overall G/T using the previously mentioned techniques in conjunction with a detailed Gregorian noise mapping. A noise temperature analysis was undertaken to provide a clear basis for a selection.

A variety of circular and coaxial horn geometries were investigated initially as possible multibeam feed elements. The initial starting point for the circular horn was the feed design for the Parkes multibeam (Bird, 1994) as the angle subtended at the tertiary reflector of the Gregorian system is almost the same as for the Parkes radio telescope. However, the Parkes multibeam was required to operate over a narrower band of frequencies (1.27–1.47 GHz) and the match of the Parkes horn design is poor outside this band, hence the need for a new design for Arecibo. The coaxial horn designs were studied for the Lovell radio telescope in the United Kingdom (Bird, 1997), although a smaller inner-to-outer conductor ratio could be used because the angle subtended at the reflector is smaller for Arecibo. The smaller inner-to-outer conductor ratio helps to achieve a wider-band match. For both circular and coaxial designs, the horns were stepped from the 150 mm diameter input waveguide up to the aperture diameter in steps designed to give a good match and pattern symmetry over the frequency band. A ground plane diameter ($2R$) of 1 m was assumed during the design phase.

After several feed designs were trialled in the Gregorian optics, it was found that the largest apertures tended to give best overall performance. The two horn designs shown in Figures 10.17 and 10.18 were obtained, and they were found to give comparable performance in the radio telescope. The input match of these horns is shown in Figures 10.19 and 10.20. Although the TE_{11}-mode stepped horn has slightly better return loss, lower variation of pattern beamwidth and lower cross-polarization, the coaxial horn is smaller and can be packed closer together in the array which reduces the reflector scan loss.

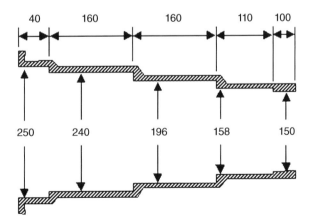

Figure 10.17 Stepped feed horn geometry (dimensions in mm)

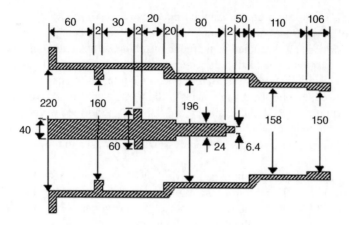

Figure 10.18   Coaxial horn design (dimensions in mm)

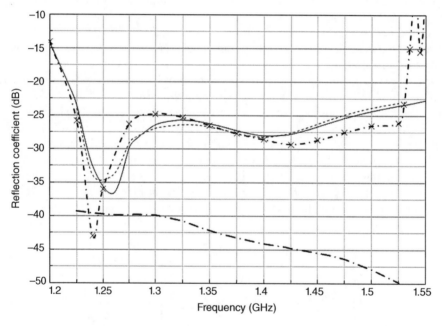

Figure 10.19   Reflection coefficient of the stepped circular horn shown in Figure 10.17. Solid curve: isolated single feed; short dash: theory – element #1 in the centre of the array; dash x: experiment – element #1 of array at output of OMT (Bird et al., 2003); and long dash – dot: coupling coefficient $|S_{21}|$ of element #2

   A detailed study concluded that the TE_{11}-mode stepped horn in Figure 10.17 offered the best compromise in performance in terms of spillover efficiency, antenna noise temperature and sensitivity and was simple to fabricate. The performance of the stepped horn is very satisfactory as shown in Table 10.5. Radiation pattern cuts of the isolated horn are given in Figures 10.21

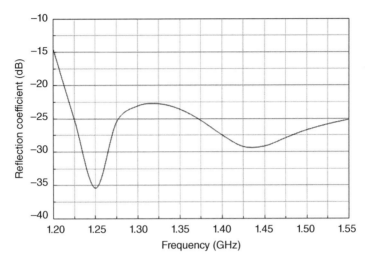

Figure 10.20   Reflection coefficient of the isolated single coaxial horn shown in Figure 10.18

Table 10.5   Summary of multibeam feed performance

Frequency (GHz)	1.225	1.375	1.525
Return loss (dB)	23.5	27.2	23.5
Edge illumination (dB)	10.9	13.5	15.9
Peak horn cross-polar (dB)	−24.4	−30.7	−33.7
Spillover efficiency (%)	92.7	95.7	97.9
Antenna system temperature (K)	36	30	26

and 10.22 at the low and mid-frequencies in the operating band. Figure 10.23 shows the corresponding two-dimensional patterns for the isolated horn. The patterns at the highest frequency in the band are even better than at the middle of the band.

The array centre is offset 27 mm in the X-direction of the focal plane, and this limits the scan loss at an outer horn location to about 1.3 dB (see Figure 10.15). For the chosen horn size and aperture spacing, it was found that the beam spacing of the radio telescope is about 1.66 times larger than the Nyquist sample spacing. Also, it was determined that the path followed by the beams in the sky, as the array rotates in the focal plane, is elliptical. The size of this ellipse depends linearly on the feed spacing in the focal plane, but the eccentricity is almost constant.

The performance of the horn in the array was predicted including the effects of mutual coupling and also a finite ground plane. Figure 10.24 shows the predicted radiation patterns of three elements in the array for the three independent beams at the mid-frequency for vertical (Y) polarization. These patterns should be compared with the corresponding pattern for the

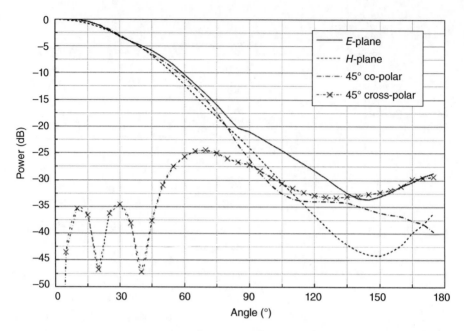

Figure 10.21   Principal radiation patterns of the stepped horn at 1.225 GHz. Solid line: long dash: short dash: 45° plane co-polar; and dash x: 45° plane cross-polar.

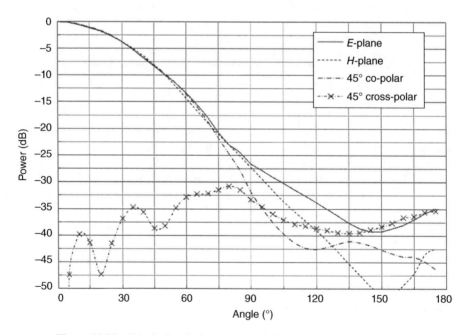

Figure 10.22   Principal radiation patterns of the stepped horn at 1.375 GHz

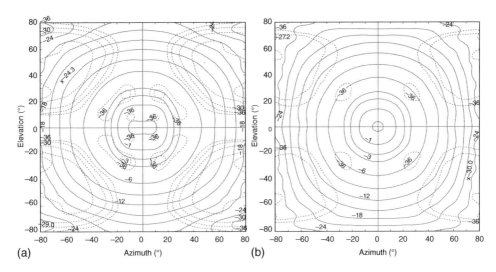

Figure 10.23 Radiation patterns of the isolated TE$_{11}$-mode stepped horn. (a) 1.225 GHz and (b) 1.375 GHz. Solid contours are co-polar patterns, dashed contours are cross-polar patterns

isolated horn in Figure 10.23b. It is found that, due to mutual coupling, the peak cross-polarization can be about 5 dB higher than that for the isolated horn at the low frequency end of the band. This is because of the high levels of the TE$_{21}$ mode excited in the aperture of the array. The result is summarized in Figure 10.25 for vertical polarization (Y-direction). Excitation of this unwanted TE$_{21}$ mode is difficult to control at low frequencies in a close-packed array of relatively small diameter horns, and tuning of the waveguide steps has little or no effect on the peak level. However, mutual coupling has a small effect on the return loss, which remains >23 dB across the entire frequency band, as shown in Figure 10.19.

The design was performed assuming the array was located in a large ground plane. With a smaller ground plane, the radiation patterns are more greatly impacted by the edge diffraction. Thus a finite ground plane should be included in the design. The size of the ground plane does not have a significant effect on the individual feed pattern; provided the ground plane is reasonably large (typically $R > 3.4a$, where a is the radius of the horn aperture in a seven-element hexagonal array), there is minor effect on the secondary radiation patterns and there is only a small change in feed spillover if the size is varied slightly from the specified diameter.

Measured results are shown in Figure 10.26 for the central element. The patterns shown are for the co-polar and cross-polar components in the 45° plane. Due to the geometrical symmetry, the pattern is almost symmetric. There is good agreement between these results and theoretical values obtained from the mode matching method which includes mutual coupling between the elements. It is seen that the peak cross-polar level is comparable to its level in the isolated horn.

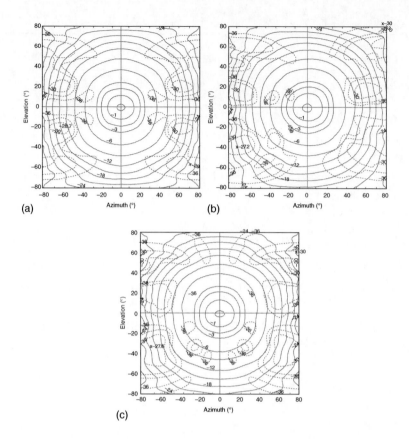

Figure 10.24 Radiation patterns of stepped horn feed at 1.375 GHz. (a) Element 1, (b) Element 2 and (c) Element 3. Vertical polarization. Solid contours are co-polar patterns, dashed contours are cross-polar patterns

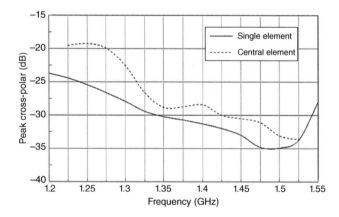

Figure 10.25 Peak cross-polar level of stepped horns. Solid curve: single element; and broken curve: element 1 of array for vertical polarization

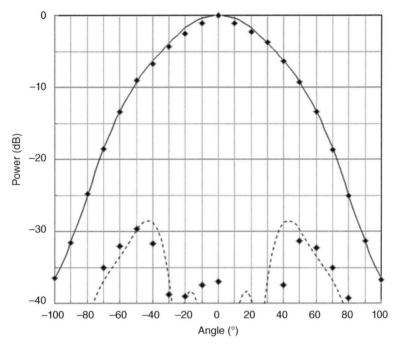

Figure 10.26 Radiation patterns in a 45° plane of element 1 of the feed array at a frequency of 1.375 GHz. Solid line: co-polar computed; dashed line: cross-polar computed; diamond: experiment

10.4 Problems

P10.1 A metal is spluttered onto a lossy dielectric. If the metal has a high conductivity that is close to aluminium, what thickness of metal coating is required to ensure the loss due to the dielectric is negligible at a frequency of 30 GHz.

P10.2 A planar near-field measurement obtains data sets v_0 and v_{90} at two probe orientations $0°$ and $90°$ for scans extending over $|x,y| \le 100\lambda$ with sample spacing of $\lambda/2$. The data obtained for the probe oriented at $\alpha = 0°$ is given by $jH(|x|-50\lambda)H(|y|-50\lambda)$, where $H()$ is the Heaviside step function. The data obtained for the probe at $90°$ is of the same functional form but with an amplitude 1/100 less. Assume the probe is a corrugated waveguide with a radius of 1λ and operating at the balanced hybrid condition.

    a. Transform the data to obtain the far-field functions.

    b. Determine the E- and H-plane patterns and also the peak cross-polarization level.

P10.3 If the measurement error is ε_i where $i = 1,...3$ for each of the three separate measurements in the three antenna method as described in Section 10.2.1 and given by Eq. 10.3, what is the total error for the method and the error in the gain of each antenna?

P10.4 Verify that Eq. 10.4 is an alternative representation of the spherical field is $\mathbf{E}(\theta,\phi) = \hat{\theta}E_\theta(\theta,\phi) + \hat{\phi}E_\phi(\theta,\phi)$.

P10.5 Design an aperture antenna for full earth coverage from a low-earth orbit (LEO) satellite, which is at an altitude of 800 km above the earth (earth radius 6378 km). Determine the required half-power beamwidth. Suggest options for aperture antennas to provide the specified half-power beamwidth (Hay et al., 1999).

P10.6 Simplify Eq. 10.8 for a probe with an axisymmetric pattern. Determine the formula for cross-polarization with this probe for an aperture antenna with a general pattern.

P10.7 Verify that Eq. 10.12 for an intermediate field measurement becomes a far-field measurement as the radius $r_b \to \infty$.

P10.8 Design a phased array feed consisting of a rectangular array of circular horns for a parabolic reflector with diameter $D = 55\lambda$ and $f/D = 0.38$. The antenna should be able to view a source at ± 2 HPBWs from boresight in any direction with minimum loss of gain. What is the size of the array and how many elements are required?

References

Bird, T.S. (1994): 'A multibeam feed for the Parkes radio-telescope', IEEE Antennas & Propagation Society Symposium, Seattle, WA, June 19–24, pp. 966–969.

Bird, T.S. (1997): 'Coaxial feed array for a short focal-length reflector', IEEE Antennas & Propagation Society Symposium, Montréal, Canada, July 13–18, pp. 1618–1621.

Bird, T.S. and Granet, C. (2013): 'Profiled horns and feeds', L. Shafai, S.K. Sharma and S. Rao (eds.) 'Vol. II: Feed systems' of 'Handbook of reflector antennas', Artech House, Norwood, MA, pp. 123–155.

Bird, T.S., LI, L. and Barker, S.J. (2003): 'Measurement of the Arecibo multibeam feed array', CSIRO Technical Report, No. TIPP 1920, Sydney, Australia.

Brigham, E.O. (1974): 'The fast Fourier transform', Prentice-Hall Inc., Eaglewood Cliffs, New Jersey.

Burnside, W.D., Gupta, I.J. and Lee, T.-H.. (2007): 'Indoor antenna measurements', In Volakis, J.L. (ed) 'Antenna engineering handbook', 4th ed., McGraw-Hill, New York.

Granet, C., Bird, T.S. and James, G.J. (2000): 'Compact multimode horn with low sidelobes for global earth coverage', IEEE Trans. Antennas Propag., Vol. **AP-48**, No. 7, pp. 1125–1133.

Hay, S.G., Barker, S.J., Granet, C., Forsyth, A.R., Bird, T.S., Sprey, M.A. and Greene, K.J. (2001): 'Multibeam earth station antenna for a European teleport application', IEEE Antennas & Propagation Society Symposium, Boston, MA, July 8–13, Vol. II, pp. 300–303.

Hayman, D.B., Bird, T.S. and James, G.C. (1998): 'Fresnel-zone measurement and analysis of a dual-polarized meteorological radar antenna', AMTA'98, Montréal, Canada, October 26–30, pp. 127–132.

Hay, S.G., Bateman, D.G., Bird, T.S. and Cooray, F.R. (1999): 'Simple Ka-band earth coverage antennas for LEO satellites', IEEE Antennas & Propagation Society Symposium, Orlando, FL, 11–16 July, pp. 708–711.

Hodges, D.J. and Greenwood, J.B. (1971): 'Optical distance measurement', Butterworths, London, UK.

Hollis, J.S., Lyon, T.J. and Clayton, L. (1970): 'Microwave antenna measurements', Scientific Atlanta, Atlanta, GA.

Keen, K.M. (1978): 'Interference-pattern intermediate-distance antenna measurement technique', IEE J. Microw, Optics Acous, Vol. **2**, No. 4, pp. 113–116.

Keenan, T., Glasson, K., Cummings, F., Bird, T.S., Keeler, J. and Lutz, J. (1998): 'The BMRC/NCAR C-band polarimetric (C-POL) radar system', J. Atmos. Oceanic Technol., **15**, pp. 871–886.

Kommineni, P.R., Hollandsworth, P.E. and JONES, J.W. (1988): 'A method of shaping an antenna panel', US Patent # 4,731,144, March 15.

Kraus, J.D. and Carver, K.R. (1973): 'Electromagnetics', 2nd ed., McGraw-Hill, International Student Edition, Kagakuska Ltd., Tokyo, Japan.

Leach, W.N. and Paris, D.T. (1973): 'Probe compensated near-field measurements on a cylinder', IEEE Trans. Antennas Propag., Vol. **AP-21**, No. 4, pp. 435–445.

Levy, R. (1996): 'Structural engineering for microwave antennas', IEEE Press, Piscataway, NJ.

Paris, D.T., Leach, W.M. and Joy, E.B. (1978): 'Basic theory of probe compensated near-field measurements', IEEE Trans. Antennas Propag., Vol. **AP-26**, No. 3, pp. 373–379.

Parsons, B.F. and Yabsley, D.E. (1985): 'The Australia telescope antennas: development of high-accuracy low-cost surface panels', Convention Digest, IREECON'85, Melbourne, Australia, October, pp. 716–719.

Poulton, G.T. (1983): 'Microwave holography for antenna measurements', Convention Digest, IREECON'83, Sydney, Australia, September 5–9, pp. 256–258.

Schofield, W. and Breach, M. (2007): 'Engineering surveying', 6th ed. Elsevier Ltd., Oxford, UK.

Staveley-Smith, L., Wilson, W.E., Bird, T.S., Disney, M.J., Ekers, R.D., Freeman, K.C., Haynes, R.F., Sinclair, M.W., Vaile, R.A., Webster, R.L. and Wright, A.E. (1996): 'The Parkes 21 cm multibeam receiver', Publ. Astron. Soc. Aust., Vol. **13**, pp. 243–248.

Weinstein, L.A. (1969): 'The theory of diffraction and the factorization method', The Golem Press, Boulder, CO.

Appendix A

Useful Identities

A.1 Vector Identities

Some useful vector identities and geometrical transformations are summarized below (Harrington, 1961; Hildebrand, 1962; Gradshteyn et al., 1994).

A vector quantity is defined in bold lettering, for example, $\mathbf{A}$. A unit vector is shown as a standard variable with a caret ('hat') symbol, for example, $\hat{a}$:

$$(\mathbf{A} + \mathbf{B}) \times \mathbf{C} = \mathbf{A} \times \mathbf{C} + \mathbf{B} \times \mathbf{C}$$

$$\mathbf{A} \cdot (\mathbf{B} \times \mathbf{C}) = \mathbf{B} \cdot (\mathbf{C} \times \mathbf{A}) - \mathbf{C} \cdot (\mathbf{A} \times \mathbf{B})$$

$$\mathbf{A} \times (\mathbf{B} \times \mathbf{C}) = (\mathbf{A} \cdot \mathbf{C}) \mathbf{B} - (\mathbf{A} \cdot \mathbf{B}) \mathbf{C}$$

$$\nabla \cdot (\mathbf{A} + \mathbf{B}) = \nabla \cdot \mathbf{A} + \nabla \cdot \mathbf{B}$$

$$\nabla \times (\mathbf{A} + \mathbf{B}) = \nabla \times \mathbf{A} + \nabla \times \mathbf{B}$$

$$\nabla \cdot (\phi \mathbf{A}) = \phi \nabla \cdot \mathbf{A} + \mathbf{A} \cdot \nabla \phi$$

$$\nabla \times (\phi \nabla \alpha) = \nabla \phi \times \nabla \alpha$$

$$\nabla \times \nabla \alpha = 0$$

$$\nabla \cdot (\mathbf{A} \times \mathbf{B}) = \mathbf{B} \cdot \nabla \times \mathbf{A} - \mathbf{A} \cdot \nabla \times \mathbf{B}$$

$$\nabla \cdot (\nabla \times \mathbf{A}) = 0$$

$$\nabla \times \nabla \times \mathbf{A} = \nabla (\nabla \cdot \mathbf{A}) - \nabla^2 \mathbf{A}.$$

Fundamentals of Aperture Antennas and Arrays: From Theory to Design, Fabrication and Testing,
First Edition. Trevor S. Bird.
© 2016 John Wiley & Sons, Ltd. Published 2016 by John Wiley & Sons, Ltd.
Companion website: www.wiley.com/go/bird448

Note that

$$\mathbf{A} \times \mathbf{B} = \begin{vmatrix} \hat{x} & \hat{y} & \hat{z} \\ A_x & A_y & A_z \\ B_x & B_y & B_z \end{vmatrix}$$

$$= \hat{x} \begin{vmatrix} A_y & A_z \\ B_y & B_z \end{vmatrix} - \hat{y} \begin{vmatrix} A_x & A_z \\ B_x & B_z \end{vmatrix} + \hat{z} \begin{vmatrix} A_x & A_y \\ B_x & B_y \end{vmatrix}$$

$$= \hat{x} \left(A_y B_z - A_z B_y \right) - \hat{y} \left(A_x B_z - A_z B_x \right) + \hat{z} \left(A_x B_y - A_y B_x \right),$$

where $|\mathbf{X}|$ is the determinant of matrix $\mathbf{X}$.

A.2 Geometric Identities

The three main co-ordinates systems used here are rectangular (x, y, z), cylindrical (ρ, ϕ, z) and spherical (r, θ, ϕ) co-ordinates as shown in Figure A.1. The relationship between the co-ordinates is

$$x = \rho \cos \phi = r \sin \theta \cos \phi$$
$$y = \rho \sin \phi = r \sin \theta \sin \phi$$
$$z = \rho \cos \theta$$
$$\rho = \sqrt{x^2 + y^2} = r \sin \theta$$
$$\phi = \tan^{-1} \left(\frac{y}{x} \right)$$
$$r = \sqrt{x^2 + y^2 + z^2} = \sqrt{\rho^2 + z^2}$$
$$\phi = \tan^{-1} \left(\frac{\sqrt{x^2 + y^2}}{z} \right) = \tan^{-1} \left(\frac{\rho}{z} \right).$$

Transformations between the co-ordinate vector components are given by the relationships

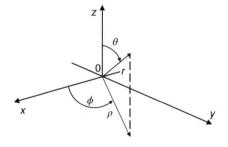

Figure A.1 Co-ordinate definitions

$$A_x = A_\rho \cos\phi - A_\phi \sin\phi$$
$$= A_r \sin\theta \cos\phi + A_\theta \cos\theta \cos\phi - A_\phi \sin\phi$$
$$A_y = A_\rho \sin\phi + A_\phi \cos\phi$$
$$= A_r \sin\theta \sin\phi + A_\theta \cos\theta \sin\phi + A_\phi \cos\phi$$
$$A_z = A_r \cos\theta - A_\theta \sin\theta$$
$$A_\rho = A_x \cos\phi + A_y \sin\phi$$
$$= A_r \sin\theta + A_\theta \cos\theta$$
$$A_\phi = -A_x \sin\phi + A_y \cos\phi$$
$$A_r = A_x \sin\theta \cos\phi + A_y \sin\theta \sin\phi + A_z \cos\theta$$
$$= A_\rho \sin\theta + A_z \cos\theta$$
$$A_\theta = A_x \cos\theta \cos\phi + A_y \cos\theta \sin\phi - A_z \sin\theta$$
$$= A_\rho \cos\theta - A_z \sin\theta.$$

The differential surface and volume elements are, respectively,

$$d\mathbf{S} = \hat{x}dydz + \hat{y}dxdz + \hat{z}dxdy$$
$$= \hat{\rho}\rho d\phi dz + \hat{\phi}d\rho dz + \hat{z}\hat{\rho}d\rho d\phi$$
$$= \hat{r}r^2 \sin\theta d\theta d\phi + \hat{\theta}r \sin\theta dr d\phi + \hat{\phi}\, r dr d\theta$$
$$dV = dxdydz = \rho d\rho d\phi dz = r^2 \sin\theta dr d\theta\, d\phi.$$

A.3 Transverse Representation of the Electromagnetic Field

In two-dimensional field problems and also in uniform cylindrical structures, it is convenient to use a transverse field representation. Thus, three-dimensional fields are represented as

$$\mathbf{A}(x,y,z) = \mathbf{A}_t(x,y,z) + \hat{z}A_z(x,y,z)$$
$$= \mathbf{A}_t(\rho,\phi,z) + \hat{z}A_z(\rho,\phi,z).$$

The gradient vector is similarly represented

$$\nabla = \nabla_t + \hat{z}\frac{\partial}{\partial z},$$

where

$$\nabla_t = \hat{x}\frac{\partial}{\partial x} + \hat{y}\frac{\partial}{\partial y}$$
$$= \hat{\rho}\frac{\partial}{\partial \rho} + \hat{\phi}\frac{1}{\rho}\frac{\partial}{\partial \phi}$$

and

$$\nabla_t^2 = \frac{\partial^2}{\partial x^2} + \frac{\partial^2}{\partial y^2}$$

$$= \frac{1}{\rho}\frac{\partial}{\partial \rho}\left(\rho\frac{\partial}{\partial \rho}\right) + \frac{1}{\rho^2}\frac{\partial^2}{\partial \phi^2}.$$

A.4   Useful Functions

Some functions that are used throughout are listed below:

$$H(u) = \begin{cases} 0; & u < 0 \\ 1; & \text{otherwise} \end{cases}$$

$$\text{Int}(x) = \text{Integer value of } x \leq x$$

$$C(x) = \frac{\cos(x)}{1 - (2x/\pi)^2} = \text{cosine function}$$

$$S(x) = \frac{\sin(x)}{x} = \text{sinc function}$$

A recursion formula for the Chebyshev polynomial of order n, $T_n(z)$, is as follows:

$$T_n(z) = 2zT_{n-1}(z) - T_{n-2}(z),$$

where $T_0(z) = 1$, $T_1(z) = z$, $T_2(z) = 2z^2 - 1$, etc. Also,

$$T_n(z) = \begin{cases} \cos[n\cos^{-1}(z)]; & |z| < 1 \\ \cosh[n\cosh^{-1}(z)]; & |z| > 1 \end{cases}.$$

References

Gradshteyn, I.S., Ryzhik, I.M. and Jeffrey, A. (1994): 'Table of integrals, series, and products', 5th ed., Academic Press Inc., London, UK.

Harrington, R.F. (1961): 'Time-harmonic electromagnetic fields', McGraw-Hill, New York.

Hildebrand, F.B. (1962): 'Advanced calculus for applications', Prentice Hall, Englewood Cliffs, NJ.

Appendix B

Bessel Functions

B.1 Properties

The Bessel equation of integer order n is given by

$$x\frac{d}{dx}\left(x\frac{d\psi}{dx}\right) + \left(x^2 - n^2\right)\psi = 0. \tag{B.1}$$

The solutions are expressed in several forms (McLachlan, 1934; Abramowitz & Stegun, 1965). Two possible solutions to this equation are called Bessel functions of first and second kind. These are $J_n(x)$ and $Y_n(x)$. A series expansion of the former type is

$$J_n(x) = \sum_{k=0}^{\infty} \frac{(-1)^k}{k!(n+k)!}\left(\frac{x}{2}\right)^{n+2k}.$$

For a negative integer order, the expansion shows that $J_{-n}(x) = (-1)^n J_n(x)$ and x real $J_n(-x) = -J_n(x)$. In addition,

$$Y_n(x) = \lim_{\nu \to n} \frac{J_\nu(x)\cos\nu\pi - J_{-\nu}(x)}{\sin\nu\pi}.$$

For small arguments,

$$J_n(x) \to \frac{1}{\Gamma(n+1)}\left(\frac{x}{2}\right)^n, \quad Y_n(x) \to -\frac{\Gamma(n)}{\pi}\left(\frac{2}{x}\right)^n,$$

where for the latter when $n = 0$, $Y_0(x) \to 2/\pi \ln(x)$.

Fundamentals of Aperture Antennas and Arrays: From Theory to Design, Fabrication and Testing,
First Edition. Trevor S. Bird.
© 2016 John Wiley & Sons, Ltd. Published 2016 by John Wiley & Sons, Ltd.
Companion website: www.wiley.com/go/bird448

For large arguments (i.e. $|z| \gg n$),

$$J_n(x) \sim \sqrt{\frac{2}{\pi x}} \cos\left(x - \frac{1}{2}n\pi - \frac{\pi}{4}\right)$$

$$Y_n(x) \sim \sqrt{\frac{2}{\pi x}} \sin\left(x - \frac{1}{2}n\pi - \frac{\pi}{4}\right).$$

On the other hand for large orders (i.e. $n \to \infty$),

$$J_n(x) \sim \frac{1}{\sqrt{2\pi n}} \left(\frac{xe}{2n}\right)^n$$

$$Y_n(x) \sim -\sqrt{\frac{2}{\pi n}} \left(\frac{xe}{2n}\right)^{-n}$$

and for positive values of n

$$J_n(n \sec h\alpha) \sim \frac{\exp(n \tan h\alpha - \alpha)}{\sqrt{2\pi n \tan h\alpha}}$$

and also

$$J'_n(n \sec h\alpha) \sim \sqrt{\frac{\sin h 2\alpha}{4\pi n}} \exp(n \tan h\alpha - \alpha).$$

The ratio of the derivative of J'_n and J_n for large orders can be expressed as

$$\frac{J'_n(z)}{J_n(z)} \sim \sqrt{\left(\frac{n}{z}\right)^2 - 1}.$$

Zeros of the ordinary Bessel function and its derivative can be approximated by the first two terms of McMahon's expansion. For n fixed and s large, an estimate for the zero of the Bessel function is

$$j_{n,s} \sim \beta - \frac{4n^2 - 1}{8\beta},$$

where $\beta = (n/2 + s - 1/4)\pi$, and for the derivative of the Bessel function,

$$j'_{n,s} \sim \beta' - \frac{4n^2 + 3}{8\beta'},$$

where $\beta' = (n/2 + s - 3/4)\pi$. These formulae give adequate first estimates to commence a search for more accurate values. The exception is $j'_{1,1} \sim 1.98$ compared with $j'_{1,1} = 1.84118$.

The Hankel functions of first and second kinds are defined by

$$H_n^{(1)}(x) = J_n(x) + jY_n(x)$$
$$H_n^{(2)}(x) = J_n(x) - jY_n(x),$$

and for large arguments the Hankel functions and their derivative are

$$H_n^{(1,2)}(x) \sim \sqrt{\frac{2}{\pi x}} \exp\left[\pm j\left(x - \frac{1}{2}n\pi - \frac{\pi}{4}\right)\right]$$

$$H_n^{(1,2)'}(x) \sim \pm j\sqrt{\frac{2}{\pi x}} \exp\left[\pm j\left(x - \frac{1}{2}n\pi - \frac{\pi}{4}\right)\right].$$

Approximations with greater degree of accuracy than given above are available in the references (e.g. Abramowitz & Stegun, 1965).

Recurrence relations for any of the Bessel functions, $Z_n(x)$, which stands for any one of the ordinary Bessel functions or Hankel functions, J_n, Y_n or $H_n^{(1,2)}$, are as follows:

$$Z_n(x) = \frac{x(Z_{n-1}(x) + Z_{n+1}(x))}{2n}$$

$$Z_n'(x) = \frac{Z_{n-1}(x) - Z_{n+1}(x)}{2}$$

$$= \frac{Z_{n-1}(x) - nZ_n(x)}{x} \tag{B.2}$$

$$= -\frac{Z_{n+1}(x) + nZ_n(x)}{x}.$$

A useful series expansion for $Y_0(x)$ is

$$Y_0(x) = \frac{2}{\pi}\left[\ln\left(\frac{x}{2}\right) + \gamma\right]J_0(x) - 2\sum_{k=1}^{\infty}(-1)^k\frac{J_{2k}(x)}{k},$$

where $\gamma = 0.5772156649053$ is Euler's number.

The modified Bessel functions of first and second kind, namely, $I_n(x)$ and $K_n(x)$, are solutions of the Bessel equation, Eq. B.1, with an imaginary argument. They can be expressed as

$$I_n(x) = \exp\left(\frac{-jn\pi}{2}\right)J_n(jx); \quad -\pi < \arg(x) \le \frac{\pi}{2}$$

$$= \exp\left(\frac{j3n\pi}{2}\right)J_n(-jx); \quad \frac{\pi}{2} < \arg(x) \le \pi$$

$$K_n(x) = \left(\frac{j\pi}{2}\right)\exp\left(\frac{jn\pi}{2}\right)H_n^{(1)}(jx); \quad -\pi < \arg(x) \le \frac{\pi}{2}$$

$$= -\left(\frac{j\pi}{2}\right)\exp\left(\frac{-jn\pi}{2}\right)H_n^{(2)}(jx); \quad \frac{\pi}{2} < \arg(x) \le \pi$$

with

$$K_0(x) = -\left[\ln\left(\frac{x}{2}\right) + \gamma\right] I_0(x) + 2\sum_{k=1}^{\infty} \frac{J_{2k}(x)}{k}.$$

For large arguments $K_n(x)$ and its derivative are

$$K_n(x) \sim \sqrt{\frac{2}{\pi x}} \exp(-x)$$

$$K_n'(x) \sim -\sqrt{\frac{2}{\pi x}} \exp(-x)$$

Recurrence relation Eq. B.2 applies also for $I_n(x)$ or $K_n(x)$ with the following changes in the expressions $Z_n(x) \Rightarrow Z_n(x)$, $Z_{n+1}(x) \Rightarrow (-1)Z_{n+1}(x)$ and $Z_{n-1}(x) \Rightarrow Z_{n-1}(x)$.
Some Wronskian relations for the Bessel functions are

$$J_{n+1}(x)Y_n(x) - J_n(x)Y_{n+1}(x) = \frac{2\pi}{x}$$

$$J_n'(x)Y_n(x) - J_n(x)Y_n'(x) = \frac{2\pi}{x}$$

$$H_{n+1}^{(1)}(x)H_n^{(2)}(x) - H_n^{(1)}(x)H_{n+1}^{(2)}(x) = \frac{-4j}{\pi x}$$

$$H_n^{(1)'}(x)H_n^{(2)}(x) - H_n^{(1)}(x)H_n^{(2)'}(x) = \frac{4j}{\pi x}$$

$$J_n'(x)H_n^{(2)}(x) - J_n(x)H_n^{(2)'}(x) = \frac{2j}{\pi x}.$$

In coaxial structures it is convenient to adopt compound Bessel functions of the form

$$Z_n(x,y) = J_n(x) - \frac{J_n'(y)}{Y_n'(y)} Y_n(x)$$

$$\Lambda_n(x,y) = J_n(x) - \frac{J_n(y)}{Y_n(y)} Y_n(x)$$

$$Z_n'(x,y) = J_n'(x) - \frac{J_n'(y)}{Y_n'(y)} Y_n'(x)$$

$$\Lambda_n'(x,y) = J_n'(x) - \frac{J_n(y)}{Y_n(y)} Y_n'(x)$$

$$U_n(x,y) = J_n(x) - \frac{J_n(y)}{H_n^{(2)}(y)} H_n^{(2)}(x)$$

$$V_n(x,y) = J_n(x) - \frac{J_n'(y)}{H_n^{(2)'}(y)} H_n^{(2)}(x),$$

where the derivative indicated by $'$ refers to the derivative with respect to the first argument only.

Useful series of Bessel functions are

$$\exp(jx\cos\theta) = J_0(x) + 2\sum_{k=1}^{\infty} j^k J_k(x)\cos(k\theta)$$

$$\exp(jx\sin\theta) = \sum_{k=-\infty}^{\infty} \exp(jk\theta)J_k(x)$$

$$1 = J_0(x) + 2\sum_{k=1}^{\infty} J_{2k}(x)$$

$$\exp(z) = I_0(x) + 2\sum_{k=1}^{\infty} I_{2k}(x)$$

$$1 = I_0(x) + 2\sum_{k=1}^{\infty} (-1)^k I_{2k}(x)$$

$$J_p(az)J_q(z) = \frac{(\tfrac{1}{2}az)^p(\tfrac{1}{2}bz)^q}{\Gamma(q+1)} \sum_{r=0}^{\infty} (-1)^r \frac{{}_2F_1\left(-r,-p-r;q+1;(b/a)^2\right)(az/2)^r}{r!\Gamma(p+q+1)}$$

${}_2F_1(s,t;u;z)$ is the hypergeometric function (Watson, 1962; Abramowitz & Stegun, 1965) given by

$$_2F_1(\alpha,\beta;\gamma;z) = \frac{\Gamma(\gamma)}{\Gamma(\alpha)\Gamma(\beta)} \sum_{n=0}^{\infty} \frac{\Gamma(\alpha+n)\Gamma(\beta+n)}{\Gamma(\gamma+n)} \frac{z^n}{n!}.$$

Some useful integrals involving Bessel functions in closed form are (McLachlan, 1934; Luke, 1962; Gradshteyn et al., 1994)

$$\int_0^{2\pi} \left\{\begin{matrix}\cos\\\sin\end{matrix}p\phi\right\} e^{jz\cos(\phi-\phi')}d\phi' = 2\pi j^p J_p(z)\left\{\begin{matrix}\cos\\\sin\end{matrix}p\phi\right\}. \tag{B.3}$$

$$\int^z z^p J_{p-1}(z)dz = z^p J_p(z)$$

$$\int_0^a [J_p(\alpha z)]^2 z\,dz = \frac{a^2}{2}\left\{[J_p(\alpha a)]^2 - J_{p-1}(\alpha a)J_{p+1}(\alpha a)\right\} \tag{B.4}$$

$$\int_0^a J_p(\alpha z)J_p(\beta z)z\,dz = \frac{a}{\alpha^2-\beta^2}\left(\alpha J_{p+1}(\alpha a)J_p(\beta a) - \beta J_p(\alpha a)J_{p+1}(\beta a)\right)$$

$$= \frac{a}{\alpha^2-\beta^2}\left(\beta J_p(\alpha a)J_{p-1}(\beta a) - \alpha J_{p-1}(\alpha a)J_p(\beta a)\right) \tag{B.5}$$

$$\int_0^a J_p(\alpha z)J_p(\beta z)\exp\left(-j\gamma z^2\right)z\,dz = \left(\frac{\alpha\beta}{4}a^2\right)^p \frac{a^2}{2\Gamma(p+1)}\sum_{m=0}^{\infty}(-1)^m\left(\frac{\alpha}{2}a\right)^{2m}\times$$

$$\frac{{}_2F_1\left(-m,-m-p;p+1;(\beta/\alpha)^2\right)}{m!\Gamma(p+m+1)}\sum_{\nu=0}^{\infty}\frac{(-j\gamma a^2)}{\nu!(p+m+\nu+1)}$$

(B.6)

$$\int_0^a dw\,w^{\alpha-1}\left(1-w^2\right)^{\beta-1}\frac{J_p(Aw)J_q(Bw)}{\left(w^2-u_m^2\right)^{\lambda}\left(w^2-u_n^2\right)^{\tau}}J_\nu(Cw) =$$

$$\frac{\Gamma(\beta)}{2}\Gamma\left(\frac{A}{2}\right)^p\Gamma\left(\frac{B}{2}\right)^q\sum_{\sigma=0}^{\infty}(-1)^\sigma Q_\sigma(p+\sigma,q+1,A,B)\left\{\sum_{\mu=0}^{\infty}\frac{(-1)^\mu}{\mu!}\left(\frac{C}{2}\right)^{2\mu}\right.$$

$$\times\left[\Gamma\left(\frac{C}{2}\right)^\nu\frac{\Gamma(\chi)}{\Gamma(\nu+\mu+1)\Gamma(\beta+\chi)}g_\nu + j\,\mathrm{sgn}(1-\beta)\Gamma\left(\frac{C}{2}\right)^{2(\lambda-\beta-\sigma+1)-(p+q+\alpha)}\frac{\Gamma(\theta)}{\Gamma(\nu-\theta+1)\Gamma(\beta-\mu)}h_\nu\right]$$

(B.7)

where $A,B<C$, $u_m=x_m/A$ and $u_n=x_n/B$ in which $J_p(x_m)=0$,

$$Q_m(s,t,x,y) = \frac{((x/2))^{2m}{}_2F_1\left(-m,-s;t;(y/x)^2\right)}{m!\Gamma(s+1)\Gamma(t)},$$

$\chi=(p+q+\nu+\alpha)/2+(\sigma+\mu)$, $\theta=(p+q+\nu+\alpha)/2+(\sigma-\mu+\beta-\lambda-1)$ and

$$g_\nu = \begin{cases} 1;\ \lambda=0=\tau \\ \frac{-1}{(u_m^2-u_n^2)}\left[u_{m1}^{-2}F_2\left(1,\chi;\beta+\chi;u_m^{-2}\right)-u_{n1}^{-2}F_2\left(1,\chi;\beta+\chi;u_n^{-2}\right)\right];\ \lambda=1=\tau \\ u_{m1}^{-2\lambda}F_2\left(\lambda,\chi;\beta+\chi;u_m^{-2}\right);\ \lambda>0,\tau=0 \end{cases}$$

$$h_\nu = \begin{cases} 1;\ \lambda=0=\tau \\ \frac{1}{(u_m^2-u_n^2)}\left[{}_1F_2\left(1,-\mu;\beta-\mu;u_m^2\right)-{}_1F_2\left(1,-\mu;\beta-\mu;u_n^2\right)\right];\ \lambda=1=\tau \\ {}_1F_2\left(\lambda,-\mu;\beta-\mu;u_m^2\right);\ \lambda>0,\tau=0 \end{cases}.$$

In the functions g_ν and h_ν above, when $\lambda=0$ and $\tau>0$, the roles of λ and τ reverse as are those of u_m u_n.

B.2 Computation of Bessel Functions

Several applications call for a sequence of Bessel functions with the same argument z but of different orders. An accurate method of generating the desired sequence of functions up to order N is to commence with $M>N$ (typically $M=N+10$) such that

$$j_M(z)=0 \quad \text{and} \quad j_{M-1}(z)=\varepsilon \sim 10^{-8},$$

where j_k are unscaled and commencing members of the sequence and z may be complex. Then use the following recurrence relation to compute other members down to $k=0$. Thus,

$$j_k(z)=\frac{2(k+1)}{z}j_{k+1}-j_{k+2}; k=M-2, M-1, \ldots, 1, 0.$$

At the same time, compute a normalization constant C from

$$C=2\sum_{k=1}^{M}j_{2k}(z)+j_0(z).$$

Also, if the Y Bessel functions are needed at the same time, compute

$$\Sigma=\sum_{k=1}^{M}\frac{(-1)^k}{2k}j_{2k}(z).$$

The ordinary Bessel functions are given by

$$J_k(z)=\frac{j_0(z)}{C}$$

and

$$Y_0(z)=\frac{2}{\pi}\left[\ln\left(\frac{z}{2}\right)+\gamma\right]J_0(z)--\frac{4\Sigma(z)}{C}.$$

Afterwards use the first Wronskian listed above to compute $Y_1(z)$. Thus,

$$Y_1(z)=\frac{J_1(z)Y_0(z)-2\pi/z}{J_n(z)}.$$

An ascending recurrence relation is used to compute the remaining members of the sequence

$$Y_{k+1}(z)=\frac{2k}{z}Y_k(z)-Y_{k-1}(z); \quad k=1,2,\ldots,N.$$

The modified Bessel functions are computed in the same way except that the relevant recurrence relations and summation formulae should be used.

References

Abramowitz, M. and Stegun, I.A. (1965): 'Handbook of mathematical functions', Dover Inc., New York.

Gradshteyn, I.S., Ryzhik, I.M. and Jeffrey, A. (1994): 'Table of integrals, series, and products', 5th ed., Academic Press Inc., London, UK.

Luke, Y.L. (1962): 'Integrals of Bessel functions', McGraw-Hill, New York.

Mclachlan, N.W. (1934): 'Bessel functions for engineers', Oxford University Press, London, UK.

Watson, G.N. (1962): 'A treatise on the theory of Bessel functions', 2nd Edition, Cambridge University Press, Cambridge, UK.

Appendix C

Proof of Stationary Behaviour of Mutual Impedance

A reason for relatively accurate results being possible with simple current approximations is that mutual impedance expression, Eq. 7.27, is stationary with respect to small variations in currents on the antennas. To prove this, substitute Eq. 7.32 in Eq. 7.27 and consider deviations in $\mathbf{J}_1$ and $\mathbf{J}_2$ expressed as $\boldsymbol{\delta}\mathbf{J}_1$ and $\boldsymbol{\delta}\mathbf{J}_2$. If the errors are small, the change in Z_{21} (given by δZ_{21}) is, to first order, given by

$$\delta Z_{21} I_1 I_2 = -Z_{21}(I_1 \delta I_2 + \delta I_1 I_2) - \iint_{S_2} dS \iint_{S_1} dS' \boldsymbol{\delta}\mathbf{J}_2 \cdot \mathbf{G}^{(e)}(\mathbf{R}|\mathbf{R}')\mathbf{J}_1$$

$$- \iint_{S_2} dS \iint_{S_1} dS' \mathbf{J}_2 \cdot \mathbf{G}^{(e)}(\mathbf{R}|\mathbf{R}')\boldsymbol{\delta}\mathbf{J}_1.$$

(C.1)

To interpret the terms on the right side of Eq. C.1, consider the scalar product of $\mathbf{E}_{21}$ and $\boldsymbol{\delta}\mathbf{J}_2$ and integrate over the surface of antenna 2 to give

$$\iint_{S_2} dS \boldsymbol{\delta}\mathbf{J}_2 \cdot \mathbf{E}_{21} = -\iint_{S_2} dS \iint_{S_1} dS' \boldsymbol{\delta}\mathbf{J}_2 \cdot \underset{=}{\mathbf{G}}^{(e)} \cdot \mathbf{J}_1$$

$$= V_{21}\delta I_2 = Z_{21} I_1 \delta I_2.$$

(C.2)

Fundamentals of Aperture Antennas and Arrays: From Theory to Design, Fabrication and Testing,
First Edition. Trevor S. Bird.
© 2016 John Wiley & Sons, Ltd. Published 2016 by John Wiley & Sons, Ltd.
Companion website: www.wiley.com/go/bird448

Similarly, the integral of the scalar product of $\mathbf{E}_{12}$ and $\delta\mathbf{J}_1$ on S_1 gives

$$\iint_{S_2} dS\,\delta\mathbf{J}_1\cdot\mathbf{E}_{21} = -\iint_{S_2} dS \iint_{S_1} dS'\,\delta\mathbf{J}_1\cdot\underset{=}{\mathbf{G}}^{(e)}\cdot\mathbf{J}_2$$
$$= V_{12}\,\delta I_1 = Z_{12}\,I_2\,\delta I_1.$$

(C.3)

Under conditions where the dyadic Green's function is symmetric and the dyadic products are symmetric, then from Eqs. C.2 and C.3,

$$\iint_{S_1} dS' \iint_{S_2} dS'\,\delta\mathbf{J}_1\cdot\mathbf{G}^{(e)}(\mathbf{R}'|\mathbf{R})\cdot\mathbf{J}_2 = \iint_{S_2} dS \iint_{S_1} dS'\,\mathbf{J}_2\cdot\underset{=}{\mathbf{G}}^{(e)}(\mathbf{R}|\mathbf{R}')\cdot\delta\mathbf{J}_1.$$

(C.4)

Using Eqs. C.2 to C.4 and also the reciprocity property that $Z_{12} = Z_{21}$, the terms on the right-hand side of Eq. C.1 cancel, giving $\delta Z_{21} = 0$. Thus the mutual impedance Eq. C.4 is stationary and is accurate to second order for an assumed current distribution that may be correct to only first order. This conclusion applies also for the special case of the self-impedance, corresponding to when antennas 1 and 2 are coincident, and also for the dual quantity mutual admittance given by Eq. 7.31.

Appendix D

Free-Space Dyadic Magnetic Green's Function

The magnetic Green's function dyadic is derived from the wave equation for a magnetic dipole in free-space (Felsen & Marcuvitz, 1973).

Commencing with the rotational Maxwell equations, the vector wave equation is formed for the magnetic field in the absence of electric currents

$$\nabla \times \nabla \times \mathbf{H}(\mathbf{r}) - k^2 \mathbf{H}(\mathbf{r}) = j\omega\varepsilon \mathbf{M}(\mathbf{r}),$$

where $\mathbf{M}(\mathbf{r})$ is the magnetic current source. Now make use of the curl-curl vector identity in Appendix A.1. With a combination of Gauss' law and the continuity equation for magnetic sources, it is found that

$$\nabla^2 \mathbf{H}(\mathbf{r}) + k^2 \mathbf{H}(\mathbf{r}) = -j\omega\varepsilon \left(\mathbf{I} + \frac{\nabla \cdot \nabla}{k^2} \right) \cdot \mathbf{M}(\mathbf{r}),$$

where $\mathbf{I}$ is the unit dyadic. If $\mathbf{M}(\mathbf{r})$ lies on a planar surface S, then a solution to the above equation is

$$\mathbf{H}(\mathbf{r}) = -j\omega\varepsilon \iint_{S'} dS' \mathbf{M}(\mathbf{r}') \cdot \left(\mathbf{I} + \frac{\nabla_t \cdot \nabla_t}{k^2} \right) G_o(|\mathbf{r} - \mathbf{r}'|)$$

$$= -j\omega\varepsilon \iint_{S'} dS' \mathbf{M}(\mathbf{r}') \cdot \underline{\underline{\mathbf{G}}}^{(h)}(\mathbf{r}, .\mathbf{r}'),$$

Fundamentals of Aperture Antennas and Arrays: From Theory to Design, Fabrication and Testing,
First Edition. Trevor S. Bird.
© 2016 John Wiley & Sons, Ltd. Published 2016 by John Wiley & Sons, Ltd.
Companion website: www.wiley.com/go/bird448

where ∇_t is the gradient operator in the transverse plane. The two-dimensional Greens dyadic is given by

$$\underset{=}{\mathbf{G}}^{(h)}(\mathbf{r}, .\mathbf{r}') = \frac{-jk}{2\pi\eta_\mathrm{o}}\left(\mathbf{I} + \frac{\nabla_t \cdot \nabla_t}{k^2}\right)G_\mathrm{o}(|\mathbf{r} - \mathbf{r}'|),$$

where $G_\mathrm{o}(R) = \exp(-jkR)/R$ is the scalar free-space Greens function, and

$$|\mathbf{r} - \mathbf{r}'| = \sqrt{(x - x')^2 + (y - y')^2 + (z - z')^2}.$$

Reference

Felsen, L.B. and Marcuvitz, N. (1973): 'Radiation and scattering of waves', Prentice-Hall, Englewood Cliffs, NJ.

Appendix E

Complex Fresnel Integrals

Several forms of Fresnel integrals have been defined by various authors, and as a result, the definition often depends on the application. In this text, a complex Fresnel integral is defined as

$$\mathcal{K}(z) = \int_0^z \exp\left(-j\xi^2\right) d\xi, \tag{E.1}$$

where z is generally complex also. Note that

$$-\mathcal{K}(z) = \mathcal{K}(-z) \text{ and } \mathcal{K}(\infty) = \frac{\sqrt{\pi}}{2} \exp\left(-j\left(\frac{\pi}{4}\right)\right).$$

The complex Fresnel integral is related to the cosine and integral functions, which are defined here as (Abramowitz & Stegun, 1965, p. 300)

$$Ci(z) = \int_0^z \cos\left(\frac{\pi}{2}t^2\right) dt \text{ and } Si(z) = \int_0^z \sin\left(\frac{\pi}{2}t^2\right) dt.$$

Thus, $Ci(z) = (\sqrt{\pi}/2)\operatorname{Re}\left\{\mathcal{K}\left(\sqrt{\pi/2}z\right)\right\}$ and $Si(z) = (\sqrt{\pi}/2)\operatorname{\Im}m\left\{\mathcal{K}\left(\sqrt{\pi/2}z\right)\right\}$ or $\mathcal{K}(z) = \sqrt{\pi/2}\left[Ci\left(\sqrt{(2/\pi)}z\right) + jSi\left(\sqrt{(2/\pi)}z\right)\right]$.

Fundamentals of Aperture Antennas and Arrays: From Theory to Design, Fabrication and Testing,
First Edition. Trevor S. Bird.
© 2016 John Wiley & Sons, Ltd. Published 2016 by John Wiley & Sons, Ltd.
Companion website: www.wiley.com/go/bird448

Series expansions for the cosine and sine integrals are given by

$$\mathrm{Ci}(z) = \sum_{k=0}^{\infty} \frac{(-1)^k (\pi/2)^{2k}}{(2k)!(4k+1)} z^{4k+1} \tag{E.2a}$$

$$\mathrm{Si}(z) = \sum_{k=0}^{\infty} \frac{(-1)^k (\pi/2)^{2k+1}}{(2k+1)!(4k+3)} z^{4k+3}. \tag{E.2b}$$

These series are quite accurate for moderate values of the argument (typically $|z| < 3$). As well, there are accurate Padé polynomial approximations to these integrals that can be used to compute the complex Fresnel integral (refer to Luke, 1969). For small arguments, $C(z) \approx z$ and $S(z) \approx \pi z^3/6$ and, therefore,

$$\mathcal{K}(z) \approx z\left(1 - j\frac{z^2}{3}\right). \tag{E.3}$$

Another variant is the one-sided Fresnel integral that has an infinite upper limit arises in some diffraction problems. It is defined by the integral

$$F_{\pm}(x) = \int_{x}^{\infty} \exp\left(\pm j\xi^2\right) d\xi, \tag{E.4}$$

where x is real. Thus,

$$F_{\pm}(x) = \left(\int_{0}^{\infty} - \int_{0}^{x}\right) \exp\left(\pm j\xi^2\right)$$

$$= \frac{\sqrt{\pi}}{2} \exp\left(\pm j\frac{\pi}{4}\right) - \begin{cases} \mathcal{K}(z); + \\ \mathcal{K}(z)^*; - \end{cases}$$

By a similar argument, when x is negative,

$$F_{\pm}(-x) = \left(\int_{-\infty}^{\infty} - \int_{|x|}^{\infty}\right) \exp\left(\pm j\xi^2\right)$$

$$= \sqrt{\pi} \exp\left(\pm j\frac{\pi}{4}\right) - F_{\pm}(|x|) \tag{E.5}$$

For large arguments, the Fresnel integral is approximately

$$F_{\pm}(x) \approx \frac{1}{2x} \exp\left(\pm j\left(x^2 + \frac{\pi}{2}\right)\right)\left(1 \pm j\frac{1}{2x^2}\right).$$

A large argument approximation to the complex Fresnel integral can be obtained from this relationship:

$$\mathcal{K}(z) = \mathcal{K}(\infty) - F_{-}(z)$$

$$\approx \frac{\sqrt{\pi}}{2}\exp\left(-j\left(\frac{\pi}{4}\right)\right) - \frac{1}{2z}\exp\left(-j\left(z^2 + \frac{\pi}{2}\right)\right)\left(1 - j\frac{1}{2z^2}\right).$$

Another function used in diffraction theory is the modified Fresnel integral (James, 1986) which is defined as

$$M_{\pm}(x) = \frac{\exp(\mp j(x^2 + \pi/4))}{\sqrt{\pi}}\int_{x}^{\infty}\exp\left(\pm jxt^2\right)dt$$

$$= \frac{\exp(\mp j(x^2 + \pi/4))}{\sqrt{\pi}}F_{\pm}(x).$$

(E.6)

Special cases of the modified Fresnel integral are

$$M_{\pm}(-x) = \exp(\mp jx^2) - M_{\pm}(x),$$

$$M_{\pm}(0) = \frac{1}{2},$$

and

$$M_{\pm}(x)\big|_{x\to\infty} \sim \frac{\exp(\pm j\pi/4)}{2x\sqrt{\pi}}.$$

The modified Fresnel integral can be evaluated numerically by means of the results given above or approximated quite accurately by the formula (James, 1979), which is often sufficiently accurate for many applications:

$$M_{\pm}(x) \approx \frac{1}{2}\frac{\exp(\pm j(\tan^{-1}(x^2 + 1.594x + 1) - \pi/4))}{\sqrt{\pi x^2 + 0.54\sqrt{x} + \exp(-0.2x^4)}}.$$

(E.7)

References

Abramowitz, M. and Stegun, I.A. (1965): 'Handbook of mathematical functions', Dover Inc., New York.

James, G.L. (1979): 'An approximation to the Fresnel integral', IEEE Proc., Vol. **67**, No. 4, pp. 677–678.

James, G.L. (1986): 'Geometrical theory of diffraction for electromagnetic waves', 3rd ed., Peter Peregrinus Ltd., London, UK.

Luke, Y.L. (1969): 'The special functions and their approximations', Vol. **2**, Academic Press, New York, pp. 422–435.

Appendix F

Properties of Hankel Transform Functions

Closed-form solutions to the integral of products of Bessel functions have been described by several authors (Nicholson, 1920; de Hoop, 1955; Luke, 1962; Watson, 1962). In coupling computations involving circular apertures (Bird, 1996b), products of up to three Bessel functions can occur in the mutual admittance expressions. A general Hankel transform involving triple products of Bessel functions is defined by

$$
C_{p,q,\nu}^{(1)}(\alpha,\beta;u_m,u_n;s) = \int_0^\infty dw \frac{w^3}{\sqrt{1-w^2}} \frac{\left[J_p(ka_iw)-\alpha J_p(kb_iw)\right]}{\left(w^2-u_m^2\right)}
$$
$$
\times \frac{\left[J_q(ka_jw)-\beta J_q(kb_jw)\right]}{\left(w^2-u_n^2\right)} J_\nu(ksw),
\tag{F.1}
$$

and a second type involving derivatives of the two compound Bessel functions is defined by

$$
C_{p,q,\nu}^{(2)}(\alpha,\beta;u_m,u_n;s) = \int_0^\infty dw \; w\sqrt{1-w^2} \frac{\left[J_p'(ka_iw)-\alpha J_p'(kb_iw)\right]}{\left(w^2-u_m^2\right)}
$$
$$
\times \frac{\left[J_q'(ka_jw)-\beta J_q'(kb_jw)\right]}{\left(w^2-u_n^2\right)} J_\nu(ksw)
.
\tag{F.2}
$$

Fundamentals of Aperture Antennas and Arrays: From Theory to Design, Fabrication and Testing,
First Edition. Trevor S. Bird.
© 2016 John Wiley & Sons, Ltd. Published 2016 by John Wiley & Sons, Ltd.
Companion website: www.wiley.com/go/bird448

Both transforms can be expressed in terms of the following general integral:

$$E_{p,q,\nu}(\mu,\sigma,\lambda,\tau;x,y,s) = \int_0^\infty dw \; w^\mu (1-w^2)^{\sigma-1} \frac{J_p(kxw)}{(w^2-u_m^2)^\lambda} \frac{J_q(kyw)}{(w^2-u_n^2)^\tau} J_\nu(ksw), \qquad (F.3)$$

where μ, σ, λ and τ are integers, and kx, ky and ks are normalized lengths. Also u_m and u_n are the m-th and n-th zeros of J_p and J_q. Expanding $C_{p,q,\nu}^{(1)}$ in terms of Eq. F.3, it is found that

$$C_{p,q,\nu}^{(1)}(\alpha,\beta;u_m,u_n;s) = E_{p,q,\nu}\left(3,\frac{1}{2},1,1;a_i,a_j,s\right) - \alpha E_{p,q,\nu}\left(3,\frac{1}{2},1,1;b_i,a_j,s\right)$$
$$- \beta E_{p,q,\nu}\left(3,\frac{1}{2},1,1;a_i,b_j,s\right) + \alpha\beta E_{p,q,\nu}\left(3,\frac{1}{2},1,1;b_i,b_j,s\right). \qquad (F.4)$$

In addition, since $J_p'(z) = -J_{p+1}(z) + (p/z)J_p(z)$, $C_{p,q,\nu}^{(2)}(\alpha,\beta;u_m,u_n;s)$ can be expanded similarly into 16 terms. In the special case when $\alpha = \beta = 0$, Eq. F.2 becomes

$$C_{p,q,\nu}^{(2)}(0,0;u_m,u_n;s) = E_{p+1,q+1,\nu}\left(2,\frac{3}{2},1,1;a_i,a_j,s\right) - \left(\frac{q}{ka_j}\right) E_{p+1,q,\nu}\left(1,\frac{3}{2},1,1;a_i,a_j,s\right)$$
$$- \left(\frac{p}{ka_i}\right) E_{p,q+1,\nu}\left(1,\frac{3}{2},1,1;a_i,a_j,s\right) + \left(\frac{pq}{ka_i ka_j}\right) E_{p,q,\nu}\left(0,\frac{3}{2},1,1;a_i,a_j,s\right). \qquad (F.5)$$

The real and imaginary parts of the integral $E_{p,q,\nu}(\mu,\sigma,\lambda,\tau;x,y,s)$ can each be expanded as two double series of simple integrals (Watson, 1962). It can be shown that various infinite integrals involving Bessel products can be simplified by substituting series for the Bessel functions under the integral sign and expressing the result in terms of other functions such as the hypergeometric function. In other cases, these series can be expressed in terms of hypergeometric functions (see Appendix B.1). A closed-form solution is available for all integers p, q and ν for some cases such as when there are no poles (i.e. $\lambda = 0 = \tau$) (Bird, 1996a). In another case, a closed-form solution to Eq. F.3 in the form of series is available under limited conditions (Bird, 1996b), namely, for the real part of $E_{p,q,\nu}(\mu,\sigma,\lambda,\tau;x,y,s)$, when $u_{m,n} > 1$, and also for the imaginary part, when $u_{m,n} < 1$. In the case of mode coupling contributions to the self-admittance (when s = 0), the integral transform function becomes $E_{p,q,0}(\mu,\sigma,\lambda,\tau;x,y,0)$, which now involves only a double product of Bessel functions as can be seen from Eq. F.3.

References

Bird, T.S. (1996a): 'Cross-coupling between open-ended coaxial radiators', IEE Proc., Microwaves, Antennas Propag., Vol. **143**, No. 4, pp. 265–271.

Bird, T.S. (1996b): 'Improved solution for mode coupling in different-sized circular apertures and its application', IEE Proc., Microwaves, Antennas Propag., Vol. **143**, No. 6, pp. 457–464.

De Hoop, A.T. (1955): 'On integrals occurring in the variational formulation of diffraction problems', Proc. Konink. Nederl. Akad. Wetensch., Series B, Vol. **58**, pp. 325–330.

Luke, Y.L. (1962): 'Integrals of Bessel functions', McGraw-Hill, New York.

Nicholson, J.W. (1920): 'Generalization of a theorem due to Sonine', Q. J. Pure Appl. Math., Vol. **48**, pp. 321–329.

Watson, G.N. (1962): 'A treatise on the theory of Bessel functions', 2nd ed., Cambridge University Press, Cambridge, UK.

Appendix G

Properties of Fock Functions for Convex Surfaces

The Fock functions occur in the asymptotic representation of the circular cylinder and are also used in the description of general convex surfaces (Bowman et al., 1963; Fock, 1965). These functions are integrals in the complex τ-plane of the Airy integral $w_2(\tau)$ or its derivative $w_2'(\tau)$. The Airy integral is defined by

$$w_2(\alpha) = \frac{1}{\sqrt{\pi}} \int_{\Gamma_2} d\tau \exp\left(\alpha\tau - \frac{\tau^3}{3}\right). \tag{G.1}$$

where the contour Γ_2 in the τ-plane is shown in Figure G.1.

G.1 Surface Fock Functions

A surface Fock function of the n-th kind, order m, is defined by

$$L_m^{(n)}(z) = \int_{\Gamma_1} d\tau \ \tau^n \left(\frac{w_2'(\tau)}{w_2(\tau)}\right)^m \exp(-jz\tau), \tag{G.1}$$

where m, n are integers and the contour Γ_1 in the complex plane is shown in Figure G.1. Two families of functions have been defined. These are known as soft and hard surface functions, depending on $m > 0$ or $m < 0$, respectively.

Fundamentals of Aperture Antennas and Arrays: From Theory to Design, Fabrication and Testing,
First Edition. Trevor S. Bird.
© 2016 John Wiley & Sons, Ltd. Published 2016 by John Wiley & Sons, Ltd.
Companion website: www.wiley.com/go/bird448

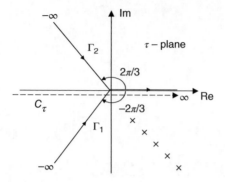

Figure G.1   Contours in the complex τ-plane. Crosses indicate zeros of w_2 or w_2' depending on the integrand of the soft or hard Fock functions

G.1.1   Soft Surface Functions (m > 0)

For convenience these are defined as

$$u_n^{(m)}(z) = \frac{\exp(j(n+3/2)\pi/2)z^{n+3/2}}{2\Gamma(n+3/2)} L_m^{(n)}(z); \quad m,n = 0,1,\dots \tag{G.2}$$

$$u_n(z) \equiv u_n^{(0)}(z) \quad \text{and} \quad u(z) \equiv u_0(z) \equiv u_0^{(0)}(z).$$

The first derivative is given by

$$u_n^{(m)'}(z) = (n+3/2)\left[u_n^{(m)}(z) - u_{n+1}^{(m)}(z)\right]\Big/z.$$

A residue series that is sufficiently accurate for $z > 1$ is found to be

$$u_n(z) = \frac{\pi \exp(j(n+3/2)\pi/2)z^{n+3/2}}{2\Gamma(n+3/2)} \sum_{p=1}^{\infty} (t_p)^n \exp(-jzt_p), \tag{G.3}$$

where t_p ($p = 1,2,\dots,\infty$) are the zeros of w_2 and are related to the zeros α_p of the Airy function $Ai(t)$ (Abramowitz & Stegun, 1965) as follows:

$$t_p = |\alpha_p|\exp\left(-\frac{j\pi}{3}\right),$$

where α_p are available in tables (Abramowitz and Stegun, 1965) for $p = 1$ to10.
For $p > 10$ a highly accurate approximation to $|\alpha_p|$ is

$$|\alpha_p| \approx \chi^{2/3}\left[1 + 5/(48\chi^2)\right],$$

where $\chi(p) = \frac{3\pi}{4}(p-1/4)$.

A small argument approximation can be found from Eq. G.1. It can be shown that

$$u_n(z) \approx 1 + (-1)^n \sum_{m=1}^{\infty} \exp\left(-\frac{j3m\pi}{4}\right) a_m(n) z^{3m/2}. \tag{G.4}$$

The first five coefficients of Eq. G.4, $a_m(n)$ ($m = 1, 2, \ldots, 5$), are given by

$$a_1(n) = \frac{\pi/4}{\Gamma(1-n)\Gamma(n+3/2)},$$

$$a_2(n) = \frac{5\pi/32}{\Gamma(5/2-n)\Gamma(n+3/2)},$$

$$a_3(n) = \frac{15\pi/64}{\Gamma(4-n)\Gamma(n+3/2)},$$

$$a_4(n) = \frac{1105\pi/1024}{\Gamma(11/2-n)\Gamma(n+3/2)}, \quad \text{and}$$

$$a_5(n) = \frac{1695\pi/1024}{\Gamma(7-n)\Gamma(n+3/2)}.$$

A list of the first 20 coefficients of the small argument approximation is given in the references (Bird, 1985).

G.1.2 Hard Surface Fock Functions (m < 0)

The n-th order hard surface Fock functions are defined as

$$v_n^{(m)}(z) = \frac{\exp(j(n+1/2)\pi/2) z^{n+1/2}}{2\Gamma(n+1/2)} L_{-m-1}^{(n)}(z); \quad m, n = 0, 1, \ldots, \tag{G.5}$$

The first derivative is

$$v_n^{(m)'}(z) = \left(n + \frac{1}{2}\right) \frac{v_n^{(m)}(z) - v_{n+1}^{(m)}(z)}{z}.$$

It is conventional for $m = 0$ to let

$$v_n(z) \equiv v_n^{(0)}(z)$$

and the zeroth-order function to be given by

$$v(z) \equiv v_0 \equiv v_0^{(0)}(z).$$

As for the soft surface Fock functions, a residue series is available for $z > 1$. In this instance

$$v_n(z) = \frac{\pi \exp(j(n+1/2)\pi/2) z^{n+1/2}}{2\Gamma(n+1/2)} \sum_{p=1}^{\infty} \left(t_p'\right)^{n-1} \exp\left(-jzt_p'\right), \qquad (G.6)$$

$t_p'(p = 1, 2, \ldots, \infty)$ are the zeros of w_2' which are related to the zeros α_p' of the derivative of the Airy function $Ai'(t)$ as follows:

$$t_p' = \left|\alpha_p'\right| \exp\left(-\frac{j\pi}{3}\right),$$

where α_p' are also tabulated in the references (Abramowitz and Stegun, 1965) for $p = 1$ to 10. For $p > 10$ a highly accurate approximation to $|\alpha_p'|$ is

$$\left|\alpha_p'\right| \approx \chi^{2/3}\left[1 - 7/\left(48\chi^2\right)\right],$$

where this time $\chi(p) = (3\pi/2)(p - 3/4)$.

A small argument approximation can be found by returning to the definition in Eq. G.1. It can be shown that

$$v_n(z) \approx 1 + (-1)^n \sum_{m=1}^{\infty} \exp\left(-\frac{j3m\pi}{4}\right) b_m(n) z^{3m/2}, \qquad (G.7)$$

The first five coefficients of this series, $b_m(n)$ ($m = 1, 2, \ldots, 5$), are as follows:

$$b_1(n) = \frac{\pi/4}{\Gamma(2-n)\Gamma(n+1/2)},$$

$$b_2(n) = \frac{7\pi/32}{\Gamma(7/2-n)\Gamma(n+1/2)},$$

$$b_3(n) = \frac{21\pi/64}{\Gamma(5-n)\Gamma(n+1/2)},$$

$$b_4(n) = \frac{1463\pi/2048}{\Gamma(13/2-n)\Gamma(n+1/2)} \quad \text{and}$$

$$b_5(n) = \frac{2121\pi/1024}{\Gamma(8-n)\Gamma(n+1/2)}.$$

The first 20 coefficients of the small argument approximation are given in the references (Bird, 1985).

A zero-order hard surface Fock function of the first kind is also required for an asymptotic expansion described in Chapter 8. This function is defined

$$v_0^{(1)}(z) = \frac{\exp(j\pi/4) z^{1/2}}{2\Gamma(1/2)} K_{-2}^{(0)}(z). \qquad (G.8)$$

Its residue series is

$$
v_0^{(1)}(z) = \sqrt{\pi} \exp(j3\pi/4) z^{1/2} \left(1 + \sum_{p=1}^{\infty} \left(1 + jzt_p' \right) \frac{\exp\left(-jzt_p'\right)}{\left(t_p'\right)^3} \right)
$$

$$
= \sqrt{\pi z} \exp(-j\pi/4) \left(1 + \sum_{p=1}^{\infty} (1+w) \frac{\exp(-w)}{\left|\alpha_p'\right|^3} \right),
$$

where $w = z\alpha_p' \exp(j\pi/6)$.

A small argument approximation is

$$
v_0^{(1)}(z) = \sqrt{\pi} \exp(j\pi/4) z^{1/2} \left(1 + \sum_{m=1}^{\infty} c_m(0)(-z)^{3m/2} \right). \tag{G.9}
$$

The first five coefficients of this series are as follows:

$$
c_1(0) = \frac{1/2}{\Gamma(5/2)},
$$

$$
c_2(0) = \frac{1/2}{\Gamma(4)},
$$

$$
c_3(0) = \frac{49/64}{\Gamma(11/2)},
$$

$$
c_4(0) = \frac{105/64}{\Gamma(7)}, \quad \text{and}
$$

$$
c_5(0) = \frac{19019/4096}{\Gamma(17/2)}.
$$

A list of the first 10 coefficients is available (Bird, 1985). Functions of higher order but still of the first kind are calculated from the following recursion relation

$$
v_n^{(1)}(z) = \frac{z^2}{(n-1/2)(n-3/2)} \left[(n-1)v_{n-2}^{(1)}(z) - (n-3/2)v_{n-1}^{(1)}(z) \right].
$$

G.2   Acoustic Fock Functions

These functions arise in the asymptotic representation of the field scattered by or radiated from a curved surface. Two basic forms are used in Chapter 8.

G.2.1   Soft Acoustic Fock Function

The lowest order of these functions are defined by (Logan, 1959)

$$f(z) = \frac{1}{\sqrt{\pi}} \int_{\Gamma_1} d\tau \frac{\exp(-jz\tau)}{w_2(\tau)}. \tag{G.10}$$

When $z = 0$, $f(0) = 0.776\exp(j\pi 3)$. When $x > 0$, $f(z)$ is accurately represented by the residue series

$$f(z) = \exp(j\pi/3) \sum_{p=1}^{\infty} \frac{\exp\left(z|\alpha_p|e^{-j5\pi/6}\right)}{Ai'\left(-|\alpha_p|\right)} \tag{G.11}$$

and for $z < -1$,

$$f(z) \approx -2j \exp\left(jz^3/3\right) \left[1 + j\frac{1}{4z^3} + \frac{1}{2z^6} - j\frac{175}{64z^9} - \frac{395}{16z^{12}} + j\frac{318175}{1024z^{15}} + \cdots \right].$$

A Taylor series expansion of $f(z)$ is frequently adopted in the intermediate region $-1 < z < 1$. This representation is

$$f(z) = \sum_{n=0}^{\infty} (\gamma_n - j\delta_n) \frac{z^n}{n!}.$$

The first 11 coefficients of this series are listed in Table G.1 (Logan, 1959, Table 24).

A polynomial approximation for the range $-3 < z < 0$ is also useful. To establish this, it is convenient to define a new function that is continuous, namely,

$$F(z) = f(z) \begin{cases} \exp(-jz^3/3); & z < 0 \\ 1; & z > 0 \end{cases}.$$

Table G.1   Coefficients of Taylor series for $f(z)$ in the range $-1 < z < 1$

n	γ_n	δ_n
0	3.879110E−01	−6.718810E−01
1	0.000000E+00	1.146730E+00
2	−4.314790E−01	−7.473430E−01
3	−1.748730E+00	−1.009630E+00
4	9.977770E+00	0.000000E+00
5	−1.264780E+01	7.302190E+00
6	−2.453740E+01	4.250000E+01
7	0.000000E+00	−3.594720E+02
8	3.602850E+02	6.240320E+02
9	2.711050E+03	1.565220E+03
10	−2.910550E+04	0.000000E+00

Table G.2   Coefficients of 7th-order polynomial approximations of amplitude and phase of $F(z)$ in the range $-3 < x < 0$

n	b_n	ϕ_n
0	7.757384E–01	–1.047185E+00
1	–9.668020E–01	7.394584E–01
2	4.029465E–01	4.664478E–01
3	–6.896238E–02	1.573962E–01
4	–1.256473E–01	2.612710E–02
5	–5.095698E–02	6.100471E–04
6	–9.824301E–03	–4.277579E–04
7	–7.800576E–04	–4.556784E–05

The amplitude and phase of this function is approximated by a 7th-order polynomial, the coefficients of which are listed in Table G.2. The approximation is valid for the range $-3 < x < 0$

$$B(z) = \sum_{n=0}^{7} b_n z^n \left(\text{error} \le |1| \times 10^{-4} \right)$$

$$\Phi(z) = \sum_{n=0}^{7} \phi_n z^n \left(\text{error} \le |2| \times 10^{-5} \text{radians} \right),$$

where $F(z) \approx B(z) \exp(-j\Phi(z))$. An estimate of the error of the polynomial approximations is shown in brackets. The modified soft acoustic Fock function $F(z)$ is plotted in Figure 8.6 over the argument range $-5 < z < 10$.

G.2.2   Hard Acoustic Fock Function

This is defined by

$$g(z) = \frac{1}{\sqrt{\pi}} \int_{\Gamma_1} d\tau \frac{\exp(-jz\tau)}{w_2'(\tau)}, \tag{G.12}$$

where $g(0) = 1.399$. When $z > 0$, $g(z)$ is accurately represented by the residue series

$$g(z) = \sum_{p=1}^{\infty} \frac{\exp\left(z|\alpha_p'|e^{-j5\pi/6}\right)}{|\alpha_p'| \left|Ai\left(-|\alpha_p'|\right)\right|} = \sum_{p=1}^{\infty} \frac{\exp\left(-z|\alpha_p'|(\sqrt{3}+j)/2\right)}{|\alpha_p'| \left|Ai\left(-|\alpha_p'|\right)\right|}. \tag{G.13}$$

While for $z < -1$,

$$g(z) \approx 2 \exp\left(jz^3/3\right) \left[1 - j\frac{1}{4z^3} - \frac{1}{z^6} + j\frac{469}{64z^9} + \frac{5005}{64z^{12}} - j\frac{31\,122\,121}{1024z^{15}} + \cdots \right].$$

Table G.3   Coefficients of Taylor series for $g(z)$ in the range $-1 < z < 1$

n	α_n	β_n
0	1.399380E+00	0.000000E+00
1	−6.472530E−01	3.736920E−01
2	−3.431040E−01	5.942730E−01
3	0.000000E+00	−2.949540E+00
4	1.741350E+00	3.016110E+00
5	7.740490E+00	4.489700E+00
6	−5.619460E+01	0.000000E+00
7	8.458020E+01	−4.883240E+01
8	1.852550E+02	−3.206110E+02
9	0.000000E+00	3.083790E+03
10	−3.451710E+03	−5.978540E+03

Table G.4   Coefficients of 7th-order polynomial approximations of the amplitude and phase of $G(z)$ in the range $-3 < z < 0$

n	a_n	θ_n
0	1.399427E+00	1.633610E+00
1	−6.450223E−01	2.674878E−01
2	−1.056456E−01	3.376310E−01
3	1.453212E−01	1.654477E−01
4	7.815282E−02	2.993846E−02
5	9.556067E−03	−3.475570E−03
6	−1.858469E−03	−2.159459E−03
7	−4.039875E−04	−2.410199E−04

The following Taylor series expansion of $g(z)$ is usually used in the range $-1 < z < 1$:

$$g(z) = \sum_{n=0}^{\infty} (a_n - jb_n) \frac{z^n}{n!}.$$

The first 11 coefficients of this series are listed in Table G.3 (Logan, 1959, Table 25). A useful polynomial approximation for the range $-3 < z < 0$ has also been obtained. For this, a new function was defined, namely,

$$G(z) = g(z) \begin{cases} \exp(-jz^3/3); & z < 0 \\ 1; & z > 0 \end{cases}.$$

The amplitude and phase of this function is approximated by a 7th-order polynomial, which has coefficients listed in Table G.4. The approximation is valid for the range $-3 < z < 0$

$$A(z) = \sum_{n=0}^{7} a_n z^n \left(\text{error} \le |5| \times 10^{-4}\right)$$

$$\Theta(z) = \sum_{n=0}^{7} \theta_n z^n \left(\text{error} \le |2| \times 10^{-5} \text{rad}\right),$$

where $G(z) \approx A(z) \exp(-j\Theta(z))$. An estimate of the error of these approximations is shown in brackets beside the polynomials. The modified hard acoustic Fock function $G(z)$ is plotted in Figure 8.2 for the argument range $-5 < z < 10$.

References

Abramowitz, M. and Stegun, I.A. (1965): 'Handbook of mathematical functions', Dover Inc., New York.

Bird, T.S. (1985): 'Accurate asymptotic solution for the surface field due to apertures in a conducting cylinder' IEEE Trans. Antennas Propagat., Vol. **AP-33**, pp. 1108–1117.

Bowman, J.J., Senior, T.B.A. and Uslenghi, P.L.E. (1963): 'Electromagnetic and acoustic scattering by simple shapes', North-Holland Publishing Company, Amsterdam, the Netherlands.

Fock, V.A. (1965): 'Electromagnetic diffraction and propagation problems', Pergamon Press, London, UK.

Logan, N.A. (1959): 'General research in diffraction theory', Missiles and Space Division, Lockheed Aircraft Corporation, Sunnyvale, CA. Vol. 1: Rep. LMSD-288087 and Vol. 2: Rep. I. MSC-288088.

Index

Fundamentals of Aperture Antennas and Arrays: From Theory to Design, Fabrication and Testing,
First Edition. Trevor S. Bird.
© 2016 John Wiley & Sons, Ltd. Published 2016 by John Wiley & Sons, Ltd.
Companion website: www.wiley.com/go/bird448

Printed and bound by CPI Group (UK) Ltd, Croydon, CR0 4YY

22/04/2025

14659800-0001